T0406086

Geological Storage of Highly Radioactive Waste

Roland Pusch

Geological Storage of Highly Radioactive Waste

Current Concepts and Plans for Radioactive Waste Disposal

Prof. Dr. Roland Pusch
Geodevelopment International AB
Ideon Science Park
SE-223 70 Lund
Sweden
pusch@gedevelopment.ideon.se

ISBN: 978-3-540-77332-0 e-ISBN: 978-3-540-77333-7

Library of Congress Control Number: 2008929595

Cover design: deblik, Berlin

Printed on acid-free paper

9 8 7 6 5 4 3 2 1

springer.com

Preface

A major part of this book is based on work performed by several of the national organizations that are responsible for disposal of radioactive waste from nuclear reactors, with the Author involved in the research as well in the reporting. He is greatly indebted to the organizations and to their representatives that were engaged in the projects, and to the European Commission, represented by Mr Christophe Davies, that supported the work financially and otherwise. Mr Davies' services are gratefully acknowledged. The author also expresses his thanks to the following persons who assisted in various ways in the preparation of the book: Christer Svemar, Swedish Nuclear Fuel and Waste Management Co (SKB), Sweden; Wolf Seidler, Agence National pour la gestion des Dechets Radioactifs (ANDRA), France; Jan Verstricht, Studiecentrum voor Kernenergie-Centre d'Etude de l'énergie Nucleare (SCK-CEN), Belgium; and Tilmann Rothfuchs, Gesellschaft für Anlagen- und Reaktorsicherheit GmbH (GRS), Germany.

Lund, January 2008 *Roland Pusch*

Acknowledgment

The author expresses his thanks to the following persons who assisted in various ways in the preparation of the book: Christer Svemar, Swedish Nuclear Fuel and Waste Management Co (SKB), Sweden; Wolf Seidler, Agence National pour la gestion des Dechets Radioactifs (ANDRA), France, and Tilmann Rothfuchs, Gesellschaft für Anlagen- und Reaktorsicherheit GmbH (GRS), Germany.

Contents

List of Abbreviations*

ACE	Acoustic emission
AECL	Atomic Energy of Canada Limited
AGP	Almacenamiento Geológico Profundo (Spanish for crystalline geological formation)
Andra	Agence Nationale pour la question des Déchets Radioactifs, France
BAMBUS	Backfill and Material Behaviour in Underground Repositories in Salt
BCE	Buffer/container experiment
BEM	Boundary Element Method
BMT	Buffer mass test
BWR	Boiling water reactor
CEC	Cation exchange capacity
CH-TRUW	Contact-handled transuranic radioactive waste
CMI	Computational Mechanics Centre, UK
CROP	Cluster Repository Project (EU project)
DAS	Data Acquisition System
DBE	Deutsche Gesellschaft zum Bau und Betrieb von Endlagern für Abfallstoffe
D&D	Decontamination and decommissioning
EBS	Engineered barrier system
EC	European Commission
EDZ	Excavation disturbed zone (here both damaged by excavation and disturbed by stress redistribution)
Enresa	Empresa Nacional de Residuos radiactivos, S.A., Spain
FEP	Features, Events and Processes
FEBEX	Full-scale engineered barriers experiment in crystalline host rock
FEM	Finite Element Method

* Only major and frequently used abbreviations are listed.

FMT	Fracture-Matrix Transport
GRS	Gesellschaft für Anlagen- und Reaktorsicherhait mbH, Germany
HE	Heater Experiment
HLW	High-level radioactive waste
IAEA	International Atomic Energy Agency
ICRP	International Committee on Radiation Protection
ILW	Intermediate level radioactive waste
IPC	Inorganic Phosphate Cement
ITT	Concrete plug interaction test
KBS-3	Kärnbränslesäkerhet Concept 3 (Swedish for "nuclear fuel safety 3")
LANL	Los Alamos National Laboratory
LBL	Lawrence Berkeley Laboratories
LLW	Low-level waste
LOT	Long Term Test of Buffer Material (Äspö field test)
LRDT	Low Risk Deposition Technology (EU project)
MLW	Medium Level Waste
Nagra	Nationale Genossenshaft für die Lagerung radioaktiver Abfälle, Switzerland
NBS	Natural Barrier System
NIST	US National Institute for Standards and Technology
OPC	Ordinary Portland Cement
OPG	Ontario Power Generation Inc., Canada
PA	Performance Assessment
POSIVA	POSIVA Oy, Finland
PWR	Pressurised Water Reactor
QA	Quality Assurance
RH TRUW	Remote handled transuranic radioactive waste
SA	Sensitivity Analysis
SCK-CEN	Studiecentrum voor Kernenergie/Centre d'étude de l'Energie Nucléaire'. Nationaal onderzoekcentrum in Mol, Belgium (België).
SF	Spent Fuel
SFR	Swedish final repository for low-and intermediate level radioactive waste
SKB	Svensk Kärnbränslehantering AB (Swedish Nuclear Fuel and Waste Management)
SNF	Spent nuclear fuel
SNL	Sandia National Laboratories
TBM	Tunnel boring machine
TRUW	Transuranic radioactive waste
TSDE	Thermal Simulation of Drift Emplacement
URL	Underground research laboratory
USDOE-CBFO	USDOE Carlsbad, USA
WIPP	Waste Isolation Pilot Plant
WIT	Wessex Institute of Technology, UK

Symbols and Parameters*

Mathematical	Exponential functions expressed as E+ or E- (e.g. E-6 = 10^{-6}).
Geometrical	
d	Grain diameter (µm, mm).
H	Height (m, km).
L	Length (nm = E-9, Å = E7 mm, m, km).
Soil physico/ chemical	
e	Void ratio (ratio of pore volume and volume of solids).
g	Gravity (m/s^2).
i	Hydraulic gradient (m/m).
k	Permeability (m^2).
n	Porosity (ratio of pore volume and total volume), (%).
m	Mass (kg).
t	Time (s).
w	Water content (ratio of mass and solids), (%).
w_L	Atterberg Liquid Limit (%).
w_P	Atterberg Plastic Limit (%).
CEC	Cation exchange capacity mE/100 g solids).
D	Diffusivity (m^2/s), Diameter (m).
D_a	Apparent diffusivity (m^2/s).
D_e	Effective diffusivity (m^2/s).
D_p	Pore diffusivity (m^2/s).
K	Hydraulic conductivity (m/s).
K_d	Sorption factor (m^3/kg).
S	Sievert, radiation dose.
S_r	Degree of fluid saturation (ratio of volume of satur. Voids and total volume, (%).
T	Temperature (°C, K).
η	Viscosity (Ns/m^2).

* Only major and frequently used abbreviations, symbols and parameters are listed.

ρ Density at actual degree of fluids saturation (kg/m^3).
ρ_d Dry density (kg/m^3). Mass of solids divided by total time.
ρ_{sat} Density at fluid-saturation (kg/m^3).

Grain size
Clay Particle size smaller than 2μm, soil with at least 20% particles of this size.
Silt Particle size ranging from 2 to 60 μm, with at least 20% particle of this size.
Sand Particle size ranging from 60 to 200 μm, with at least 40% particle of this size.
Gravel Particle size from 2 to 20 mm, with at least 40% particle of this size.
Cobble Particle size ranging from 60 to 600 mm, with more than 40% cobbles.
Boulder Particle size exceeding 600 mm, soil with more than 40% boulders and cobbles.

Soil and rock mechanical
c Cohesion (kPa, MPa). Shear strength for zero effective normal effective stress.
p Pressure (kPa, MPa).
p_s Swelling pressure (kPa, MPa).
q Load intensity (N/m^2).
u Porewater pressure (kPa, MPa).
E Modulus of elasticity (kPa, MPa, GPa).
G Shear modulus (kPa, MPa, GPa).
M Odeometer modulus.
K Compression modulus (N/m^2).
P Force (N, kN, MN, tons).
ε Strain (m/m).
Φ Angle of international friction.
σ Stress (kPa, MPa).
Effective stress (kPa, MPa).
$(\sigma_1 - \sigma_3)_o$ Reference deviator stress (kPa, MPa).
$(\sigma_1 - \sigma_3)_f$ Deviator stress of failure.
τ Sheare stress (kPa, MPa).
υ Poisson's ratio.

Thermal
C Heat capacity (Ws/kg, K).
λ Heat conductivity (W/m, K).
α Coefficient of linear thermal expansion ($1/C^\circ$) or (1/K).
α_L Theoretical lower bound of linear thermal expansion ($1/C^\circ$) or (1/K).

α_U	Theoretical upper bound of linear thermal expansion ($1/C^{\circ}$) or (1/K).
β	Coefficient of volume expansion ($1/C^{\circ}$) or (1/K).
γ	Grüneisen parameter.
γ_{th}	Grüneisen parameter: thermodynamic gamma.
δ_r	Anderson-Grüneisen parameter.
ε	Strain.
Θ_D	Debye temperature (K).
ρ	Resistivity (Ωm).
ρ	Specific mass (kg/m^3).
σ	Stress (Pa).
A	Area (m^2).
c	Stifness tensor.
C_P	Specific heat constant pressure ($J\ kg^{-1}\ K^{-1}$).
C_V	Specific heat in constant volume ($J\ kg^{-1}\ K^{-1}$).
D	Diameter (m).
E	Modulus of elasticity (Young's modulus) (N/m^2).
E	Energy of lattice vibrations (J).
GF	Gauge factor.
K	Bulk modulus (compressibility).
K_T	Isothermal bulk modulus (Pa).
L	Length (m).
n	Number of atoms in chemical formula (mol).
N	Axial force (N).
P	Pressure (Pa).
R	Residence (Ω).
R	Gas constant, 8.314*($J\ mol^{-1}\ K^{-1}$).
S	Entropy ($J\ mol^{-1}\ K–^{1}$).
T	Temperature (°C) or (K).
THMCBR	Thermal (T), Hydraulic (H), Mechanical (M), Chemical (C), Biological (B), Radiological (R) processes.
U	Internal energy (J).
V	Volume (m^3).

Introduction

National and International Work

Nuclear energy appears to be needed for a long time in many countries and the issue of safe disposal or reprocessing of the highly radioactive waste emanating from it becomes an increasingly important matter. Organizations like IAEA and the European Commission watch this and help to develop a basis for selecting safe methods for disposal of spent and reprocessed reactor fuel, collectively termed high level radioactive waste – HLW – in this book, as well as of intermediate (ILW) and low-level (LLW) radioactive waste, but the national governments take the decisions. IAEA's regulations form the basis of designing and constructing safe repositories [1].

For assisting the individual countries utilizing nuclear power in developing and refining their concepts for disposal of highly radioactive waste and waste with long-lived radionuclides, the EU initiated and partly financed a comprehensive project in year 2000, "Cluster Repository Project" (CROP), Contract FIR1-CT-2000-2003, for summarizing the experience from the national programs concerning repository design and construction as well as modelling and performance assessment. This project was co-ordinated by SKB through the head of the Äspö HRL,[1] Mr Christer Svemar, who also edited and authored the final report assisted by the present author [2]. It outlined how further development of technically/economically optimal concepts for disposal of HLW can be achieved and included attempts to improve the concepts. The numerous reports that have been published by the respective national organizations form the basis of the report and of further national and co-operative international work.

The CROP report gave detailed descriptions of the national concepts focusing on construction, design and performance with special reference to the results of experiments performed in underground laboratories. The present book is primarily a synthesis of the national contributions to the report but widens the matter by

[1] SKB's current underground laboratory at Äspö, "**H**ard **R**ock **L**aboratory".

R. Pusch, *Geological Storage of Highly Radioactive Waste*,
DOI: 10.1007/978-3-540-77333-7_1, © Springer-Verlag Berlin Heidelberg 2008

describing and assessing special concepts, like deposition of HLW in very deep boreholes. Disposal of HLW in crystalline, salt and argillaceous rock, and in clastic clay was treated in both CROP and LRDT described later, and they are considered also in this book but focus is on crystalline rock.

Nine end-user organisations took part in CROP representing Europe as well as North America. They will be termed the Organizations in this book and were, in alphabetical order:

1. Agence National pour la gestion des Dechets Radioactifs (ANDRA), France.
2. Empresa Nacional de Residuos Radiactivos SA (ENRESA), Spain.
3. Gesellschaft für Anlagen- und Reaktorsicherheit GmbH (GRS), Germany.
4. Nationale Genossenschaft für die Lagerung radioaktiver Abfälle (NAGRA), Switzerland.
5. Ontario Power Generation Inc (OPG), Canada, supported by G.R. Simmons & Associates Consulting Services Ltd.
6. Posiva Oy, (POSIVA), Finland.
7. SKB (Svensk Kärnbränslehantering AB), Sweden, supported by Geodevelopment AB.
8. Studiecentrum voor Kernenergie-Centre d Etude de l'energie Nucleare (SCK-CEN), Belgium.
9. United States (U.S.) Department of Energy Carlsbad Field Office (CBFO), USA, supported by GRAM, Inc. and Sandia National Laboratories (SNL).

NAGRA, OPG and U.S. CBFO were self-funded Participants. Deutsche Gesellschaft zum Bau und Betrieb von Endlagern für Abfallstoffe (DBE) participated as an observer since 2002. The CROP participants run underground rock laboratories (URLs) in several different geological media, such as crystalline rock in Sweden (Äspö), Finland (Olkiluoto), Switzerland (Grimsel), and Canada (URL, Pinawa), bedded salt in the United States of America (WIPP) and a salt dome in Germany (Asse), and sedimentary clay formations in Belgium (Mol), France (Bure) and Switzerland (Mt Terri).

Basic Principle of Final Storage of Hazardous Waste

The basic, common principle of isolating HLW from the surroundings is that in Fig. 1.

The CROP project demonstrated that the man-made engineered barriers, termed EBS, have similar functions in the different repository host media and many of the solutions and techniques documented are believed to be applicable to disposal concepts in various rock types. The results from tests conducted in many different geological media and involving a large number of EBS components were expected to be valuable to all organisations engaged in development of repository design and construction. In particular, it was expected that improved solutions of a number of not yet solved technical problems in disposal of hazardous waste would emanate

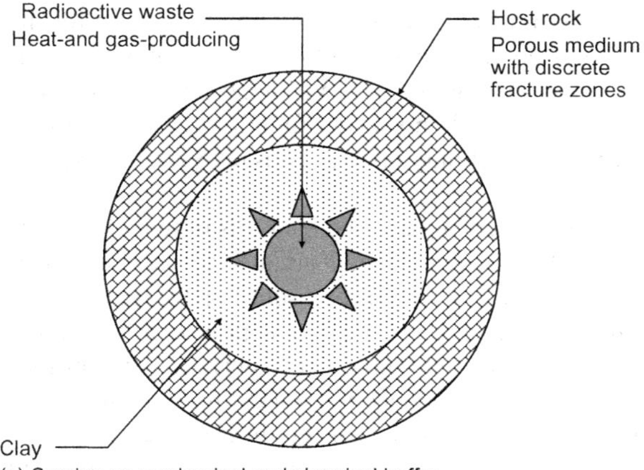

Fig. 1 Isolation of HLW. The containers are embedded in clay in crystalline and argillaceous rock, as in the picture, or in crushed salt in rock salt [3]

from the work performed in the CROP and LRDT projects. Some changes in design and construction have been proposed in later years, hopefully resulting from the discussions and presentations in the two projects. The present book is hoped to serve as catalyzer for further development of repository concepts.

The CROP Project

The general conclusions from the CROP project were:

- The existing underground laboratories (URLs) for testing the performance of rock and EBS components have performed according to plans and represent a large enough variety of geological conditions for providing credible generic information on repository concepts in all geological media of consideration in the world today. However, they are not capable of providing the site-specific information that the disposal programmes require for final verification and fine-tuning of the safety case and the detailed construction/operation parameters.
- The proposed location of repositories with respect to depth is in the interval 200–600 m and all concepts are of one-level type. Access to the repositories is through shafts for salt while both ramps and shafts are planned for crystalline rock and clay.
- The use of natural and engineered barrier systems for retarding possible release of radionuclides to acceptable levels in the biosphere requires prediction of the evolution and performance of the repository system, including degradation processes. The temperature in the near-field during the heat pulse is shown to be an important factor since it could strongly affect the chemical processes in the EBS components, and groundwater flow.

- SKB, POSIVA and OPG are designing very corrosion-resistant canisters for the spent fuel, while less durable canisters represent an option for several of the other organisations. For long-lived transuranic waste (TRUW) disposed in the WIPP, a licensed American repository in bedded salt, no barrier function is ascribed to the steel containers, and a canister lifetime of only up to a thousand years is assumed for several other concepts.
- The embedment of HLW canisters, called buffers, and backfills of tunnels, rooms and shafts, are of different types. For the two concepts related to salt rock, magnesium oxide, crushed salt and/or salt blocks will be used. For those related to crystalline rock and clay sediments, dense smectitic clay is proposed for preparing buffer and backfill, either alone or mixed with other materials. Cementitious seals and concrete are options in ANDRA's concept intended for crystalline and argillaceous rock, and for SCK-CEN's repository concept in clastic clay.
- There is need for structurally sound, low-permeable plugs in tunnels, drifts and shafts as well as in boreholes. The isolation provided by them is closely related to the extension and properties of the excavation-disturbed zone (EDZ) at the tunnel periphery in all geological media, and to the capability of the plug to seal or cut off the EDZ in a long-term perspective.
- Instrumentation of a real repository should be left out for avoiding radionuclide migration along cables and tubings. However, in underground laboratories where whole concepts or parts thereof are investigated, properly selected and placed instrumentation is required for testing predictions made on the basis of theoretical models. Moisture gauges, like psychrometers, are required for recording the rate of wetting and drying of EBS components but they are concluded to be the least reliable of all instruments and the first to fail according to the experience gained in several URLs. Each type of sensor serves accurately only up to a certain degree of fluid saturation, and combination of different types of sensors to cover the full range of expected fluid saturation is recommended. The number of gauges that can be placed in a large-scale demonstration test is limited by the space they require and the need to avoid physical interaction of adjacent sensors.
- The conditions for hydration of clay components of EBS in crystalline rock are not well known and the role of the rock structure, including the EDZ, in providing water needs to be clarified. A further question that also has to do with inflow of water is how one can stop excessive water inflow into deposition holes or drifts in the construction and waste placement phases, and during the placement of clay buffers and backfills.
- Maturation of buffer clay in salt rock and argillaceous rock is intimately connected to the rheology of the host medium and the rate of convergence will need further study for developing complete models. The understanding of creep processes in argillaceous rock and clastic clay as well as in salt buffer and rock should benefit from additional study of the mechanisms on the micro-structural scale.
- Gas release from canisters and migration through the buffer require more attention in HLW repositories although conceptual understanding of the processes involved has been developed.

- Modelling of chemical processes in clay buffer is an issue that requires more work, especially concerning complexation and cementation. A remaining task is to develop more accurate theoretical models for cementation by precipitation of dissolved elements, and conversion of smectite to nonexpendable minerals.

For the various types of geological media the following conclusions were drawn in the CROP project.

Crystalline Rock

- Crystalline rock has brittle characteristics and contains fractures, which may form patterns of groundwater transport pathways from a repository to the ground surface. The existence of these fracture systems and their transport characteristics are the major factors that influence both the construction conditions and the long-term safety case.
- Canister design and manufacturing, as well as the performance of different canister materials in repository environment, require more work for finding optimum solutions.
- The most important single parameter for the design of the repository is the maximum temperature allowed in the near-field.
- The selection of tunnel construction method needs consideration with respect to the repository design and operation and to the consequences of the presence and extension of the EDZ.
- Grouting as a means for sealing drift walls against inflowing water needs to be developed as current experience shows limited success in grouting unsupported rock surfaces. Low-pH cements should be used because ordinary cementitious materials (high pH) are potentially detrimental to the properties of the clay-based EBS.
- Swelling clay alone or mixed with other materials is an outstanding material for use in buffer, backfill and plugs and has been thoroughly investigated for more than 20 years. Production of highly compacted blocks of various size has been demonstrated.
- Methods have been tested on a full scale for vertical in-hole emplacement, but emplacement in other configurations must be demonstrated on full scale.
- Backfilling of deposition drifts in the KBS-3 V vertical in-hole emplacement method has been tested on full scale for low-electrolyte-water but is not verified for salt water.

Salt Rock

- Crushed salt makes a good backfill as stress and creep-induced room closure (convergence) ultimately leads to consolidation of crushed-salt backfill and complete encapsulation of waste canisters.

- With given weight and size of waste packages, the repository design is a country- and site-specific issue, of which the US DoE's WIPP repository is an example. The design of seals and plugs, which are the main engineered barriers in salt rock, are also site-specific.
- For a future German domal salt repository, work remains to be conducted for verification/optimisation of design details.
- Good capability exists for modelling of excavation-induced effects, e.g., EDZ generation, but adequate prediction of EDZ healing in salt rock needs further investigation. For argillaceous rock and clastic clay it is required to find out whether the EDZ really self-seals.
- Final confirmation of the technical emplacement system for Cogéma canisters, as well as the testing of the feasibility of the emplacement of alternative canisters for spent nuclear fuel into 300-m-deep boreholes, is pending.
- In situ testing of suggested drift seal design under representative conditions is pending.
- The physical behaviour of crushed salt backfill is understood in principle. The effectiveness of including geochemical additives in the EBS to increase sorption of special radionuclides in the near-field has not been tested adequately.

Argillaceous Rock and Clastic Clay

- The disposal concepts for argillaceous rock (clay shale, clay stone or argillite) depend on the usually limited thickness and homogeneity of the layers. In-room emplacement of canisters is favoured and not in-hole emplacement (vertical mode). The alternative is several tens of meter long horizontal large diameter boreholes or micro-drifts.
- Excavation is more complex than in salt or crystalline rock, but proven technology can be applied. Verification of optimum techniques has been achieved for soft clay in the Mol URL, where the rock needs permanent support of liners to prevent creep-induced time-dependent deformation of excavated openings.
- The rock is the main barrier in the multi-barrier system, but the near-field of the repository has to be designed and constructed so that geological features and construction of the repository do not jeopardise long-term safety.
- Design is difficult, as for salt, with respect to the existence of unexpected structural elements/discontinuities in the rock, drift convergence, and EDZ formation and de-saturation.

The Low Risk Deposition Technology Project

Additional research and development that is relevant to the issue of disposing HLW and that was conducted in approximately the same time period as CROP and supported by the European Commission (Contract EVG1-CT-2000-00020) concerned disposal of very hazardous chemical waste. It dealt with disposal of very hazardous waste in abandoned mines and was termed "Low-Risk Deposition Technology" (LRDT), [4]. It was performed by the following parts:

1. Geodevelopment AB,[2] Lund, Sweden
2. DURTEC GmbH, Neubrandenburg, Germany
3. NTUA, Technical University of Athens, Greece
4. Computational Mechanics International Ltd, Southampton, UK
5. Wessex Institute of Technology, Southampton, UK

The LRDT project, which considered the same rock types as CROP and, in addition, porous limestone of coral reef type, gave the following major conclusions:

- In *crystalline rock* transport of toxic elements takes place solely by flow in major conductive discontinuities, primarily fracture zones, representing discontinuities of low orders according to the categorization principle adopted in the project. Focus has been on transport in minor (3rd order) zones of dissolved species and clay colloids that can bring such species attached to them all the way to wells for drinking water but the sorbing potential of weathered rock in the fracture zones and of fracture coatings make such transport insignificant. It may be important, however, in major fracture zones like those of 1st and 2nd orders and particularly in the "channels" formed by their intersections. Flow analysis must include numerical modelling of groundwater contamination (Fig. 2). The LRDT project showed that the rock structure is of great importance but since it is nearly impossible to derive reliable structure models that are valid for long periods of time, the EBSs is of even greater importance.
- *Bedded salt and salt domes* (NaCl or KCl) have rather low strength and a substantial creep potential. The excavation-disturbance (EDZ) may therefore be significant but after emplacing the waste and backfilling the mine with crushed salt,

Fig. 2 Example of predicted concentration plots during the transient stage of release of contaminants from a repository in rock with orthogonal major fracture zones. Flow takes place from left to right [4]

[2]Now Geodevelopment International AB.

creep of the host rock will compress the EDZ and backfill and the entire repository will ultimately become completely homogeneous. No free water exists in salt and toxic elements are therefore effectively retained. Local brine zones exist but they are easily identified, drained or avoided.
- *Argillaceous rock and limestone* have low strength even if they are of high density, which makes it difficult to locate disposal rooms at larger depths than 200–300 m. The expected frequent failure in the form of an extensive EDZ will be a major transport route in the repository and it needs to by sealed by special grouting or be cut off by plugs that are keyed into the surrounding undisturbed rock. Self-sealing of clay rock by creep can not be assured.
- A general conclusion is that salt rock provides excellent isolation of hazardous waste although the matter of gas release needs further consideration: the extreme tightness of salt rock can lead to enclosed gas under very high pressure that can cause catastrophic breakage. Argillaceous rock has several disadvantages, primarily a low mechanical strength that puts a limit to the size of storage rooms, and special steps have to be taken to seal off the richly fractured surrounding of waste-filled rooms. Crystalline rock has excellent mechanical strength and provides stable rooms for waste isolation but the content of significantly active hydraulic conductors is high and one has to rely on effective engineered barriers. In this respect salt is a particularly good candidate rock type except that gas emanating from the waste may cause difficulties.

Like CROP, the LRDT project was focused on investigating if and how very hazardous waste can be isolated from the biosphere for very long periods of time, the latter project dealing with non-radioactive waste represented by mercury batteries and organic pesticides. It went deeper into the role of rock structure and related rock stability and groundwater flow than the first mentioned project, which will be referred to in the present book. The reasonings and discussions are based on deepened analyses and the promises and limitations considered from a somewhat different view of angle. Above all, it puts emphasis on the importance of the evolution of the earth crust to the long-term performance of HLW repositories, a matter that makes large-scale geophysical and geochemical processes more important than in the EU projects.

Use of mines for disposal of HLW may, as a first glance, be unsafe but does not have to be unacceptable if certain fundamental conditions are fulfilled. One is that no exploitable ore is left so that the risk of future renewed mining is eliminated, and another that the rock must be sufficiently tight and stable. There are examples of concepts that imply that waste is stored and suitably isolated in exploited parts of a big mine while mining continues in the opposite end. Such rational use of the often very large volumes of open space in remote mine districts might well be worth considering. The issue is further discussed in Chap. 6 in this book.

The ESDRED Project

A further example of the co-operative work that has been made for solving common problems is the EU project ESDRED (Contract FI6 W-CT-2004-508851) that was conducted somewhat later but yielded the same major conclusion as the CROP project. It was performed by the same consortium that conducted the ESDRED but with ANDRA as co-ordinator.

The ESDRED project demonstrated that techniques are available for handling and placement of canisters with HLW and clay buffers and backfills. It has also shown that temporary plugs for supporting clay-based backfills can be made.

Options for Disposal of HLW

The highly radioactive waste representing spent reactor fuel or vitrified material, both being termed HLW in this book, is the most hazardous rest product of the nuclear energy cycle. Historically, we have seen some concepts for its disposal that appear odd today like sending it off to the sun by rockets for getting rid of it once and for all. Another idea implied disposal of it in very deep trenches in the oceans where there are no living species. The reason for abandoning the firstmentioned idea is the lack of confidence in rocket performance – the deadly load can come back – while the other was unacceptable because the heat production would cause convective water flow from the depth, bringing with it released radionuclides to shallow, organic-rich levels in the sea.

A third, not finally deemed unacceptable, was the idea of an open, windblown cemetery for piles of canisters on top of a mountain like Parthenon. Thick granite or diabase slabs as roof and large-diameter columns of such rock would serve as construction elements and arranged so that the HLW can be effectively isolated from the surroundings and so that rain and meltwater are drained off, leaving the canisters exposed only to air for cooling (Fig. 3). The persistence of such a mausoleum would last for tens of thousands of years or even much longer in a climate like that in present desert areas.

Naturally, there are considerably uncertainties in such a project. Climate can change and even a very solid rock construction will ultimately collapse but the most challenging matter is the persistence of the canisters. They can be shielded from eroding windblown sand but not from chemical attack by oxygen, sulphur and

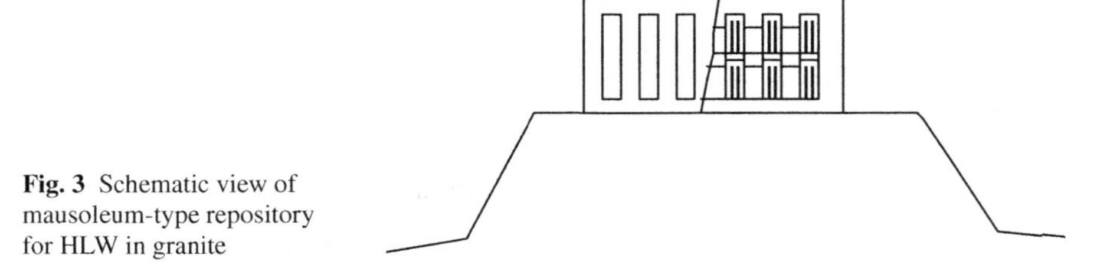

Fig. 3 Schematic view of mausoleum-type repository for HLW in granite

chlorides, which will cause corrosion of the initially very hot and strongly radiating canisters. Easy access to the canisters can not be avoided, which implies risk of removal of the canisters for illegal use and risk of injury by released radioactivity. Bomb attack by terrorists would only be hindered by very comprehensive fence arrangements.

Moving to the idea of on-ground landfill-type disposal one can image placement of HLW in concrete vaults covered by a clay-based top liner with overlying erosion-resistant friction soil and stone fill (Fig. 4). This solution is being proposed for disposal of low- and intermediate-level waste and can theoretically be used also for HLW although the heat production is expected to speed up degradation of the concrete and clay cover.

As for any other hazardous waste deep geological disposal offers very good possibilities for effective isolation from the biosphere. This is because the average hydraulic conductivity of crystalline and argillaceous rock drops with depth, which makes migration of radionuclides released from the waste by groundwater flow extremely slow and nearly negligible at a depth of several hundred meters in fracture-poor rock. However, intersection of a repository by water-bearing fracture zones, termed low-order discontinuities in this book, can cause quick transport of radionuclides to the ground surface. The condition for such flow is the hydraulic conductivity of these discontinuities, and the magnitude and orientation of the regional hydraulic gradient, which depends on the topography.

In comparing the function of different geological media one finds that crystalline rock can have a rather high hydraulic conductivity because of its intrinsic content of interacting water-bearing discontinuities in the rock, while argillaceous rock is much tighter but still not as tight as salt rock. The conductivity is high in porous rock like limestone and sandstone and they are hence not suitable for disposal of hazardous waste. Crystalline rock is available in most countries and dominates in Sweden and Finland while it is not very abundant in Belgium, Holland, France, Spain and Greece where dense argillaceous rock is common. Salt rock is present in Denmark, Germany, France and a few more European countries and some deep caverns that can be used for waste disposal have been excavated in them. The limited access to suitable rock means that repositories may have to be located in relatively pervious crystalline rock in countries like Sweden and Finland, and in argillaceous rock with

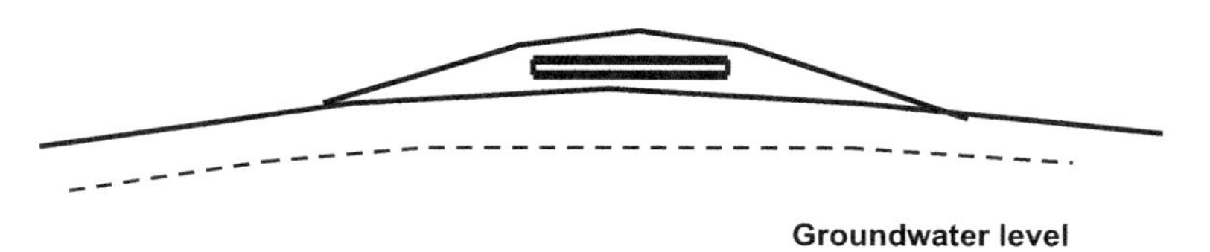

Fig. 4 Schematic view of disposal of radioactive waste in on-ground repository

a structure that is not favourable from the viewpoint of mechanical stability like in certain areas in France and Spain. This demands careful planning of the location and orientation of deposition tunnels and holes for optimal utilization of the rock. We will examine these issues in detail to see whether one can rely on the various rock types.

References

1. IAEA. 2003. Scientific and technical basis for geological disposal of radioactive wastes. International Atomic Energy Agency (IAEA), Technical Report Series, #413, Vienna.
2. Svemar Ch, 2005. Cluster Repository Project (CROP). Final Report of European Commission Contract FIR1-CT-2000-20023, Brussels, Belgium.
3. Pusch R, Yong RN, 2006. Microstructure of smectite clays and engineering performance. Taylor & Francis, London and New York. ISBN 10: 0-415-36863-4, ISBN13: 9-78-0-415-36863-6.
4. Popov V, Pusch R, (Eds), 2006. Disposal of hazardous waste in underground mines. WIT Press, Southampton, Boston. ISBN: 1-85312-750-7. ISSN: 1476-9581.

Chapter 1
The Geological Base for Developing and Assessing Concepts for HLW Disposal

1.1 Rock Types Considered for HLW Disposal

One of the most discussed items in the search for sufficiently good conditions for safe disposal of HLW is to identify a suitable rock type and get access to an acceptable site. As concluded from the CROP [1] and LRDT [2] projects effective isolation of such waste can be provided by repositories constructed in crystalline, salt, and argillaceous rock, as well as in clastic clay. The two firstmentioned normally offer very stable rooms at 400–500 m depth in contrast to the other two, which instead provide excellent tightness. Salt rock is in fact not only providing stable rooms but also better isolation than any of the other rock types. Rooms and tunnels in crystalline and argillaceous rock will be surrounded by a continuous, very permeable excavation-disturbed zone (EDZ) that may have to be cut off by constructing plugs that are keyed into the rock to a depth that depends on the excavation damage and the safety criteria since the EDZ may serve as a very effective conductor and transport path for radionuclides. Salt rock has the excellent property of undergoing self-sealing of the EDZ and any void or defect in it, but may in fact be too tight and burst if gas produced by canister corrosion or otherwise, becomes pressurized.

For all these rock types, strain caused by redistribution of the stresses in the virgin rock by creating openings leads to high hoop stresses near the periphery of the openings and these stresses determine the stability of the openings and the groundwater flow into and along them. The distribution of stress and strain is strongly influenced by the rock structure, i.e. the defects in the form of fissures and fractures, and we will start the book by briefly considering the structure of rock of the three types, keeping in mind that for each of them the present constitution represents just a stage in its transient evolution. This means that the conditions for safe isolation of HLW will change with time, a matter that will be currently referred to in the book.

R. Pusch, *Geological Storage of Highly Radioactive Waste*,
DOI: 10.1007/978-3-540-77333-7_2, © Springer-Verlag Berlin Heidelberg 2008

1.1.1 Strain in the Shallow Earth Crust

The continents' relative displacement speeded up in Cretaceous and Tertiary times, i.e. yesterday on the geological timescale, and led to splitting of the supercontinent Pangea and formation of the present continents. It is an ongoing process that may be coupled to an expansion of the globe, and associated with stress alteration in the Earth crust. The changes in direction and magnitude of the principal stresses have generated systems of major discontinuities when and where the stress level has been critically high. Because of the change in orientation of the regional major principal stress from NE/SW in Precambrian time to the present NW/SE direction in Europe, systems of major discontinuities as we see them today are oriented NE/SW, N/S, NW/SW and W/E, the firstmentioned presumably being the most ancient. The rotation of the global major stress fields in the crust is extremely slow but regional creep-related stress accumulation and release have caused and will cause more or less intermittent strain and reorganization of the stresses. These are wellknown processes as we will touch on in the subsequent chapter and in the chapter on site selection, while exogenic events that alter the stress conditions, like loading by glaciers and subsequent unloading when they melt, are less well understood and more difficult to model. Since a repository for HLW has to provide isolation of radionuclides for tens to hundreds of thousands of years both short- and long-term stress changes and their impact on the rock structure should be taken into consideration in predicting the performance of the rock hosting a repository. This does not seem to have been the case for all the Organizations.

1.1.2 Elements in the Earth Crust that can Affect Isolation of Radionuclides in HLW Repositories – Rock Structure

1.1.2.1 General

Numerous studies of the performance of fracture systems and other defects that are related to groundwater flow and mechanical stability of rock masses have been made by the Organizations as a basis of safety analyses respecting migration and concentration of radionuclides in the groundwater. There has been much debate on how to describe conductive rock elements and there is a whole range of methods and denotations, some referring to faults, fracture zones or "deformation" zones, which terms are of little use in underground construction work, and others calling them discontinuities of different orders with respect to geometrical, mechanical and hydraulic properties. The latter can be used for general description of the build-up of a rock mass and in conceptual and theoretical modelling of groundwater flow and stress and strain.

One realizes that minor fractures and fissures play a role for diffusive transport of water and radionuclides, while practically important migration of them takes place in hydraulically interacting, macroscopic fractures. This suggests that only the latter need to be considered in the "far-field", i.e. by regional transport modelling, while for the "near-field" rock surrounding deposition holes and tunnels, one would have to assume all sorts of structural elements to be present and consider migration of radionuclides both by water flow and diffusion. Water-bearing major structural elements are naturally of primary importance in the site-selection process, while the finer ones should be in focus for predicting the performance of the rock where the waste containers are placed. The major water-conducting structural elements, which represent "fracture zones", can be identified by examination of cores from deep, graded boreholes and logging of them, and by use of geophysical methods, while presently hydraulically inactive ones can hardly be identified and characterized. The problem is that the latter ones may come alive and be important conductors in the future through strain induced by changes of the stress fields generated by constructing rooms, tunnels and holes, and by the thermal pulse caused by the radioactive decay. Overlooking them can invalidate all sophistical geohydraulic modelling attempts and lead the decision-makers to select unsuitable sites. It is a dilemma that we will come back to in Chap. 6.

1.1.2.2 Categorization of Structural Elements

For geological engineers employed in design and construction of repositories, a system for simple terming and labelling of the structural features, i.e. a categorization scheme, is needed and we will use the one in Table 1.1. Most of the discontiuities of 3rd and higher orders, i.e. finer weaknesses, can not be revealed until in the construction phase.

Naturally, this way of categorizing weaknesses, having its equivalents in the programs of organizations like ENRESA and NAGRA, can only be used as guidance to designers and modellers of the performance of planned underground repositories

Table 1.1 Categorization scheme for rock discontinuities [3,4]

Geometry				Characteristic properties		
Order	Length, m	Spacing, m	Width, m	Hydraulic conductivity	Gouge content	Shear strength
Low-order (conductivity and strength refer to the resp. discontinuity as a whole)						
1st	>E4	>E3	>E2	Very high to medium	High	Very low
2nd	E3-E4	E2-E3	E1-E2	High to medium	High to medium	Low
3rd	E2-E3	E1-E2	E0-E1	Medium	Medium to low	Medium to high
High-order (conductivity and strength refer to bulk rock with no discontinuities of lower order)						
4th	E1-E2	E0-E1	<E-2	Low to medium	Very low	Medium to high
5th	E0-E1	E-1 to E0	<E-3	Low	None	High
6th	E-1 to E0	E-2 to E-1	<E-4	Very low	None	Very high
7th	<E-1	<E-2	<E-5	None	None	Very high

E denotes the log scale exponent, i.e. E4=10 000, E1=10, E-2=0.01 etc.

because of the varying geometry and properties of the respective discontinuities. It must be kept in mind that a structural feature identified in nature is termed according to what can be observed and quantified and that it may well be a component of a structural component of lower order, i.e. a larger structural element. The primary value of the scheme is that it offers a possibility also for constructing companies and laymen to understand the general structural character of a rock mass.

SKB has used a version of this categorization scheme that defines what structural elements can be accepted in repository rock. Their scheme distinguished between features that can be described as:

- D1 Very large discontinuity, fault that must not intersect a repository.
- D2 Large discontinuity, fracture zone, that may intersect a ramp or shafts of a repository but that must not intersect or interact with the parts of a repository where waste is disposed.
- D3 Moderately large discontinuity, minor fracture zone, that may intersect parts of a repository where waste is disposed but not holes or parts of tunnels where waste containers (canisters) are located.
- D4 Discrete water-bearing and mechanically active or activable fracture that is accepted anywhere in a repository including holes or parts of tunnels where waste containers (canisters) are located.

Either of these schemes can be used depending on the circumstances and competence of the users. We will see later in the book that some of the discontinuity categories agree with those used in stochastic treatment of the hydraulic and mechanical performances of rock masses.

1.1.2.3 Evolution of Rock Structure

Referring to the more complete categorization scheme in Table 1.1 we shall briefly describe how the various discontinuities evolved in nature, paying special attention to the difference between crystalline and argillaceous rock on the one hand, and salt and clastic clay on the other. This is because the stiffness and brittleness of the firstmentioned implies breakage at small strain, while the other rocks are ductile and can undergo larger strain and still maintain coherence until breaks are generated.

We will make a first estimate of what types of discontinuities that are really important for the construction and long term performance of a geological repository. When these matters began to be considered by the Organizations the major issue was the one of groundwater flow in the repository rock since this would be the primary transport mechanism in bringing possibly released radionuclides from the repository to the biosphere. The huge knowledge and experience gained earlier in mining engineering and in practical hydrology formed the basis of the more or less systematic R&D, which was naturally focused on identification and characterization of what are termed 1st, 2nd and 3rd order discontinuities in this book. This matter is still a major issue and has a very strong impact on the selection of sites of HLW repositories. In fact, it has led to abandoning crystalline rock and selection of weakly metamorphosed clayey host rock by some of the Organizations because of the much more frequent water-bearing discontinuities in the firstmentioned rock type, of which many remain unknown. We will come back to this issue in Chap. 6.

Later, and in conjunction with assessing rock also from a construction point of view, and considering the importance of diffusive transport of chemical elements including radionuclides, the properties of discontinuities of higher orders were also found to be important so the whole spectrum of weaknesses is now being taken into consideration in working out conceptual models of the evolution of the near-field, including both rock and engineered barriers. Naturally, the detailed structural constitution of the rock in the near-field can never be revealed even by very careful and comprehensive exploration, and conceptual modelling of the near-field will therefore always be largely hypothetic for crystalline rock, somewhat less uncertain for argillaceous rock and clastic clay, but realistic for salt rock.

1.2 Crystalline and Argillaceous Rock

1.2.1 Evolution of Discontinuities

Magmatic rock, like granite and all its cousins – rhyolite, porphyry, syenite, trachyte, diorite, andesite, gabbro, basalt etc – and strongly metamorphosed rock, such as gneiss, greenschist, amphibolite and leptite, represent the crystalline rock types that are considered as host media for HLW repositories. Argillaceous rock, represented by moderately metamorphosed rock like clay shales and phyllites, has become favourite in recent years in countries that have plenty of them. The two rock categories have a number of common properties, like brittleness, and we will consider them as one group here keeping in mind that weakly metamorphosed clayey rock is essentially a strongly overconsolidated,[1] very fine-grained sediment with the layering and stratification clearly visible and causing anisotropy. We will call them all crystalline rock as a technical family name.

One can illustrate the evolution of discontinuities by considering an initial 7th order discontinuity, the physical equivalent being a micro-void at the numerous junctions of three or more crystals. Assuming that it is a little elliptic with the large diameter oriented more or less in the direction of a local major principal compressive stress, it will grow in this orientation to become a discontinuity of 6th order if the stress level is high enough. It will remain an isolated, stable weakness unless the major principal stress increases or the minor stress drops, which can be caused by rotation of the stress field. The little fissure then starts to grow, in the same direction or deviated depending on the orientation and magnitude of the principal stresses,

[1] In soil mechanical sense.

by overstressing the crystal matrix at its tip and thereby become a discontinuity of 5th order. Again, it will come to a rest if the stresses remain constant but grows again if they are increased, hence developing it to become a 4th order discontinuity, i.e. a fracture with a length of meters to tens of meters. This long it will intersect other fractures of the same type and become a member of a hydraulically active network.

The 4th order discontinuities are those called individual water-bearing fractures and they represent a very important group of discontinuities in many contexts. They create the conditions for extracting raw blocks for the stone industry as we see in Fig. 1.1, and they are the ones that determine the bulk hydraulic and gas conductivities of a rock mass located between zones of densely spaced fractures.

The 4th order discontinuities in natural rock can be assumed to have existed for many millions of years; in Scandinavia many of them were probably formed by the repeated rise of the Caledonian ridge in Cambrium and Tertiary times. They have been currently kept active by the neverending massage of the earth crust due to tidal effects and have been percolated by hydrothermal solutions by which feldspars, micas and heavy minerals exposed in the discontinuities have been converted to chlorite, epidote and clay minerals. Carbonates precipitated at various times depending on the chemical constitution and temperature of the percolating solutions.

Having been formed by stress changes in the rock, the 4th order discontinuities continued to react on changes in the stress field and since they are never perfectly

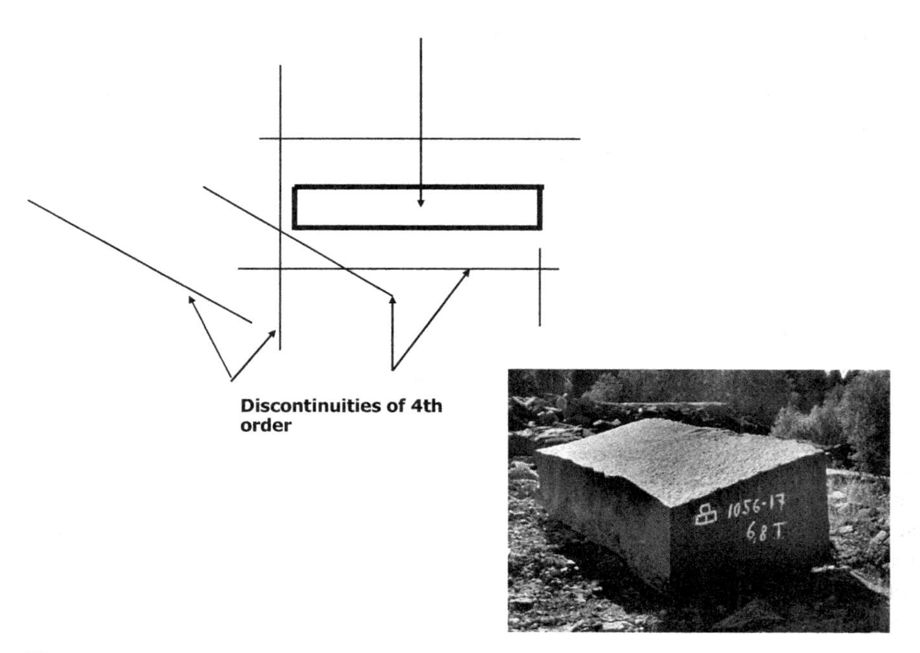

Fig. 1.1 Rock structure that determines the possibility to extract high-quality raw blocks for the stone industry. The 6.8 ton block is diabase rock

plane new breaks emerged from certain parts if the stress field changed by tectonic events, glaciation, deglaciation or temperature changes. This yielded the wellknown secondary ("splay") fractures that can make apparently intact raw blocks useless for the stone industry (Fig. 1.2).

Naturally, the evolution of weaknesses in the earth crust continues. Thus, presently identified 4th order discontinuities can grow and propagate even for small changes of the local stress field and those of 3rd, 2nd and 1st orders can do the same if the regional stress fields are changed. The larger the discontinuity the smaller the change in stress field required to cause activation. It is in fact true that the largest ones are constantly undergoing deformation.

The net effect of this is a system of large discontinuities of 1st to 3rd orders appearing in a matrix of more or less regularly spaced discontinuities of higher orders. In principle, the structural build-up major discontinuities in crystalline or argillaceous rock masses can hence be generalized as in Fig. 1.3. The examples show that there is a core that is usually very heterogeneous and often clay-bearing in larger zones of 1st and 2nd orders while secondary breaks representing local "protuberances" extend into the surroundings similar to the splay fractures grown from discrete 4th order discontinuities. They are of importance in deciding how far from the major discontinuities that one can locate holes and tunnels with HLW canisters; i.e. the "respect distance".

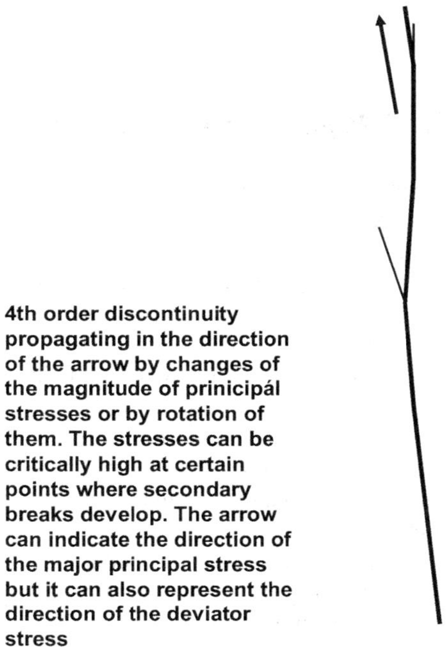

Fig. 1.2 Birth of secondary breaks at critically stressed parts of a propagating 4th order discontinuity

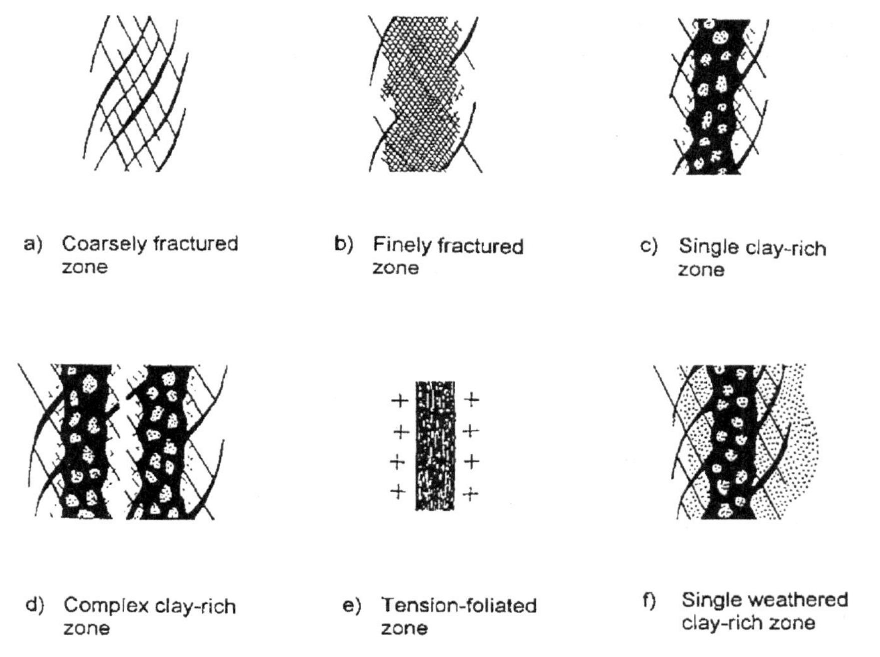

Fig. 1.3 Fracture zones representing discontinuities of 1st to 3rd orders (after E. Broch). Notice the "secondary" breaks developed from the complex core of the large discontinuities with both permeable and tight parts, the latter containing clay formed by weathering of feldspars and certain heavy minerals in conjunction with percolation of hot hydrothermal solutions

1.2.2 Regular or Random Structural Constitution

Do the low-order discontinuities form any regular geometrical patterns and are there any preferred orientations? The answer is that the rotation of the global major stress fields, associated with displacement of whole continents and relative movements within them, gave a complex and apparently arbitrary distribution of the major discontinuities. However, one can often see some general preferred orientations and it is usually possible to identify their average spacing by examining geologic and topographic maps as we will see in Chap. 6. Naturally, the information about both steep and flatlying features of this sort is limited since comprehensive geophysical investigations and boring programs are needed for deriving large-scale, reliable three-dimenisonal (3D) structural models. Water-bearing weaknesses can be found through systematic investigations while a lot of hydraulically dormant ones escape identification.

A generalized structural model that contains low-and high-order discontinuities is shown in Fig. 1.4. One recognizes a major discontinuity (1) that can represent the weaknesses in which big rivers have eroded valleys to large depths, a few 2nd- and

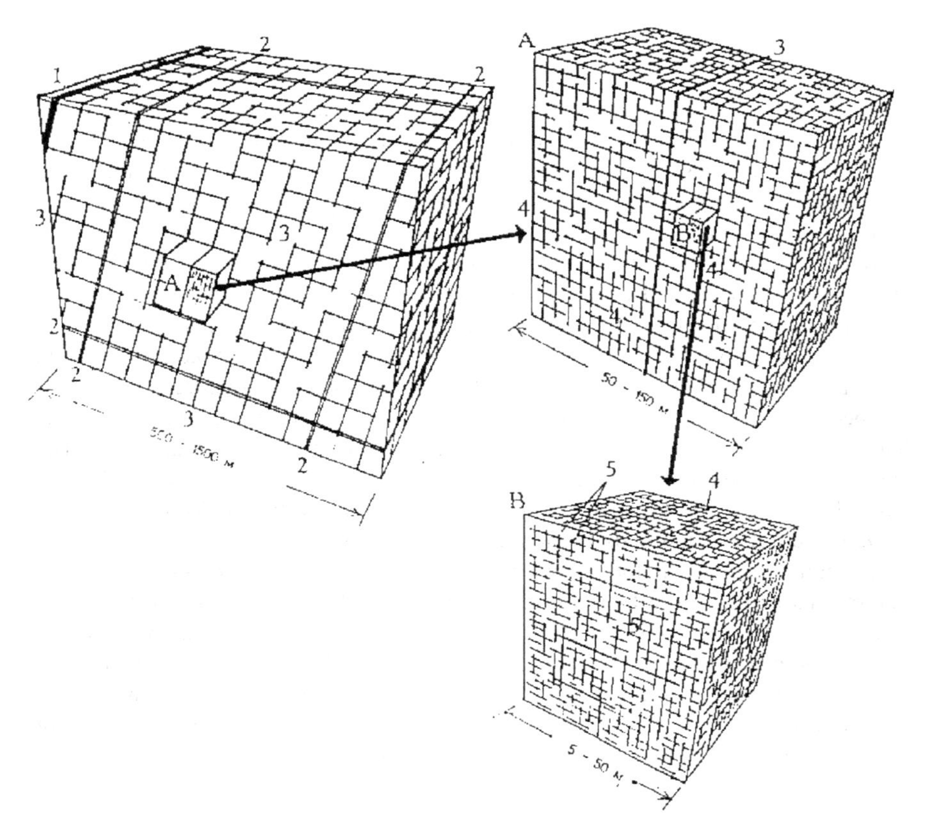

Fig. 1.4 Generalized rock structure model for crystalline rock. 1) Very large permeable fracture zone representing 1st order discontinuities, 2) Major fracture zones representing 2nd order discontinuities, 3) Minor fracture zones representing 3rd order discontinuities, 4) Persistent permeable fractures representing 4th order discontinuities, 5) Minor fractures representing 5th order discontinuities [3]. Sixth and 7th order discontinuities are represented in large to very large numbers between those of 5th order

3rd-order fracture zones, and a system of more or less interacting 4th order fractures. In the little block with 5 to 50 m edge length, 5th order discontinuities form a sub-system in the network of those of 4th order. Geometrically, the various discontinuities form fractal-like patterns but the different categories have quite different mechanical properties.

What happens with the discontinuities in the course of construction of a repository and are new ones formed? How do discontinuities affect the strength and stability, and the hydraulic and gas conductivities of virgin rock, and how will near-field repository rock with natural and neoformed discontinuities perform? We will look at these matters in the subsequent chapters after having considered the other rock types with which we are concerned.

1.3 Salt Rock

We commonly call the minerals that make up salt rock, halides, of which sodium chloride, halite, and potassium chloride, sylvite, are the most common ones. Large deposits of evaporates consisting of the latter mineral are known in Germany, with the Asse mine being a major formation and once proposed as host rock for storing highly radioactive waste, and in New Mexico, where a repository for low- and intermediate-level radioactive waste is licensed and in operation. Halite is more common than sylvite and occurs as extensive beds in Utah, Texas, New Mexico, Iraq and Iran and several other areas with similar geological histories.

The halides, which were formed by evaporation of salt water, make up stratiform deposits with beds that have a thickness from a few meters to hundreds of meters. The beds in New Mexico are typical in the sense that they were originally brine that solidified in conjunction with intermittent deposition of clay, silt and sand. The heterogeneity of the beds, which is illustrated by Fig. 1.5, can cause rock fall from clayey layers in the roof, and such layers can also serve as flow paths for brine.

Deformation of such beds by non-uniform loading of overlying sediments has occasionally resulted in local upward squeezing, leading to dome-like masses, diapirs, that can be more than a thousand meters high with their tops often being near-surface. The mechanism leading to this type of megastructure is the very significant time-dependent strain of salt rock (creep). Figure 1.6 shows a generalized section of the German Asse mine in a diapir of relatively high degree of homogeneity but which is locally interlayered by conformous, often steeply oriented layers of clay or brine, which is the term for very salt-rich fluid.

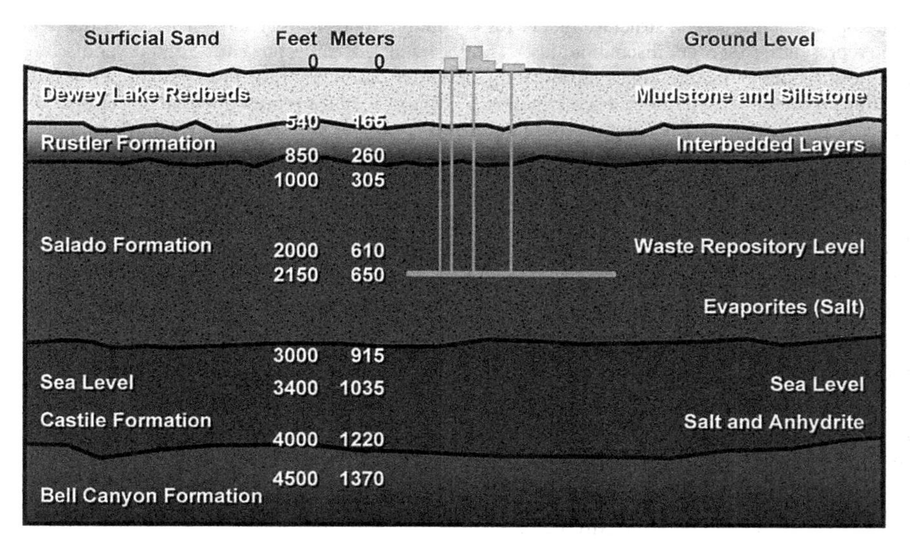

Fig. 1.5 The Salado formation with the WIPP waste repository at 650 m depth. The repository is covered by about 345 m salt and rests on very thick salt layers [1]

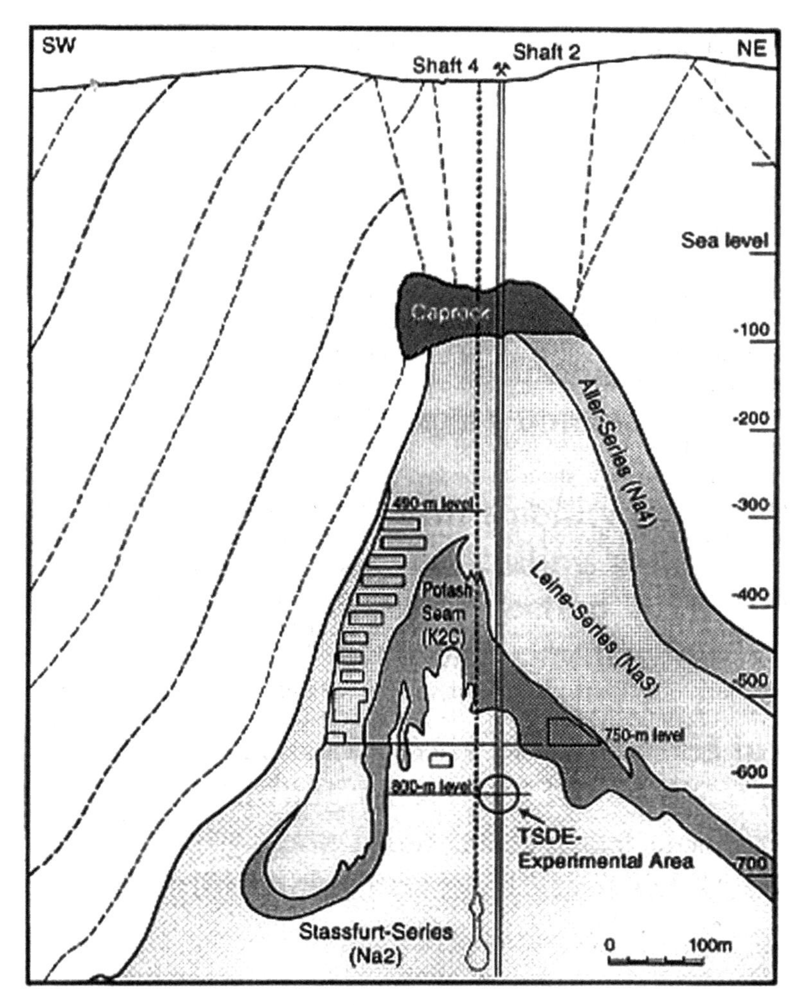

Fig. 1.6 Section through the Asse mine with drifts where salt has been mined and with an underground experimental center at about 800 m depth [1]

Salt has a significant self-healing ability that brings the porosity down to almost nil under even moderate effective stresses. It is due to the high intra- and intercrystalline slip potential and ductility of the mass of halide crystals. They combine to yield the homogeneity and lack of fractures in salt beds and diapers. In contrast to the ubiquitously fractured crystalline rock the homogeneous rock salt is therefore perfectly tight and lets no water or gas through except for the soil intercalations. The confinement of waste disposed in caverns, tunnels and holes is not only due to the impermeable nature of such rock, it is also caused by the convergence of such openings. They become smaller and smaller with time because of the very significant time-dependent creep strain.

Discrete slip planes can be formed if the stress conditions become critical but they do not persist, not even on the microstructural scale, because of the very significant self-healing potential of salt rock, provided that the failed part is effectively confined by artificial supports like concrete buttresses or backfill of salt.

1.4 Clastic Clay

While both crystalline and argillaceous rock are brittle and exhibit the whole spectrum of discontinuities, normally or weakly overconsolidated clay does not contain any equivalent features unless it has been exposed to desiccation. Stratification and lamination and all sorts of structural variations related to the sedimentation history can be found at any depth, including discrete slip planes and sets of slip planes, but no open fractures. This is of course favourable for isolation of HLW but the problem with clastic clays is that the low shear strength makes it difficult to construct a stable repository at more than 100 to 200 m depth. Fairly homogeneous clastic clay layers of postglacial clay covering glacial clay with a total thickness of 100 m are known in Sweden and Tertiary clay of similar thickness has long been exploited for use in the ceramic industry and for construction of bottom and top liners of landfills of hazardous waste. The quarry where such a clay, Friedland Ton, is exploited is shown in Fig. 1.7.

The discontinuities in clastic clay are of varying nature, they represent concordant and disconcordant individual strata or series of layers. The most important processes in the formation of clay sediments is consolidation, i.e. compression under the weight of the overburden, which leads to equilibrium with a certain porosity of the various layers, and brings about anisotropy. The fact that thick beds are formed under many thousands or even millions of years means that the source material varied from clay mineral particles of colloidal size to sand and gravel grains in the

Fig. 1.7 Mining of Tertiary clay (Friedland Ton) in Germany (Photo: Frieton AG)

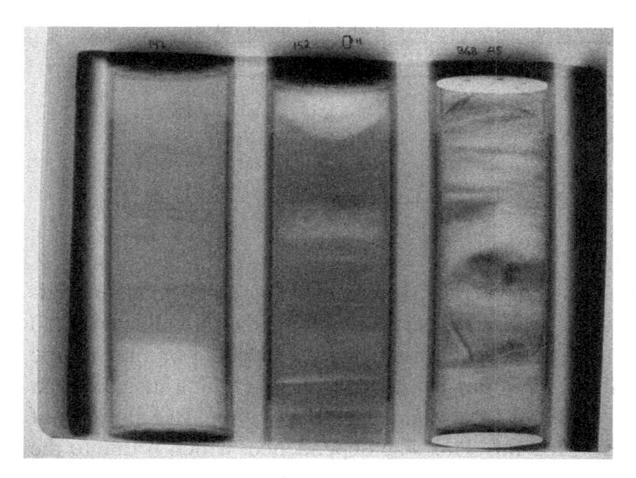

Fig. 1.8 X-ray images of sample of undisturbed clay sediment appearing to be homogenenous when examined by the unaided eye

course of sedimentation, which was often interrupted by erosion. The geotechnical properties hence vary from high strength and extreme tightness of certain layers to low strength and very significant permeability of neighbouring ones. The important thing from a practical point of view is how the complex system of layers behaves under the stress changes caused by the construction of rooms, tunnels and shafts and by the heat generated by the radioactive decay. Figure 1.8 illustrates common variations in composition and grain size distribution of apparently homogeneous clastic clay.

The difficulties with insufficient stability and anisotropic strain, and risk of quick transport of radionuclides from the HLW to the biosphere through interacting highly permeable discordant layers, makes this type of host medium moderately attractive and there is in fact only one example of a HLW disposal concept adaptated to normally consolidated clastic clay, i.e. the Belgian HADES project. We will return to this project in Chap. 4 of the book, where the various national repository concepts are described.

1.5 The Importance of Scale

1.5.1 Crystalline Rock

In general rock mechanics and in the Organizations' current work on repository siting in crystalline and argillaceous rock, as well as in repository design, the parameter rock strength is represented by the compressive strength of small specimens (core samples). It naturally reflects the average mechanical strength of the rock material but does not consider the impact of discontinuities, the frequency of which is

Table 1.2 Compressive strength of crystalline rock as a function of sample size

Volume, m^3	Unconfined compressive strength, MPa	Discontinuities
<0.001	20–100	7th
0.001–0.1	0.5–5	6th, 7th
0.1–10	0.5–2.5	5th, 6th, 7th
10–100	0.05–0.5	4th, 5th, 6th, 7th
100–10 000	0.005–0.05	3th, 4th, 5th, 6th, 7th

strongly volume-dependent. This neglicence contrasts to the carefully made investigation of the scale-dependence of the compressive strength by the Canadian (AECL) organization, which showed that this strength is strongly dependent on the volume of the compressed sample. Table 1.2 summarizes these results, which mean that an ordinary core sample with a few centimeter diameter has a compressive strength of 10–200 times that of a block with 3 m^3. This size dependence, which is due to the increased number of high-order discontinuities and appearance of bigger ones with increased rock volume, is an excellent proof of their existence. It is in fact a basic rule of the stone industry. We will return to this important matter in Chap. 2.

The practically most important discontinuities for the performance of the nearfield are those of 4th and 5th orders, which become frequent in rock volumes exceeding one or a few cubic meters. The shear strength of the firstmentioned is solely provided by friction, the friction angle ranging between 10 and 15° when smectitic clay is present as fracture minerals, and 20–35° when chlorite, epidote and carbonates dominate. For 5th order discontinuities the friction is higher because they commonly lack phyllosilicate coatings by not having been percolated by hydrothermal solutions. The friction angle has been estimated at 35–50° [4].

Enlarging the rock volume beyond a few tens to hundreds of cubic meters, we find the 3rd order discontinuities, commonly water-bearing, and they represent weaknesses that can make tunnel excavation somewhat difficult. They are not allowed to intersect a waste deposition hole or tunnels where HLW canisters are planned to be placed since they serve as megascopic hydraulic conductors and have a potential of undergoing tectonically induced shearing. Then, widening the perspective even more to consider rock volumes larger than thousands and tens to hundreds of thousands of cubic meters and beyond that, we recognize the supersized weaknesses that control regional groundwater flow and major strain in the earth crust, i.e. discontinuities of 2nd and 1st orders. This completes the spectrum of structural features that we need to keep in mind both in locating and planning repositories, and also in excavating tunnels and rooms in them.

It is appropriate here to mention that simple and adequate methods are at hand for identifying discontinuities of higher orders. Those of 4th–6th orders can be identified and examined by a lens where they appear on a free surface and their persistence from it can be estimated by small-scale geophysical methods, like the "impact echo" method that has been successfully applied in the stone industry. The technique is illustrated in Fig. 1.9 showing how testing of raw-blocks extracted from crystalline rock can be made. The technique can be used also in boreholes for determining the persistence of 4th and 5th order discontinuities.

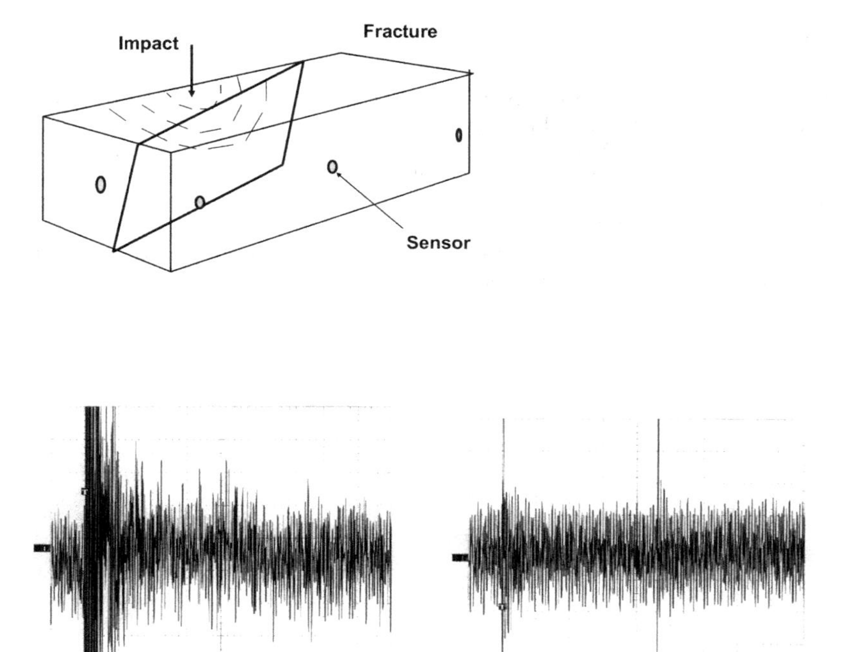

Fig. 1.9 Use of the "impact echo" method for determining the persistence of a discontinuity of 4th or 5th order. *Upper*: Placement of sensors on free surfaces of a big rock block. *Lower left*: Distorted wave pattern caused by the discontinuity. *Lower right*: Uniform wave pattern of rock containing only 6th and 7th order discontinuities

1.5.2 Salt and Argillaceous Rock, and Clastic Clay

- The mechanical strength of homogeneous salt rock is nearly independent of size.
- Argillaceous rock is similar to crystalline rock with respect to strength, except that its strongly anisotropic nature makes the compressive strength very sensitive to the orientation of the loading. The size-dependence is due both to heterogeneity caused by varying mineral and grain size compositions and varying cementing substances, and to variations in thickness and bonding of the platelets.
- In clastic clay the volume-dependence can be small if the sediment is very homogeneous, like for postglacial clay in Scandinavia, but very significant in stratified sediments containing silt lamellae and sand layers as in varved glacial clay in Scandinavia and Canada. The strength of silt and sand layers is solely due to friction and if the water pressure is high enough to reduce the effective (grain) pressure the strength vanishes, which can lead to collapse.

References

1. Svemar Ch, 2005. Cluster Repository Project (CROP). Final Report of European Commission Contract FIR1-CT-2000-20023, Brussels, Belgium.
2. Popov V, Pusch R, (Eds), 2006. Disposal of hazardous waste in underground mines. WIT Press, Southampton, Boston. ISBN: 1-85312-750-7. ISSN: 1476-9581.
3. Pusch R, 1993. Waste disposal in rock. Developments in Geotechnical Engineering, 76. Elsevier Publ. Co. ISBN: 0-444-89449-7.
4. Pusch R, 1994. Rock mechanics on a geological base. Developments in Geotechnical Engineering, 767. Elsevier Publ. Co. ISBN: 0-444-89613-9.

Chapter 2
The Rock

2.1 What is Required?

What is required from the host rock? A first and major requirement is that it must be possible to construct a repository in it, which means that the stability must be acceptable, a matter that is strongly dependent on the rock structure. A further structure-related criterion is that there must be enough space to host canisters, implying that the spacing of major weaknesses in the rock must not be too small. One more issue is that the rate of transport of water and radionuclides in the rock should be very low, which is again a rock structural matter.

One understands from the categorization of structural features that a repository will be affected in various ways by intersecting discontinuities. Where low-order discontinuities are crossed, drifts and tunnels become unstable in the construction phase and these parts will be richly percolated by flowing groundwater both in this phase and thereafter. They may also undergo tectonically induced shearing because of their low strength. The rock matrix between the major discontinuities contains 4th and higher order discontinuities and will not let much water through the rock and backfilled drifts and tunnels. However, creation of an open space, like tunnels and shafts and holes for canisters with HLW, changes the stress conditions and causes overstressing and fracturing of the rock and thereby alters its bulk behaviour, primarily the hydraulic conductivity. The natural discontinuities play a major role in this process and new generations of discontinuities can be formed from those present in virgin rock by critically high stresses.

2.2 Structure-Controlled Properties of Crystalline and Argillaceous Rock

2.2.1 General

A necessary prerequisite for describing and predicting the physical performance of the host rock is to identify relevant structural elements and quantify their hydraulic

R. Pusch, *Geological Storage of Highly Radioactive Waste*,
DOI: 10.1007/978-3-540-77333-7_3, © Springer-Verlag Berlin Heidelberg 2008

and mechanical properties. This is a very complex issue for large rock volumes realizing that the spectrum of discontinuities is very wide and that the relevance of derived results is and will always be questioned because of the difficulty to define the material properties and because only part of the spectrum can be identified and quantified.

A matter of fundamental importance is to predict how fast possibly released radionuclides from HLW can migrate through the engineered barriers and the rock and reach up to the biosphere. It can be quantified if the migration paths are identified and if their transport capacity and the driving force for transport can be determined. The rock structure in the near-field of the repository primarily controls where and how quickly radionuclides will move in the immediate vicinity of the waste while further transport in the far-field to the ground surface is determined by interacting major structural elements.

The near-field rock structure is not at all the same as in virgin rock because of the disturbance caused by the construction of the repository and by the thermal pulse that will be generated by the HLW and propagate through the engineered barriers and out into the surrounding rock. For quantifying radionuclide migration in the near-field one must therefore define a model structure and ascribe to it relevant transport properties like hydraulic conductivity and ion diffusivity. This has been made by various investigators but usually with little respect to the disturbance caused by the construction work and future tectonically generated stress changes.

A matter of equal importance is the mechanical stability of rooms, tunnels and holes. It is primarily determined by the strength of the rock material, represented by the unconfined compressive strength, but is also strongly affected by the rock structure. It is thereby of importance also to the hydraulic performance of the rock and we will therefore start by considering the strength of rock.

2.2.2 Rock Strength

Individual mineral crystals have a tremendous strength while their contacts, which contain numerous misfits between atoms in neighbouring crystal lattices, are weak. Additional weaknesses are the saltwater-filled microscopic voids at the crystal junctions. These voids, representing discontinuities of 7th order, determine the strength of small volumes. Under stable conditions they are isolated from each other (Fig. 2.1) and do therefore not contribute to a practically important extent to the bulk hydraulic and gas conductivities while they serve as diffusion paths for ions.

Discontinuities of 6th and 7th orders (Fig. 2.2) can be considered as embryonic weaknesses that can develop to lower order discontinuities under certain stress conditions. This is a first step in the evolution of the spectrum of structural features that determine the scale-dependent properties of rock. The matter is of fundamental importance for working out conceptual models of the performance of the host rock of repositories and we will return to it many times in the present book.

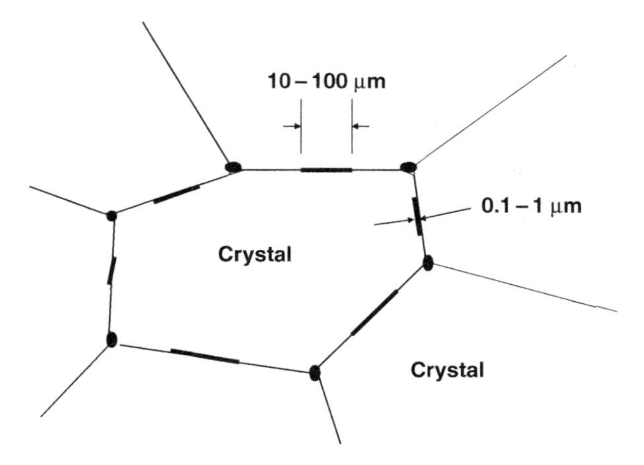

Fig. 2.1 Schematic picture of 7th order discontinuities: incomplete crystal contacts

2.2.3 Scale Dependence of Strength and Conductivity

The most important function of the microscopic and submicroscopic weaknesses is that they serve as nuclei for stress-generated formation of larger discontinuities that determine the mechanical and transport properties in bulk. The undisturbed crystal matrix of granite and similar rock types commonly has a porosity of 0.05–0.5% and is practically impervious to water and gas. Various tests have indicated that the hydraulic conductivity is on the order of E-14 to E-13 m/s. The porosity of core samples ranges between 0.1 and 1.0 % and make them low-permeable. For rock volumes larger than a few cubic decimetres the presence of discontinuities of lower orders than 6 and 7 determines the bulk conductivity.

The compressive and shear strengths of individual quartz and feldspar crystals are on the order of a few thousand MPa, while those of rock samples and larger volumes are much lower and drop with size. Table 2.1 gives typical strength data of core samples of crystalline rock and Table 2.2 indicates the scale-dependence of the mechanical strength and hydraulic conductivity using the Mohr/Coulomb failure criterion.[1] In metamorphosed clays, like clay shales and slates, precipitates like calcite or silicious compounds form bridges between the grains. They are often amorphous and represent the weakest parts of the particle network as demonstrated by numerous dispersion experiments in sedimentology. They show that boiling or ultrasonic treatment significantly reduces the average particle size by breaking the interparticle bonds.

Experimental strength data reflect the influence of all the discontinuities contained in the rock volume considered and since the failure mode is scale-dependent the strength is also depending on the rock volume. Thus, while the crystal matrix

[1] The discussion of whether one should use Hoek-Brown, Mohr/Coulomb or other failure criteria is a neverending story in rock mechanics. However, the difference between derived data is negligible in practice.

Fig. 2.2 Examples of discontinuities of 6th order. They are embryonic "dislocations" serving as effective diffusion paths and are responsible for the hydraulic and gas conductivities of the crystal matrix [1]

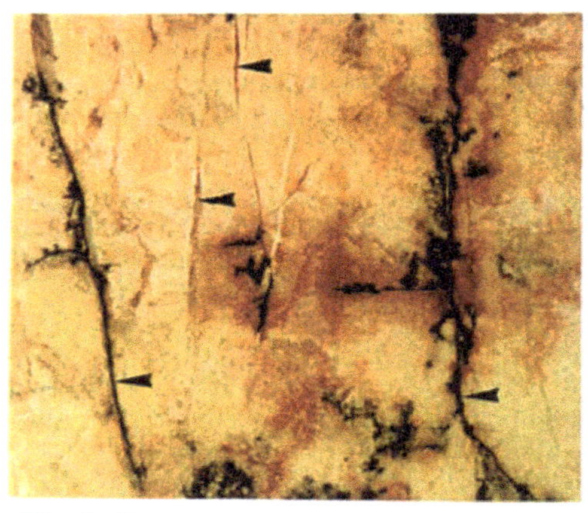

Table 2.1 Typical unconfined compression strength of core samples

Rock type	Compressive strength, MPa
Granite	150–300
Gneiss	150–300
Diabase	300–500
Quartzite	200–500
Leptite	100–300
Shale	120–200
Salt	25–35
Clay (2000–2100 kg/m^3)	0.1–2

breaks in a brittle fashion with the initial failure taking place in the form of cleavage in the load direction, i.e. in the direction of the major principal stress, larger samples break along discontinuities or by propagation of discontinuities. It is therefore not really relevant to refer to *cohesion* and *internal friction* of the rock material: for small volumes the strength is best expressed by the unconfined (uniaxial) compressive strength, while for larger volumes the fracture topography (asperities) and coatings (chlorite, micas, epidote) – and above all – the pressure normal to the fractures, determine the strength. Logically, the uniaxial compressive strength is also a function of the size of the rock sample [3], which has been validated by systematic loading tests. For granitic rock the following expression has been derived for the impact of the diameter d of cylindrical samples on the compressive strength expressed as σ_c:

$$\sigma_c = \sigma_{c50}(50/d)^{0.18} \tag{2.1}$$

where:

σ_{c50} = Uniaxial compressive strength of a core sample with 50 mm diameter and height.

This relationship implies that a 200 mm diameter sample has a strength that is only 80% of one with 50 mm diameter. Naturally, the strength of larger rock volumes drops further as vizualised in Fig. 2.3.

The figure shows the influence of the population of microscopic defects of similar type on the bulk strength. In practice, the scale dependence of strength also has another dimension, namely the appearance of discontinuities of different types when

Table 2.2 Bulk compressive strength and typical Mohr/Columb strength parameters (c and ϕ) and approximate hydraulic conductivity data of granitic rock [2]

Volume, m^3	Cohesion, (c), MPa	Angle of internal friction, °	Hydraulic conductivity, (K) m/s	Discontinuities
<0.001	10–50	45–60	<E-13	7th
0.001–0.1	1–10	40–50	E-13–E-12	6th, 7th
0.1–10	1–5	35–45	E-12–E-10	5th, 6th, 7th
10–100	0.1–1	25–35	E-11–E-9	4th, 5th, 6th, 7th
100–10 000	0.01–0.1	20–30	E-10–E-8	3rd, 4th, 5th, 6th, 7th
>10 000	<0.1	<20	E-9–E-7	All

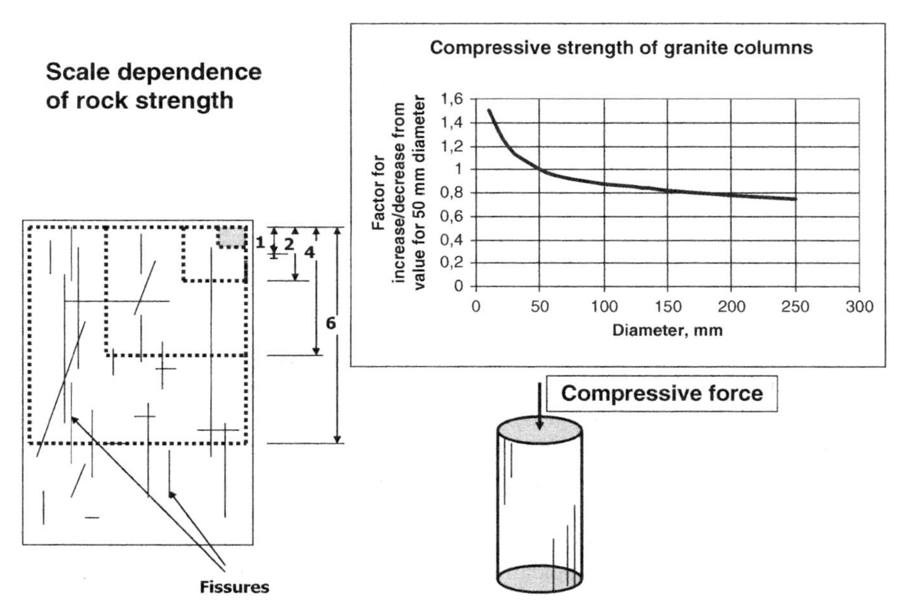

Fig. 2.3 Influence of size of sample with one type of defects on the compressive strength. The drop in strength at increased volume is explained by the increasing number of defects and the greater possibility of critical orientation and interaction of the defects

the size is increased. This is realized by considering the fact that large-diameter cores fall apart when the length exceeds a certain value, usually a few meters.

At excavation of rooms and holes for hosting canisters with HLW in crystalline and brittle argillaceous rock containing the whole spectrum of structural elements, a series of structurally related processes are initiated. The 7th order discontinuities become activated and form networks of several millimetre long fissures, which, in turn, can grow to become 6th, 5th and 4th order discontinuities in rock that is exposed to sufficiently high stresses. They are generated by mechanical agitation in the form of blasting and boring, and by stresses generated by creating the opening in question. The initial stage is exemplified by Fig. 2.4 showing breakage on the microstructural scale by stresses induced by a tunnel boring machine (TBM). It shows that disturbance in the form of microscopic breaks has taken place to about 10 mm depth from the free surface. The average porosity, i.e. the ratio of the volume of pores to the total volume, is at least 10 times higher within 3 mm distance from the surface than of undisturbed rock, and 2–5 times higher in the interval 3 to 10 mm from the surface. The hydraulic conductivity of small samples cored from the walls of TBM tunnels is two to three orders of magnitude higher than that of the unfractured crystal matrix within about 10 mm distance from the rock wall.

The small core in Fig. 2.4 broke along one of a series of inclined, more or less parallel boring-induced fissures. Most of them formed about 30–60° angle to the wall surface but some were oriented perpendicularly to this surface. All the

Fig. 2.4 Example of fracturing in the boring-disturbed wall of a 5 m diameter TBM tunnel in granite. The core has a diameter of 10 mm and a length of about 16 mm. The coring was made perpendicularly to the wall surface. Notice the axial fissures and the ones parallel to the oblique fracture surface [4]

boring-induced fissures had a spacing of a few tens of millimeters up to a few hundred millimeters according to detailed rock analyses. They reached into the crystal matrix from the wall by about 10 millimeters but some penetrated to a depth of about 50 millimeters.

A number of small rock specimens of this sort were tested in a triaxial cell for determining the hydraulic conductivity and the results are collected in Fig. 2.5, which shows that the average hydraulic conductivity within about 10 mm distance from the wall was about one order of magnitude higher than that of undisturbed rock material and that it was 2–4 orders of magnitude higher within 3 mm distance [4].

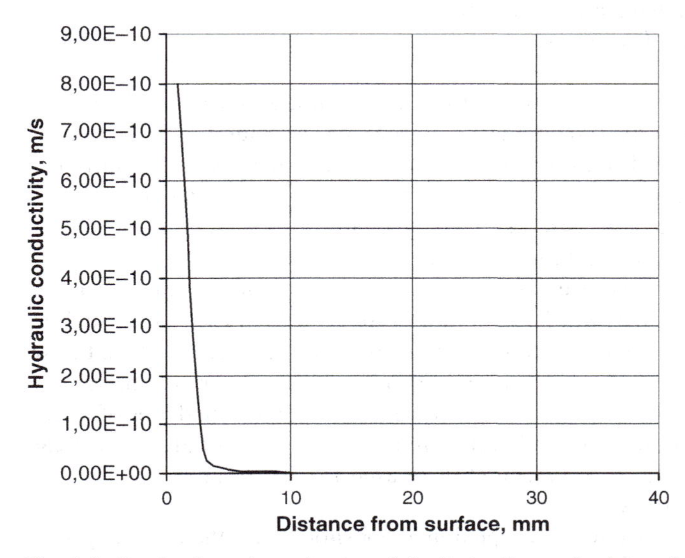

Fig. 2.5 Results from determination of the hydraulic conductivity of samples of shallow rock extracted by boring from the wall of a TBM-tunnel in Äspö granite. The conductivity was E-9 m/s for 1–2 mm depth, 2E-11 m/s at 5 mm depth, and E-13 m/s at 30 mm depth [4]

The practical significance of these findings is that the most shallow excavation-disturbed zone of rock around TBM-bored canister deposition holes and tunnels serves as a continuous thin conductor that connects all artificially created openings in a repository. Its importance for groundwater flow in the near-field is limited but water will migrate in it by diffusion ("Knudsen flow") and ions will also move readily by diffusion. A conservative estimate is that the crystal matrix around holes and tunnels bored by TBM and similar techniques will have a hydraulic conductivity that is 3 orders of magnitude higher than that of the virgin rock within 30 mm distance from the periphery.

While the natural microscopic weaknesses illustrated in Figs. 2.1 and 2.2, representing 7th and 6th order discontinuities, and those of 5th order discontinuities representing more or less isolated fissures and very fine fractures, do not contribute very much to the bulk hydraulic performance of rock, the next subgroup of discontinuities, i.e. those of 4th order by definition have a stronger influence on both transport and strength properties of the rock. Having maps worked out by direct observation in large boreholes extending from the floor of a tunnel they can be identified and their persistence and orientation estimated by examining also the floor. Working out even simple structural models requires that the nature and orientation of all important structural features, primarily discontinuities of 4th order, have been identified. This gives local structure models like those shown in the upper part of Fig. 2.6 but a fairly complete picture of the rock structure can not obtained until deposition holes or tunnels in the repository can be inspected. This can make it possible to identify water-bearing discontiuities like those in the two 9 m deep boreholes made in tonalite rock at the Finnish URL at Olkiluoto (Fig. 2.6, lower part). They have a diameter of about 1.7 m and were made by TBM-type boring technique. The mock, containing stones, was removed by applying vacuum.

Simplifying the rock to consist of a crystal matrix of high mechanical strength with intersecting 4th order discontinuities, theoretical structural models can be designed (Fig. 2.7), which makes it possible to predict changes in aperture and thereby in hydraulic conductivity of the rock caused by the changes in stress by creating the holes. In certain cases, like in stone quarries the system of high-order discontinuities appear in this fashion (Fig. 2.8).

Increasing the scale further we have to deal with discontinuities of 3rd order, i.e. minor fracture zones, which will be met with in the near-field rock of crystalline and argillaceous rock and which play a major role by forming more significant flow paths and being more apt to undergo shearing than the higher-order discontinuities. The stone industry refers to the "egg" concept meaning that the boundaries of larger blocks containing high-order discontinuities are formed by those of 3rd order. Near these boundaries the rock is too fractured to be used for extracting raw blocks for use in this industry (Fig. 2.9).

The 3rd order zones have to be accepted in a repository since their spacing is so small, i.e. 30–100 m, that intersection with them by tunnels can not be avoided. Canisters must not be positioned in them so for concepts like the KBS-3 V and horizontal deposition tunnels a number of planned canister positions must be deserted. This

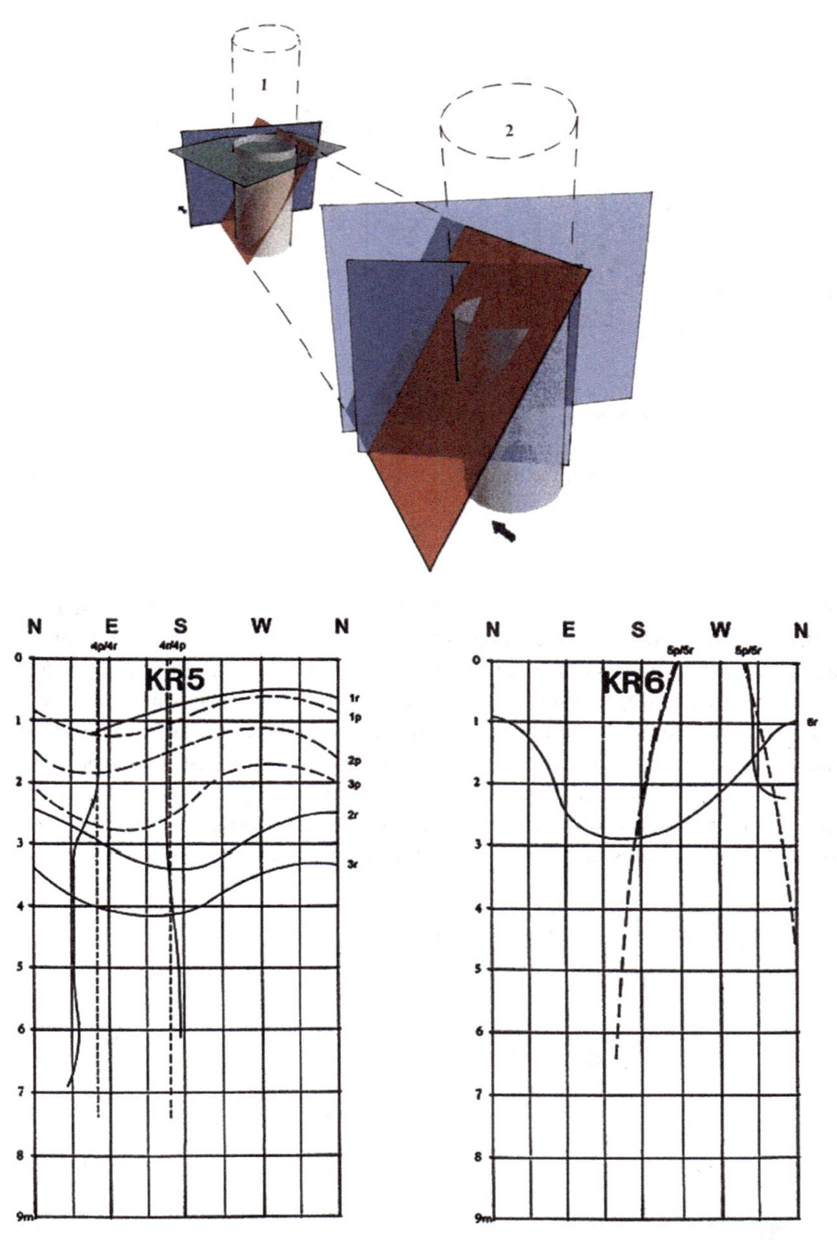

Fig. 2.6 Modelling of "near-field" rock structure. *Upper*: Local structure model derived from cored holes with water-bearing discontinuities having blue colour and tight ones marked red (Stripa BMT project). *Lower*: Mapping of full-sized deposition holes bored in POSIVA's URL at Olkiluoto. The full lines represent water-bearing 4th order fractures, some of which were generated by the blasting and hence make up part of the EDZ (Interpreted from fracture maps provided by POSIVA)

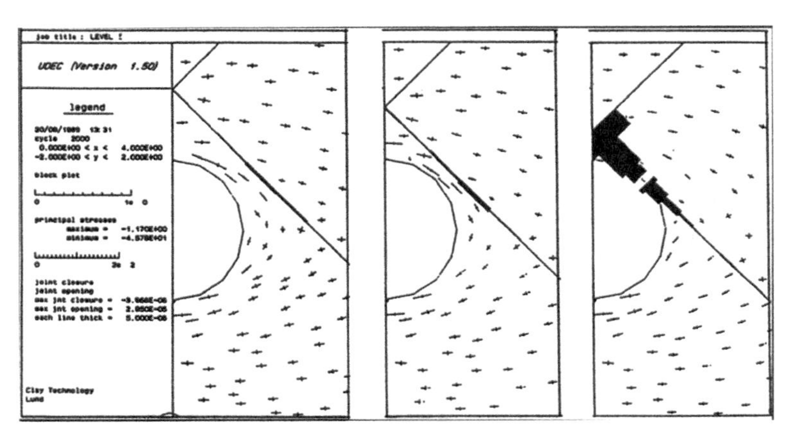

Fig. 2.7 Change in aperture of steep 4th order discontinuity (joint) located close to deposition hole. Left picture refers to the upper end of the hole while the right picture refers to the level where the joint appears in the hole [5]

criterion is not always easy to fulfil as illustrated in Fig. 2.10 showing the case of a zone of this type located a few meters below the floor of a parallel KBS-3 V tunnel. Also the cases with zones intersecting a deposition tunnel at a small angle ($<15°$) cause problems by limiting the number of acceptable canister positions. The cases with zones striking parallel to tunnels and being close to them are unsuitable since the small distance from the canisters to these, usually rather richly water-bearing zones, implies quick transport of possibly released radionuclides to the far-field. One realizes that preliminary structural investigations with respect to the presence of low-order discontinuities have to be made before decision can be taken on where

Fig. 2.8 Quarry in granite showing the typical orthogonal pattern of high-order discontinuities of granite: two sets of steep ones and one subhorizontal with water flowing in channels. The platform resulted from a subhorizontal thin 3rd order discontinuity

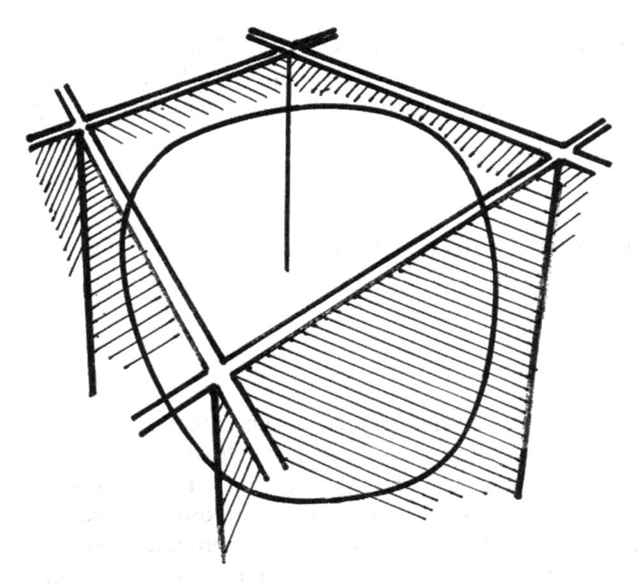

Fig. 2.9 The "egg" representing, for the stone industry, a useful part of a large rock block confined by 3rd order discontinuities in the form of minor fracture zones

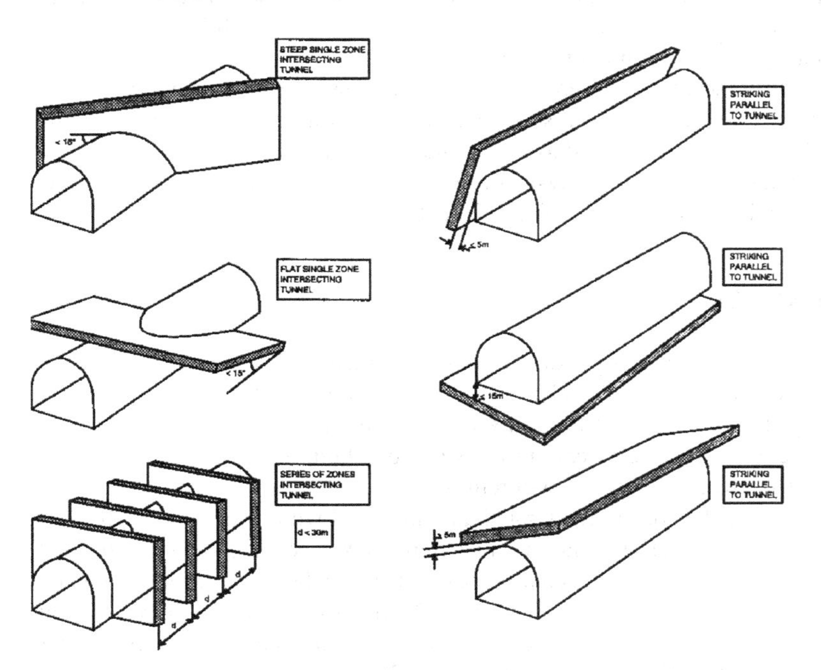

Fig. 2.10 Third order discontinuities affecting deposition tunnels. They are not allowed to intersect canister positions and great effort must be made to identify them in the rock for rational utilization of the rock [6]

and in what direction tunnels shall be positioned. However, a final decision in these respects can not be taken prior to the start of the construction work. It has to be postponed until the repository depth has been reached and subhorizontal pilot boring can begin.

2.3 Structure-Controlled Properties of Salt Rock and Clastic Clay

2.3.1 Salt Rock

Pure domal salt rock does not behave as crystalline rock because virgin rock of this type has no discontinuities of low order than 6 and 7. The mechanical strength of halide crystals is significantly lower than that of quartz, feldspars and heavy minerals, while the contacts between them is stronger in relation to those in crystalline rock. Breakage at a critical stress can therefore take place and proceed, resulting in development of microscopic breaks to become discontinuities of 6th, 5th and 4th orders. However, these self-heal under confined conditions because of the very significant creep potential of halides. This means that homogeneous salt is very tight although there may be local brine inclusions in which radionuclides can migrate. They are easily identified and done with [6].

In bedded salt the conditions are different. Here, flow of water and brine can take place in intercalated layers of soil sediments under prevailing or induced hydraulic gradients. Such layers correspond to discontinuities of 3rd to 6th order in crystalline and argillaceous rock but their physical properties are determined by the materials in the intercalations, i.e. soil and soil mixed with salt.

2.3.2 Clastic Clay

Most particles in clays with high contents of phyllosilicates like mica, chlorite, illite, and smectites, have water hulls and breakage at sufficiently high shear stresses takes place by slip at the contacts between adjacent particles. This yields slip planes along which a coherent mass of clay can move as in slope failure and as observed in the Belgian Hades URL, to which we will return.

No soil material has been investigated and modelled with respect to its physical properties as much as clastic clay. Long before rock mechanics was established as a technical discipline, soil mechanics had got its fundamental laws formulated and applied. Basic to them all are Terzaghi's effective stress concept, implying that the total pressure is the sum of the effective ("grain") pressure and the porewater pressure, and the Mohr/Coulomb failure concept that states that the shear strength is the sum of a cohesion term and the product of the effective stress and $\tan\phi$, where ϕ is the angle of internal friction. This angle has both a true mechanical and a dilatancy component. In contrast to brittle materials, like crystalline rock, the failure pattern

Fig. 2.11 Typical failure with oblique shear planes formed at uniaxial compression of dense clastic clay with a high content of clay particles $<2\mu m$

and the shear strength are the same on all scales. Thus, initiation of breakage in the form of slip along planes oriented at an angle of $(45-\phi/2)$ from the direction of the major principal stress, can be identified on the microstructural level [7], and macroscopic slip is oriented in the same way (Fig. 2.11). The development of conjugate shear planes forming the mutual angles $(45-\phi/2)$ and $(45+\phi/2)$ is typical, which is in contrast to crystalline rock (Fig. 2.12). Using soil mechanical theories and properly determined parameter values one can calculate the shear and normal stresses in clay masses in critical state for the construction phase of a repository and thereafter. In addition to analytical solutions of the equations there are a number of commercial codes for numerical treatment of problems like the stability of tunnels and holes making use of laboratory- or field-determined ϕ-values and cohesion c.

The problem with classical soil mechanical theory is that it does not include any true rheological components, except for the artificially incorporated diffusive-type pore pressure dissipation parameter, i.e. the consolidation coefficient. Thus, creep

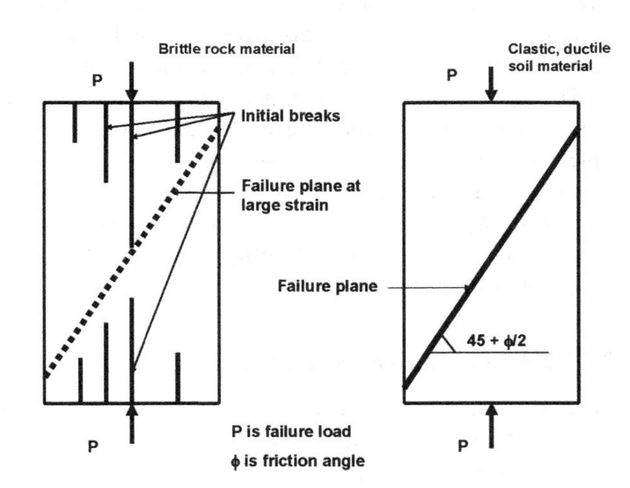

Fig. 2.12 Difference in development of failure in brittle and ductile materials at uniaxial strength testing, leading to different failure concepts [6, 7]

under constant volume conditions can not be properly modelled, which means that time-dependent strain can not be reliably predicted. We will come back to this issue when discussing the function of the Belgian repository in nearly normally consolidated Boom clay and the role of clay-based engineered barriers.

2.4 What Role do Discontinuities Play in Repository Rock?

2.4.1 Excavation-Induced Disturbance

Overstressing of several of the discontinuities in virgin rock and creation of new ones can take place when a tunnel or room is constructed and this leads to an excavation-disturbed zone (EDZ). An international study initiated by the European Commission and led by ANDRA formulated excavation disturbance in the following way:

1. In Canada the terms *Excavation Disturbed Zone* or *Excavation Damage (or Damaged) Zone* (both EDZ) are synonymous, and in the USA and Canada the term *Disturbed Rock Zone* (DRZ) has been applied.
2. In Sweden and Switzerland the *Excavation Damaged Zone* is distinguished from the *Excavation Disturbed Zone*. The *Damaged Zone* is limited to the part of the rock mass closest to the underground opening that has suffered irreversible deformation and where fracture propagation and/or the development of new fractures has occurred. This is distinguished from the *Disturbed Zone,* which is further out in the rock mass, and in which only reversible (recoverable) elastic deformation has occurred.

In this book the term EDZ will be used. As we have seen in the preceding text, stress changes can cause alterations in the performance of rock that may well be termed excavation-induced damage. It is theoretically impossible to define the outer limit of the zone because the disturbance caused by stress redistribution reaches very far from the created opening and can develop with time.

The performance of EDZ may be strongly transient. Thus, the typical construction methods in hard rocks cause instant damage to the rock from several centimeters up to several meters distance from the wall of the opening followed by creep-controlled redistribution of stresses and creation of new damage in the vicinity of the opening for months and years. Transport of fine debris in fractures opened by stress changes or created by blasting may cause clogging and successive reduction in hydraulic conductivity.

2.4.1.1 Disturbance by Stress Changes

Increasing the stress level sufficiently much in rock, 7th order discontinuities start the evolution of breakage of the virgin crystal matrix leading to macroscopic breaks. Natural discontinuities of 5th and 6th orders in the rock react earlier for the same stress rise since they are weaker, hence manifesting the scale-dependence of strength. They control the development of failure in bulk rock that lacks discontinuities of lower order. In a rock volume that is large enough to contain 4th order discontinuities these are weaker and react even earlier and they should logically be the ones that control the development of bulk failure. However, the first rock elements to react on stress changes are determined by the orientation of the principal stresses and possible restraint to movement of released blocks or fragments. The problem of predicting fracture growth – crack growth using the terminology of fracture mechanics – has been treated by numerous investigators using finite element modelling techniques (FEM) but we will use here boundary element technique (BEM) that has the advantage of not requiring remeshing of the interior domain. Another advantage is that much less computer capacity is needed (Fig. 2.13). BEM technology has been applied to provide a simple way to determine the stress state in 3D rock structure using the software BEASY for the computation [9].

Growth of 5th and 6th order discontinuities can have a very significant impact on the tightness and stability of a special part of repository rock, namely at the intersection of a TBM-drilled tunnel and a deposition hole for a HLW canister, Fig. 2.14. The primary rock stresses are 30 MPa in X-direction, 15 MPa in Y-direction and 10 MPa in Z-direction. The calculation was based on the E-modulus 105 MPa and Poisson's ratio 0.3. The calculated principal stresses at the intersection are shown in Fig. 2.15.

The graph in Fig. 2.15 shows that the highest hoop stress is about 206 MPa at the intersection, which is on the same order as the uniaxial compressive strength of good crystalline rock (Table 2.1). This would mean that there can be a risk of failure in the form of spalling and fracturing at all or most of the deposition holes in any type of rock if the primary rock stresses are high enough.

The impact of 4th, 5th and 6th order discontinuities on the stability and groundwater flow in the "near-field" is illustrated by the two cases in Figs. 2.16 and 2.17,

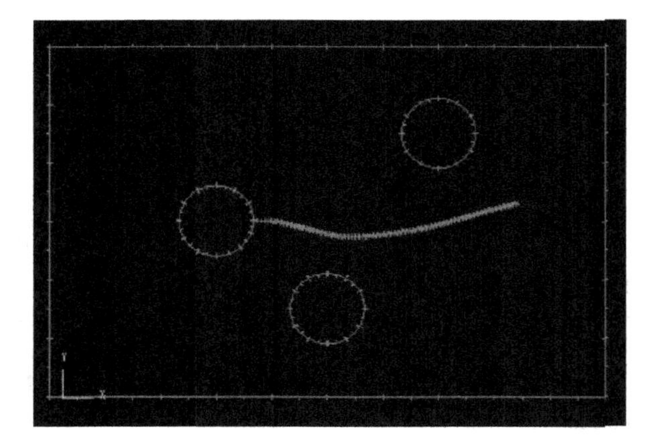

Fig. 2.13 Example of 2D boundary element model of rock with holes representing 7th order discontinuities. The fracture growth can be modelled without internal mesh generation. Note that only the boundary is defined [8]

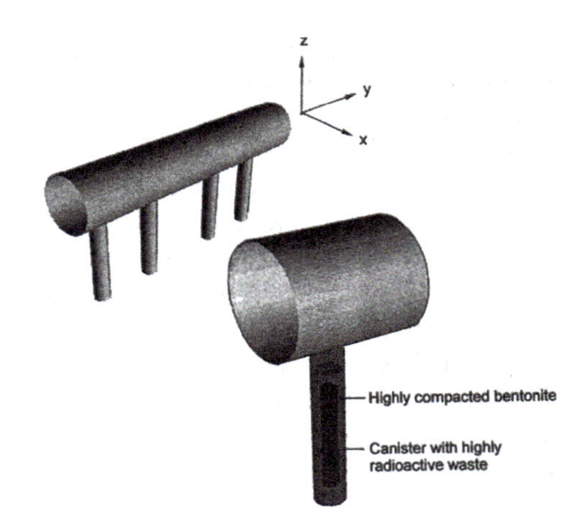

Fig. 2.14 Schematic picture of a bored repository tunnel with 5 m diameter and about 8 m deep canister deposition hole with 1.7 m diameter extending from it. *Upper*: Perspective with coordinate axes. *Lower*: Proposed design with clay-embedded canister hole

both referring to canister deposition holes like the one in Fig. 2.14 at a depth of 400–500 m. We will return to this matter several times in the book.

The stresses caused by constructing rooms, tunnels and deposition holes, and the rock structure determine where they should be located and how their orientation should be for avoiding failure. Naturally, comprehensive breakage or even a single spalling event when boring of deposition holes starts would be taken as an unacceptable risk to the workmen and an indication of improper design of the repository. The psychological effect of such events would certainly be most negative, especially since it would have to be revealed that the thermal pulse related to the radioactive decay and appearing after backfilling and closing of the tunnels would further increase the stresses and cause even more comprehensive breakage.

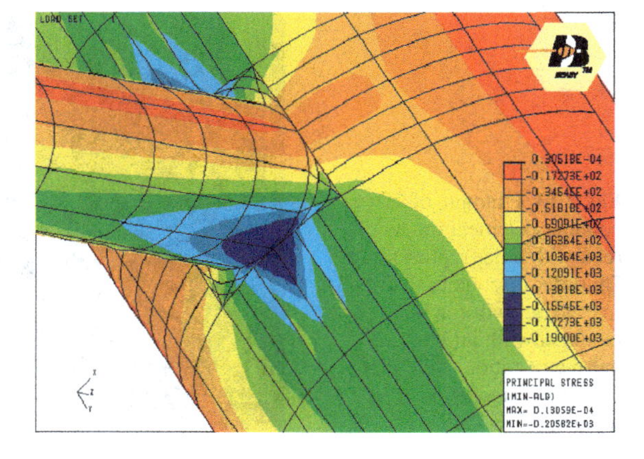

Fig. 2.15 Close-up view of the major principal stress plotted on the surface of the tunnel near the intersection. The hole is rotated from the position in Fig. 2.14 [10]

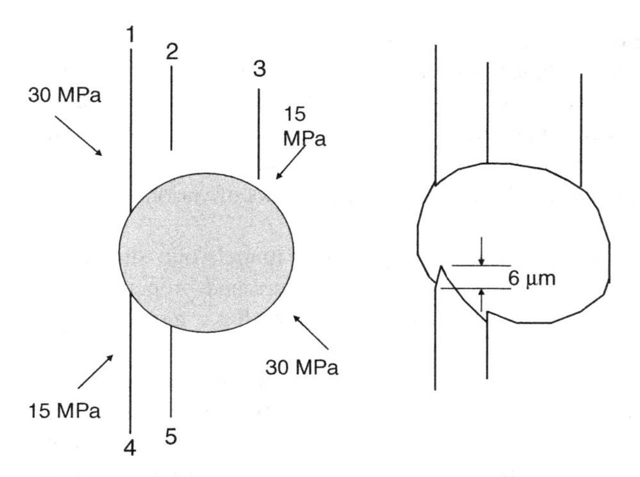

Fig. 2.16 BEASY-calculated shearing and propagation of 5th and 6th order discontinuities (1 to 5) in the near-field of a canister deposition hole with 1 m diameter. They become interconnected and hydraulically active [10]

2.4.1.2 Disturbance by Construction – Blasting

Blasting, being an alternative to TBM-boring for constructing tunnels and drifts, has strong impact on the rock. It causes stress relaxation at the strongly fractured periphery of the rooms and therefore less risk of spalling, and it increases the hydraulic conductivity very significantly to a depth of as much as one meter. Blasting was introduced in underground excavation more than 100 years ago and comprehensive experience in this field has been accumulated in most countries, which guarantees

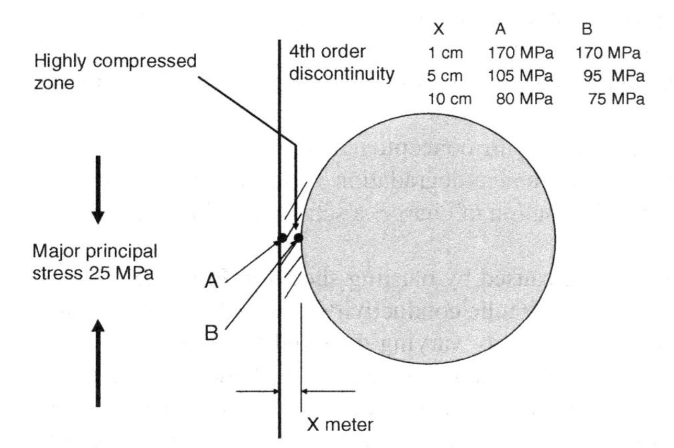

Fig. 2.17 Compression of the thin rock slab formed between a long discontinuity of 4th or 5th order and the periphery of a 1 m canister deposition hole. For distances (x) smaller than a few centimetres breakage can take place depending on the compressive strength of the rock [10]

that even complex excavation can be made with minimal impact on the remaining rock. Careful blasting can, for instance, give a depth of the EDZ of a few decimetres except in the floor where it extends to about one meter [5]. The importance of the rock structure is obvious in this context: the reflection of pressure waves and dynamic shear waves causes separation and relative movement of rock blocks along existing open fractures and latent weaknesses, by which blocks of various size may become unstable.

The fragmentation of the rock in constructing rooms, tunnels and shafts will normally cause more extensive changes of the rock structure and properties than the stress changes induced by creating the openings. Blasting has a much stronger impact on a rock mass than the abrasive effect of TBM boring because not only 7th order discontinuities but also those of 6th and 5th orders become activated and grow. The matter has been in focus for a long time among the Organizations and several field investigations have been performed for assessing the importance of the excavation-disturbed zone (EDZ), some investigators claiming that the net effect on the axial hydraulic conductivity along a blasted tunnel is negligible and others stating that it is an important water and gas conductor. We will verify here that the latter opinion is correct.

There are two ways of quantifying the hydraulic conductivity of the EDZ: (1) By evaluating the hydraulic conductivity from packer tests in a large number of holes bored into it, normally to the periphery of the room in question and mathematically integrating the results to determine the effective permeability tensor of the rock mass, and (2) By performing flow tests on a scale sufficiently large to average the individual effects of many fractures. This large-scale "macro approach" attempts to simulate the testing conditions of a porous medium, where each field measurement represents the average flow in many individual conduits. This latter approach was followed in the two tests described here. They were performed in SKB's earliest underground laboratory at Stripa, some 200 km NW of Stockholm, Sweden. The drift had been excavated by boring blast-holes with 3 m length and 0.5 m spacing, the charge being 0.5 kg Gurite per meter of the contour holes and 0.5 kg Dynamex as bottom charge. This technique can be classified as "careful production-type blasting".

The aim was to get a basis for developing conceptual models of the EDZ, an early finding being that the degree of mechanical degradation varies along the length of a blasted tunnel because of the distribution of charge, a schematic illustration being that in Fig. 2.18.

The varying degree of fracturing caused by blasting shown in Fig. 2.18 implies that spot-wise determination of the hydraulic conductivity in short boreholes drilled normal to the rock walls will give strongly varying data and that it is impossible to get definite information on the water transport capacity of the most shallow, disturbed rock over longer axial distances by performing such tests. It can only be obtained by measuring the flow and flow-driving hydraulic gradient over a sufficient length of a blasted tunnel.

The first test, called the "Macropermeability Experiment", was designed to evaluate alternative methods of characterizing rock mass hydraulics and conceived as

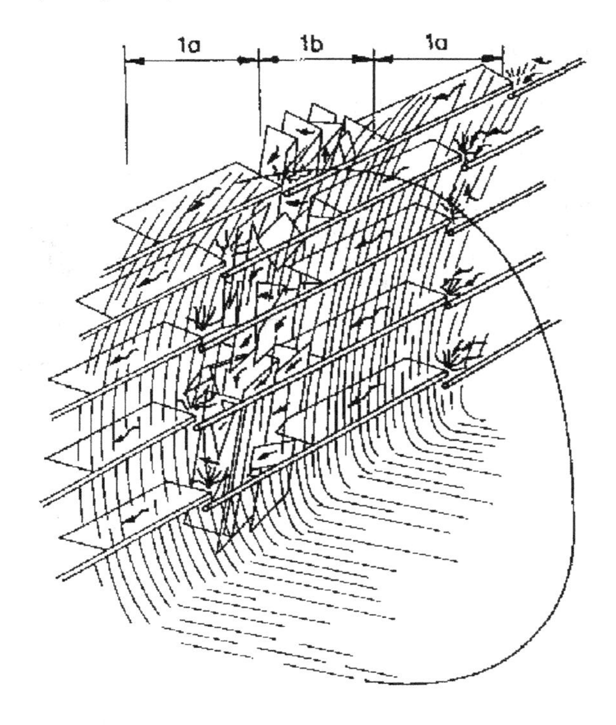

Fig. 2.18 Major types of damage by tunnel blasting. 1a-zones are characterized by regular sets of plane fractures extending radially from the central parts of the blast-holes. 1b-zones represent strongly fractured parts at the tips of the blast-holes [5]

part of the Lawrence Berkeley Laboratory fracture research program at the Stripa Mine [11]. It was a ventilation test in a 33 m long drift at about 360 m depth for measuring the inflow of water into a drift with simultaneous recording of the piezometric pressures in the rock. This test gave an average hydraulic conductivity of about E-10 m/s of the undisturbed rock mass and showed blast-disturbance within 0.5–1 m from the drift wall yielding an EDZ with an isotropic hydraulic conductivity of 5E-8 m/s. Stress-induced reduction of the radial conductivity to 4E-11 m/s, i.e. an obvious "skin" effect, was found for the surrounding 3 m thick annulus.

The second test was performed in the same drift in the Stripa Project [5,12,13] and had the form of pressuring a row of radially bored hole at the inner end of a 11 m long part of the 33 m drift and measuring the flow to a corresponding set of holes at the outer end (Figs. 2.19 and 2.20).

The drift was lined with epoxy and rubber liners and filled with bentonite slurry pressured by a 100 m^3 rubber bladder for preventing water to flow into the drift. The water pressure was measured at different distances along the drift for evaluation of the hydraulic gradient and conductivity of the rock at different distances from the drift. This was made by equipping the boreholes for injection and reception of water with packers that could be activated or closed in different positions.

Like the "Macropermeability Experiment" the subsequent test, informally called the "Macroflow test", was so arranged that it would be possible to evaluate the axial and radial hydraulic conductivity of the rock around the drift, which had a height and width of about 5 m.

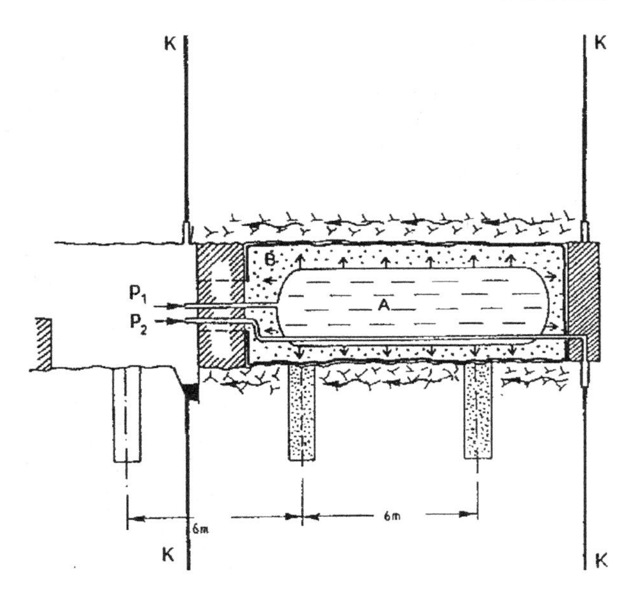

Fig. 2.19 Test set-up for determination of the arrangement for evaluating the hydraulic conductivity of the EDZ of the BMT test drift [5, 12]. A is the water-filled bladder and B the bentonite slurry that prevented water in the rock to flow into the drift. K are the curtains of radially bored holes

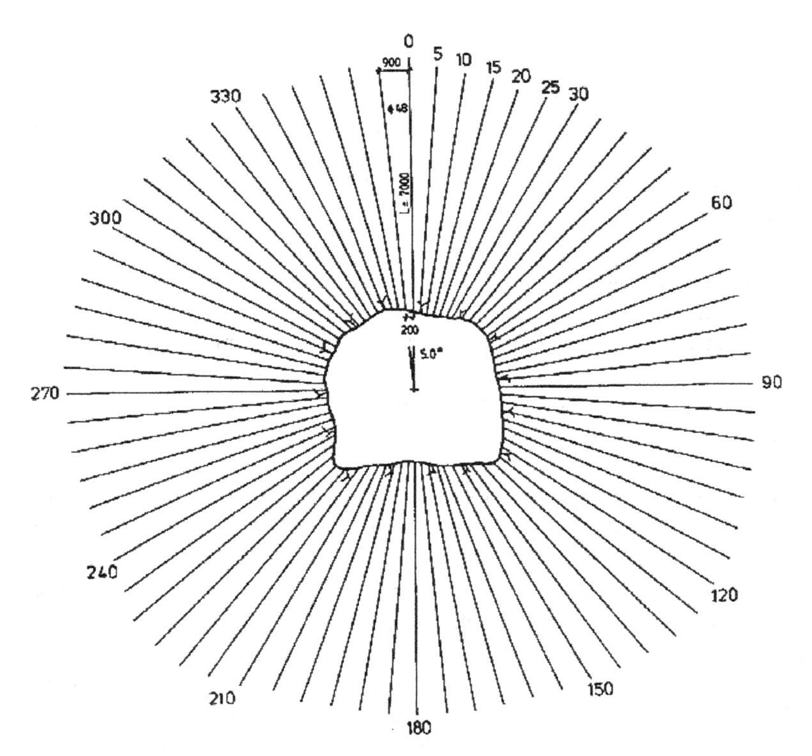

Fig. 2.20 Geometry of the borehole curtains, the diameter of the boreholes being 48 mm in the inner curtain and 35 mm in the outer [4,12]

The number of the radially oriented 7 m long″ percussion-drilled holes of each of the curtains was 76. They overlapped to about 0.75 m distance from the periphery thus forming a slot of this depth that was separated from the holes and the drift by packers. At their outer ends the boreholes had a spacing of 0.9 m. The holes extending from the slot were equipped with packers at 3 m depth and mutually connected in 4 separated sets to represent the roof, the two walls, and the floor so that in- and out-flow of the nearest 0.75 m rock annulus, the surrounding 0.75–3 m annulus, and the outer 3–7 m annulus, could be measured sector-wise. Each sector at the inner end of the drift was equally and simultaneously pressurized while measuring the inflow sector-wise into the outer slot with respect to the amount and distribution of discharged water. The bladder was pressurized by water to a level corresponding to the highest rock water pressure along the tested part.

The experiment started with inflow tests that gave a total of 62 l/day from the inner curtain and 29 l/day from the outer one. A significant number of packed-off curtain holes were used for determining the hydraulic conductivity of the rock, which was found to range between 1.7E-10 and 7E-9 m/s. Continued testing was made by stepwise increase of the pressure up to 950 kPa, keeping the bladder pressure at about the same level as the injection pressure throughout the respective tests. Stationary flow was obtained after slightly less than 2 days.

Evaluation of the test required development of a flow model with boundaries sufficiently wide away to take the investigated rock mass with its differently permeable parts as porous media. A finite element flow model was worked out and applied for evaluating the hydraulic conductivity using the continuously recorded pressure and flow data. The prerequisite for this evaluation was a linear drop in pressure from the inner to the outer galleries, which was verified by the piezometer readings at steady state. The FEM model adopted is shown in Fig. 2.21. A basic

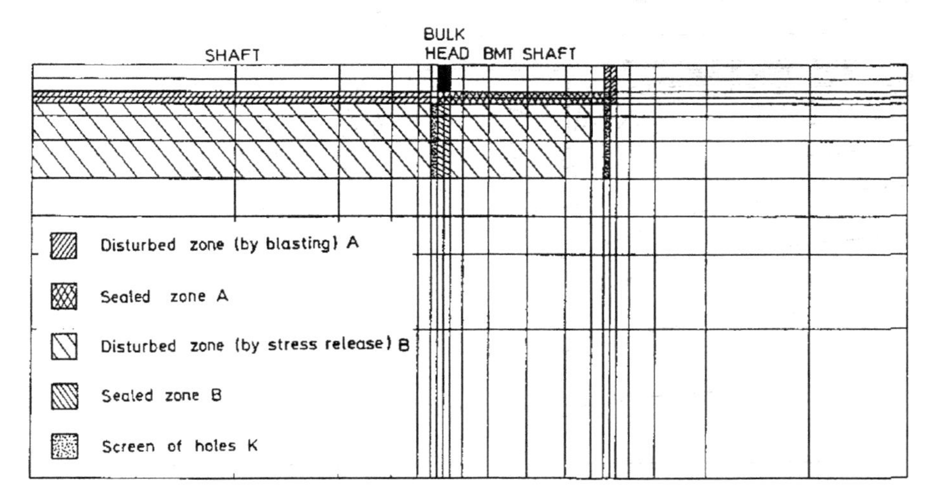

Fig. 2.21 Axisymmetric element mesh around the central axis of the drift. The solid black part is the bulkhead that separated the 11 m long test section from the rest of the drift [5,12]

Table 2.3 Evaluated hydraulic conductivity of rock zones; K_r=conductivity in radial direction, K_a=conductivity in axial direction

Rock zone	K_a, m/s	K_r, m/s	Inner radius, m	Outer radius, m
Undisturbed	3E-11 to 9E-11	3E-11 to 9E-11	5.4	40
Stress-disturbed	3E-10 to 9E-10	7.5E-12 to 2.3E-11	3.1	5.4
Blast-disturbed	1.2E-8	1.2E-8	2.4	3.1

assumption derived from the Macropermeablity Experiment and interpretation of the rock structure through a large number of short radial boreholes was that there are two EDZs, a blasting-induced zone extending to 0.7 m depth from the tunnel wall, and a stress-induced zone extending from 0.7 to 3 m depth.

The evaluated hydraulic conductivity of the various rock zones is given in Table 2.3. One concludes that the average hydraulic conductivity of the undisturbed rock is close to what the "Macropermeability Experiment" gave, i.e. on the order of E-10 m/s. The intervals given in the table represent two models that gave slightly different results, one of them implying a shift in hydraulic conductivity of the virgin rock from the inner to the outer half.

Further conclusions were (1) that the difference in radial and axial conductivity of the stress-disturbed zone and of the virgin rock clearly demonstrates a "skin" effect, and that the blast-disturbed zone that extends to 0.7 m depth from the drift wall is hydraulically isotropic with an average hydraulic conductivity that is more than 100 times higher than that of the virgin rock.

The calculated pressure distribution after drilling the curtains and with no pressure applied is shown in Fig. 2.22. It agreed acceptably well with the piezometer recordings: for the mid section of the drift the measured pressures at 0, 3 and 6 m

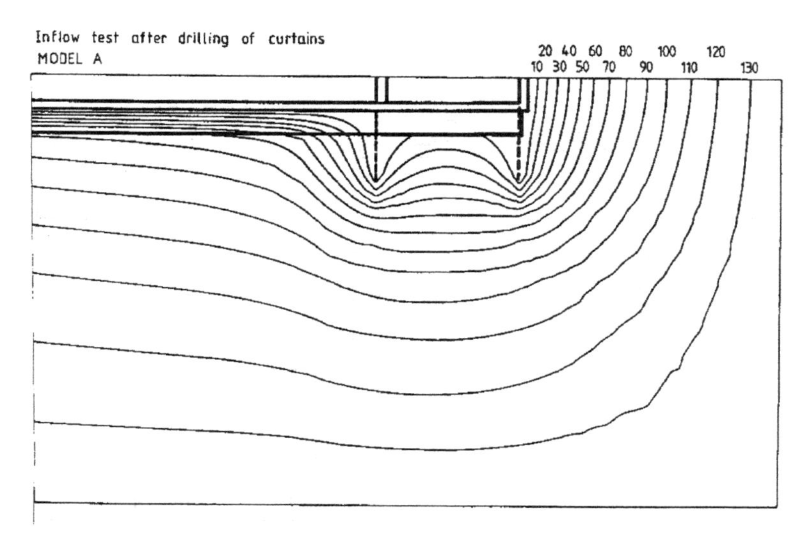

Fig. 2.22 Calculated pressure distribution after drilling of the curtains (pressures in meter water head, 100 m corresponds to 1 MPa), [12]

distance from the drift wall were 0, 30 kPa and 180 kPa, respectively, while the calculated pressures were 0 for the smaller distances and 100 kPa for the 6 m distance. In summary:

- It is obvious from both large-scale tests that there are very significant changes in the hydraulic properties of the rock around a tunnel excavated by use of relatively careful blasting technique. The "Macropermeability Experiment" showed a "skin" effect but also the draining effect of the rock close to the drift wall, i.e. the EDZ.
- The "Macroflow Test" quantified the changes in hydraulic conductivity. Thus, the tightening effect with respect to radial inflow at a few meters distance from the drift wall that was found in the preceding experiment was confirmed and found to represent a drop in radial conductivity by 50–90%. The drop in pressure in the rock closer to the drift wall than 1–3 m was interpreted to be caused by an increased hydraulic conductivity of the rock radially as well as axially within 0.7 m from the drift wall by more than 100 times.
- The "Macroflow Test" also showed that the axial conductivity within the surrounding stress-disturbed zone extending from 0.7 to 3 m depth increased by about 10 times. The straight, weak lines in Fig. 2.23 represent pressures in sets of long radial holes (R) in the "Macropermeability Experiment", verifying a skin effect, i.e. stress-induced radial tightening of the rock at about 1–3 m distance from the drift wall. The thick broken curve represents the calculated radial pressure drop using the FEM hydraulic model of the "Macroflow" concept for two of the sets that intersect the drift. They are represented by the symbols x and rings. One finds that there is very good agreement between the recorded and calculated pressures, which is a strong support of the validity of the hydraulic model and therefore also of the rock structure model.

In a separate investigation using packer tests in 80 boreholes with 0.9 m depth and oriented normally to the walls and floor of the drift the evaluated conductivity varied considerably and gave an average of E-9 m/s, i.e. somewhat lower than the values derived from the full-scale test. The outer packers in the boreholes had to be located at 15 cm depth and the packed-off part of the holes did therefore not include the most shallow, fractured and permeable part of the EDZ [12,14].

No other field study of excavation-induced disturbance of similar comprehension has been reported but several other tests on a smaller-scale have been performed and described like those conducted in SKB's present underground laboratory, the Äspö HRL,[2] and in crystalline rock on German test sites. The firstmentioned investigations were made in the so-called Zedex drift at about 300 m depth in granitic rock. Figure 2.24 is a schematic illustration of the field test that was made with the primary purpose to investigate the conductivity and rate of saturation of tunnel backfills. The innermost part of the drift was filled with very permeable crushed

[2]Hard Rock Laboratory.

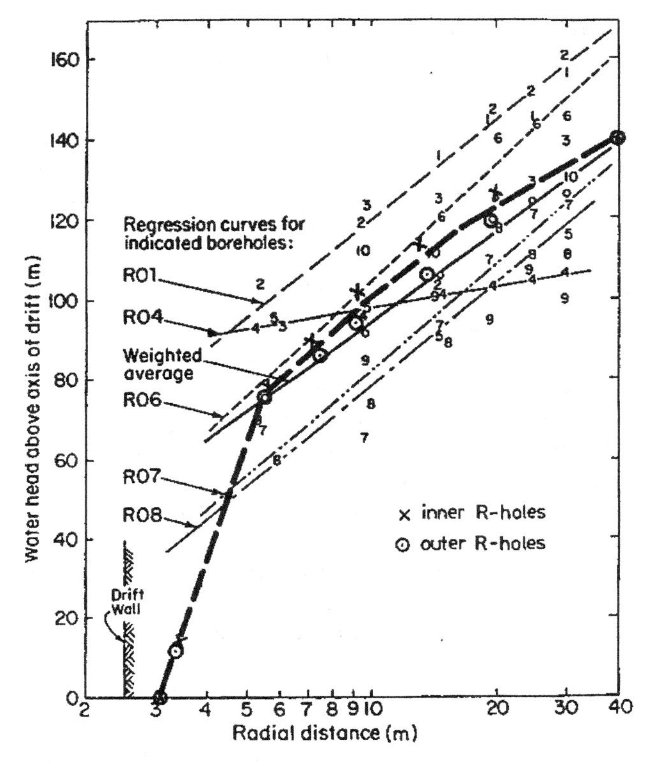

Fig. 2.23 Calculated pressure drop using the hydraulic model of the "Macroflow" concept for two of LBL's sets of holes that intersect the drift and are represented by the symbols x and rings. The calculated pressures give the thick broken curve that is nearly conformous to the pressure curves derived in the "Macropermeability Experiment" [11,14]

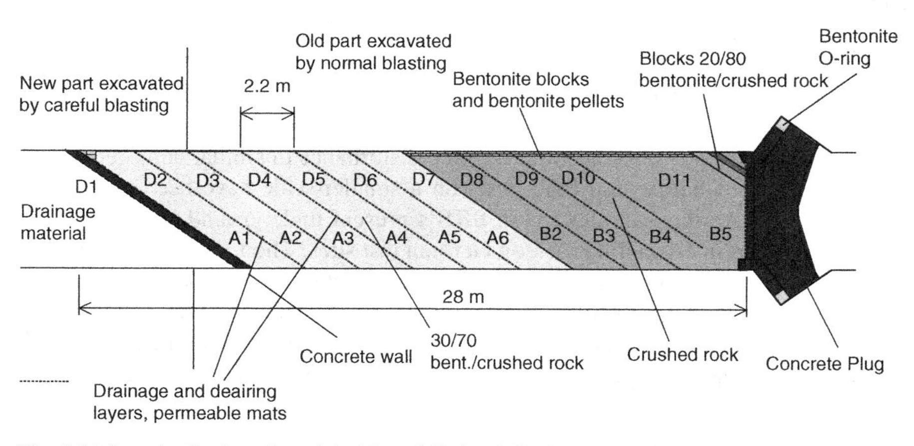

Fig. 2.24 Longitudinal section of the blasted Zedex drift [6]

rock while the rest of the drift contained inclined compacted layers of clayey material, i.e. mixed bentonite/crushed rock (30/70). Both parts were about 14 m long. The outermost backfill consisted of crushed rock over which blocks of highly compacted bentonite and pellets were placed for compensating the expected settlement of the underlying crushed rock fill that would otherwise have led to an open gap at the roof.

The Zedex drift was blasted even more carefully than the Stripa drift, i.e. by using smooth blasting technique designed by ANDRA [16]. The EDZ would hence be less conductive but measurements in short boreholes extending from the periphery of the drift indicated a hydraulic behaviour of the EDZ that was similar to that at Stripa. The general hydraulic behaviour of both rock masses was similar as indicated by the figure 6E-11 m/s for the bulk conductivity of undisturbed Äspö rock derived from large-scale flow modelling.

Repeated spot-wise determination of the conductivity of the EDZ in the Zedex drift was made in a comprehensive and careful study performed by ANDRA through Ecole de Géologie de Nancy, using the Seppi Tool [16]. The evaluated conductivity had a maximum close to the periphery – around E-10 to E-9 m/s – but the most shallow 10 cm part, which has the highest conductivity, could not be investigated because the packers were deeper than this measure from the walls. Since the rock around the Zedex drift was estimated to have an average hydraulic conductivity of no less than E-8 m/s to a depth of a few decimeters, ANDRA's study verifies that spot-wise determination cannot give a reliable value of the average conductivity over a long distance. The outcome of the Zedex experiment verifies the conclusion from the Stripa tests that the EDZ in blasted tunnels in crystalline rock is a very important water conductor over longer distances.

While full-scale testing of the entire near-field is required for adequate interpretation of the extension and nature of the EDZ in blasted crystalline rock, spot-wise determination of the hydraulic conductivity can give good information on the conductivity of the TBM-induced EDZ because of its relative homogeneity. Techniques for determining the hydraulic conductivity by spot measurements have been developed by BGR in Germany and they were applied in TBM tunnel walls at Äspö HRL and at NAGRA's URL at Grimsel [17].

In Germany, BGR's work for determining EDZ properties in blasted tunnels and drifts in crystalline rock using spot-wise testing in boreholes has given results that agree well with those from the Stripa and AEspoe experiments [17]. Thus, the EDZ of conventionally blasted tunnels in the central Aare granite area, which is characterized by a very tight, somewhat gneiss-like structure and fractures healed by biotite, chlorite or epidote and quartz, was concluded to contain induced macro- and micro-fractures to about 0.3 m depth, within which depth stress relaxation of cores was also found. Where the original discontinuities were more or less parallel to the tunnel the EDZ had a depth of more than one meter. The hydraulic characteristics of the EDZ rock were measured by use of a surface packer system and a short interval packer system developed and used for in situ tests. The conductivity within 0–0.3 m depth was found to vary between E-10 and E-3 m/s.

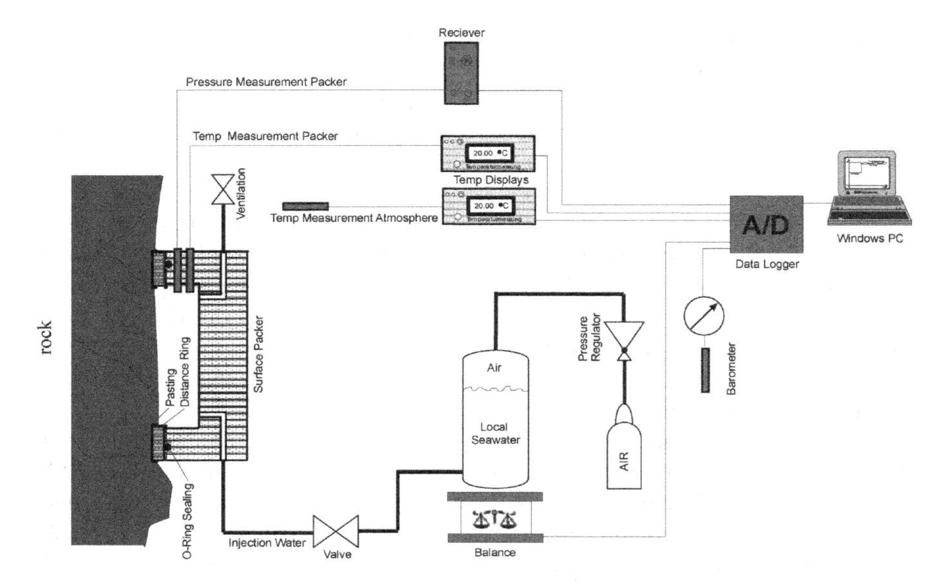

Fig. 2.25 Equipment used for measuring the hydraulic and gas conductivities in shallow rock. A plane ring-shaped surface is ground and polished for tight attachment of the surface packer which is pressurized with water. Measurement of the flow and pressure give the hydraulic conductivity surface using a flow model of porous media [17]

BGR also made field experiments at about 450 m depth in a TBM tunnel in granitic rock in SKB's underground research laboratory at Äspö, by use of an equipment consisting of a hollow metal cylinder fixed by a metal ring to the tunnel wall (Fig. 2.25). The horizontal holes were oriented perpendicularly to the tunnel wall. At each test spot experiments were made by pressurizing the packer with water and, separately, with gas. The recorded pressure drop over time in the control space was interpreted in terms of conductivity by applying "Two Phase Flow Theory" and a finite element method. This gave values in the interval E-10 to 2.5E-10 m/s for the most shallow 10 mm part of the EDZ, i.e. almost exactly the data reported earlier in this chapter. No significant differences were found between water and gas. For rock extending from the wall to 100 mm depth the average hydraulic conductivity was estimated at E-11 to E-10 m/s.

Comparison of the field measurements and the lab testing of small samples shows that those located less than 5 mm from the tunnel wall have a higher conductivity than in the field tests, which gave an average value of the conductivity of this pervious part and of the deeper, tighter rock. The reason for the higher value in the field test of this deeper rock than of samples from it, is simply that the larger volume involved in the field tests included flow in a few larger fissures, that were not represented in the small samples.

The European Commission has conducted workshops that have given a good overview of the subject [18].

2.4.1.3 Disturbance by Water-Jet Cutting

Water-jet cutting has been used for a number of years in manufacturing rock [5] and metals. AECL tested the technique for making deposition holes in granite at 240 m depth in the underground laboratory at Pinawa [18]. It gave rather rugged walls and would require a significant development program to optimise it for deposition hole drilling, but it could be considered for certain purposes in URLs like cutting of recesses and shallow troughs and trenches for cables, pipes etc.

2.4.1.4 Disturbance by Boring – TBM Technique

TBM boring has been used in many years in all sorts of rock for making discharge tunnels for hydropower plants, and transport tunnels of all kinds. The technique is well developed but can cause severe problems in rock with alternating stiff and soft parts as shown by the present tunnel excavation project in the Hallandsås horst in Sweden. The diameter of TBM tunnels can be more than 10 m and down to a couple of meters. In fact, the principle of equipping a bore head with drillbits has been used for many decades for making holes of different size in deep drilling for oil and gas prospection and production. There are varieties of the techniques, like raise-boring, in which the bore head is rotated and pulled through the rock using a pilot hole, and use of road-headers by which the bore head is moved for attacking rock similar to a dentist's drill.

For boring vertical deposition holes with a diameter of 1.5–2 m various other equipments have been developed as illustrated by Fig. 2.26, the muck being removed by vacuum technique or by using bore muds. With proper anchoring and orientation of the boring machine perfectly vertical and straight holes can be made with at least

Fig. 2.26 TBM-type equipment (Robbins) used for boring of vertical deposition holes with 1.8 m diameter and 8 m depth

Table 2.4 Approximate water flux across the assumed 80 m^2 near-field for the hydraulic gradient i = unity distributed over the various EDZ components. The backfilled tunnel is assumed to be impermeable [14,15]

Case	Permeated cross section, m^2	Hydraulic conductivity, m/s	Water flux, m^3/s
Virgin rock, no tunnel	80	E-11	8E-10
Blasted tunnel	20		2.6E-7
* Blast-EDZ	20	E-8	2E-7
* Stress-EDZ	60	E-9	6E-8
TB tunnel	80		2.6E-9
* Stress-EDZ	18	E-10	2E-9
* Virgin rock	62	E-11	6.2E-10

8 m depth, which is required for making canister deposition holes according to the Swedish and Finnish repository concepts. It should be mentioned that core drilling for preparing holes with 0.76 m diameter and very smooth walls and no abrasive disturbance of the rock was made in the Stripa Project [4]. It is possible to develop this technique so that to wider boreholes can be made.

TBM boring and related techniques, like raise-boring with similar cutting heads and use of road-headers, cause much less damage to the rock than even careful blasting. It is estimated that the average axial and radial conductivities of the EDZ zone will increase by 10 times and by about 100 times closest to the bored room, while the radial conductivity of the surrounding rock may drop to one fifth of the conductivity of the virgin rock. The effects depend on the rock structure.

The overall hydraulic effect of the EDZ is illustrated by Table 2.4 showing the estimated change in axial flux across the entire nearfield, 80 m^2, around a blasted tunnel with 25 m^2 cross section area [14]. It is composed of the blast-affected EDZ equalling 20 m^2 and of the surrounding stress-induced EDZ of 60 m^2. For TBMs with the same tunnel size, the EDZ extending to 0.1 m depth represents 18 m^2. For comparison of the two cases the net flux across the same section area, 80 m^2, is given in the table, implying that the larger part of the section is made up of virgin rock for the TBM case.

Assuming the conductivity of the virgin rock to be E-11 m/s and that of the blasted zone to be E-8 m/s and taking the axial conductivity of the stress-generated EDZ around the blast-disturbed EDZ to be E-9 m/s [14,15], one finds the total flux across the 80 m^2 near-field section area to be about 100 times higher for a blasted tunnel than for an equally large TBM tunnel, for which the conductivity of the EDZ is taken as E-10 m/s.

Comparing these data one finds the expected water transport capacity of a cross section corresponding to the total near-field of TBM tunnels to be no more than about 1% of that of blasted tunnels. The importance from the point of safety assessment of this fact is obvious.

2.5 The Integrated Performance of Host Rock

2.5.1 What is Important?

The behaviour of the rock in the construction phase is not trivial. Passing through the unevitable low order discontinuities down to the repository level, the selection of which is a first issue in a performance analysis, one will meet problems with inflowing water into ramps and adits and all tunnels and shafts in crystalline rock, and loss of stability in argillaceous rock, salt, and clastic clay. Subsequent to construction, waste emplacement and maturation of the buffers and backfills, there is a first phase that can be termed "the thermal period" of a few thousand years in which the rock is affected by a heat pulse followed by a cooling period of similar duration, the various temperature changes imposing thermal stresses and chemically induced changes of the rock. After this comes the never-ending period of apparent resting of the rock but that in fact involves a number of processes that will have an impact on it, like permafrost with cooling to below the freezing point, glaciation implying loading and unloading by glaciers, exposure to percolation of meltwater from the ice after glaciation cycles, and all sorts of tectonic events caused by altered regional stress fields.

The question is how much of all this that one needs to consider in planning and assessing of the various disposal concepts. The most important issue from a safety analysis point of view is of course the impact that the various processes will have on the ability of the host rock to provide confinement of the engineered barriers without yield, and on the flow paths represented by the virgin major discontinuities and neo-formed defects. For assessing these functions one needs to estimate what the rock strain will be in a 100 000 year perspective and what consequences it will have on the performance of the engineered barriers and on its hydraulic performance. We will deal with this matter in Chap. 6, which is devoted to site selection.

2.5.2 Overview of Rock Issues

The CROP project comprised national inventories of major issues in planning, construction closing stages and this unique work is summarized here in Tables 2.5–2.9, quoting the text in the Final report [5]. The tables illustrate the main differences between the three major rock types and give additional information on the properties and performance of rock hosting HLW repositories.

Table 2.5 Excavation stage (construction damage; stress redistribution)

	Crystalline rocks	Salt rock	Argillaceous rock[1]	Plastic clays
Processes and Events	• Excavation procedure giving rise to EDZ with high – K for drill-and-blast within 0.1–0.8 m from periphery. Significant K-increase within millimetres for TBM. – Stress redistribution due to opening, causing tension, compression and shearing in different parts around the drift.	• Effects of excavation procedures are less significant. • Stress redistribution due to opening, causing tension, compression and shear in different parts around the drift • Micro-cracking at locations where dilatancy takes place. • Permeability[2] increase with opening of micro-cracks; • EDZ extends 0.5 m into wall and 1.5 m into floor.	• Stress redistribution giving rise to strongly anisotropic, deviatoric compression and/or tensile stresses, causing (a) tensile and shear fracturing along bedding planes and (b) vertical extensional or tensile fracturing in rock near side walls • Data show vertical fracture in drift walls: 10–30/m within the first m; 5–10/m in the 1-2 m and ~0/m beyond 2 m • Generally, EDZ is one drift radius from drift wall with ΔK up to ~6 om • Transient pore-pressure dissipation	• Stress release at excavation (stress redistribution) leading to contractant and dilatant processes with induced fracturing; • Undrained behavior expected in the short term; • Dilation inducing suction and pseudo-strength increase; • Shear behavior also expected (slick and sliding); • Curved (eye-shaped) fractures with apex at front end of 4.8-m drift, extending to 7 m from rock walls; oxidation pyrite up to 2 m indicating communication with drift; but fractures closed beyond 2 m; • Evidence of extension fractures (relay fractures) for ~4 m on front of excavation; • Piezometric response far ahead of excavation front (60 m), but negligible ΔK

Table 2.5 (continued)

	Crystalline rocks	Salt rock	Argillaceous rock[1]	Plastic clays
Property and Design Parameters	• In situ stress; • Rock strength (failure initiation strength); • Initial in situ fracture network (density and orientations), • Orientation of drift relative to principal stress directions, size and shape of drifts; • Excavation sequence, quality assurance and control.	• State of stress (depth of repository); • Dilatancy and creep properties of the salt • Geometry of drift; • Procedure for rock support immediately following excavation	• Rock property parameters; anisotropy (bedding planes) • In situ stress state, including stress anisotropy • Drift orientation relative to bedding plane directions; • Drift (gallery) shape, • Methods of excavation with emplacement of support, • Moisture content in rock	• In situ stress-strength ratio; • Excavation and lining emplacement procedure to minimize wall convergence • Parameters related to shear and extensional fracturing • Presence of bedding planes or planes of weakness • Excavation shape and initial stress state

Table 2.5 (continued)

	Crystalline rocks	Salt rock	Argillaceous rock[1]	Plastic clays
Some Technical Factors	• Difference in EDZ developed along part of the drift away from and the part near to fracture zones intercepting the drift. More studies of EDZ recommended. • Some apparent differences in results from Stripa, Grimsel (FEBEX), Aspo and Pinawa URL; perhaps due to different in situ stress, drift size and drift orientation and fracture network in rock, as well as different excavation methods.	• Deviatoric stress can increase air Kup to 4 om by inducing new micro cracks.	• Medium characterized by dependence on moisture content: low moisture case corresponds to harder rock and high case corresponds to ductile (soft) behavior • Role of structures (bedding planes), weathering (drying), dissolution and oxidation; • Nonlinear stress-strain behavior dependent on stress level, suction (water content) and weathering conditions; • Stress and strain localization: onset and propagation of discontinuity; effect of heterogeneity.	• Naturally occurring fractures are found in outcrops with spacing 0.5 to several meters; mostly extensional fracture with a small part shear fractures. However, they are not seen at depth. Confining pressure, as well as moisture content, is reason for absence of fractures.

[1] In the CROP report the term was "Indurated Clays"
[2] Permeability used for hydraulic conductivity (K)
EDZ = excavation disturbed/damaged zone; ΔK = increase in permeability K; $\Delta\phi$ = change in porosity; om = orders of magnitude.

Table 2.6 Open-drift stage (ventilation; supports; EDZ cut-off and sealing)

	Crystalline rocks	Salt rock	Argillaceous rock	Plastic clays
Processes and Events	• EDZ dehydration • Air entry resulting in oxidizing conditions, two-phase flow conditions (effective water conductivity decreasing), • Potential chemical and bacterial activities with possibility of clogging against flow • Some indication from Canadian data that humidity (wetting of drift wall) increases development of induced fracturing	• Drift ventilation and salt dehydration with oxidizing condition affecting salt creep properties and hence EDZ; • Room convergence (data show convergence of 0.3 m in 10 years for a 7 × 10 m drift). Thus, EDZ is still developing unless rock support is put in • Salt creep onto support surfaces – EDZ development slows and begins to reverse	• Rock creep; and drift (gallery) wall convergence; also floor heaving due to water during construction and operation • Ventilation-induced damage; dehydration with resulting rock strengthening and contraction; • Oxidation of pyrites-production of sulfuric acid; • EDZ is evolving.	• Transition from dilatant to contractant regime; creep and possible fracture closing and drift wall moves against support • EDZ is still changing. • Suction development on the wall due to dilatant (induced consolidation); meaning suction increases without desaturation • Axial displacement of front of drift ~10 cm/yr • Drift ventilation and rock dehydration changing clay properties; ventilation retards self sealing • Oxidizing conditions – penetrating ~1 m into rock

Table 2.6 (continued)

	Crystalline rocks	Salt rock	Argillaceous rock	Plastic clays
Property and Design Parameters	• Drift humidity condition and its variation • Fracture apertures and porosity in EDZ • Fracture density in EDZ • EDZ mineralogy	• Drift humidity and its variations; • Drift support system	• Drift (gallery) humidity changes; rock pore pressure changes. • Rock support, lining settling time, • In situ stress; depth of drift (gallery), • Drift orientation with respect to natural bedding planes and in situ principal stress directions.	• Drift humidity and its variations • Suction as a function of dilatancy • Rock strength parameter changes • Support structure designs
Some Technical Factors	Need to grout fracture zones intersecting drift prior to excavation; post-excavation grouting is more difficult due to lack of back pressure	• Differential stress versus confining pressure: – Up to dilatancy limit (maybe a band), above which opening of grain boundary: Δk up to 2–3 om – Up to failure boundary, above which macro fracturing takes place • ΔK and $\Delta \phi$ relationship is an important question • Creep deformation changes with back compression from supports • Δk reversible, but full reversibility still needs study	• Process has 4 stages in stress-strain development 1. Closing of pre-existing fractures 2. Elastic deformation of material 3. Plastic deformation; occurrence and growth of micro cracks 4. Localization of stress strain and initiation and propagation of macro fractures • Peak of Δk is behind peak of stress as a function of strain • Sealing of EDZ by emplacement of backfill after removal of EDZ damage materials or by radial cuts that are filled with bentonite – yet to be studied.	• EDZ is still changing • Fractures develop several meters in rock, but if convergence is limited to fracture can be limited less than 1 or 2 m from rock wall • Ventilation drying changes EDZ properties near drift wall • Combination of low k and deformation (undrained) causes change in suction and clay properties

Table 2.7 Early closure stage (backfill; resaturation; heating)

	Crystalline rocks	Salt rock	Argillaceous rock	Plastic clays
Processes and Events	• Bentonite and backfill emplacement; EDZ resaturation • Wetting and swelling of bentonite and backfill; effect of swelling pressure on EDZ • Role of presence of faults with EDZ • Heating from waste canisters; effects of temperature gradients and moisture redistribution in bentonite and EDZ; thermal compression • chemical and bacterial processes (dissolution, precipitation, bacterial growth and clogging)	• Drift humidity building affecting rock properties; • Backfill and seal emplacement and backfill support of EDZ; • Heating from canisters and effects of temperature gradients on EDZ; • Brine pocket migration in EDZ; • Chemical and bacterial processes in micro cracks of EDZ; gas generation and pressure buildup; • EDZ is still developing • Creep onto support leading to compaction and $-\Delta K$	• Effects of swelling pressure and suction in bentonite backfill on EDZ, possibly reducing EDZ permeability; data show k in EDZ highest not at wall but at ~0.6 m into rock and is anisotropic. • Effect of transient and spatially varying temperature and saturation on rock behavior • Effect of desaturation-resaturation cycle with corresponding pore pressure changes on rock properties • Changes as bentonite and cross-cut seals swell	• Drift humidity buildup changing rock properties; • Swelling buffer materials with potential reloading of EDZ (in some designs, buffer materials without swelling capacity are chosen); • Thermal expansion in undrained conditions in the very near field near heating source; • Change in effective stress due to fracturing or sealing, • Potential increase creep rate (enhanced plasticity)

Table 2.7 (continued)

	Crystalline rocks	Salt rock	Argillaceous rock	Plastic clays
Property and Design Parameters	• Bentonite and backfill physical and chemical properties and their initial conditions • Chemically changing from oxidizing to reducing conditions • Fault or fracture zone geometry (and location) and flow properties • thermal loading and time variation; thermal conductivities of EDZ and backfill as a function of saturation	• Backfilling and sealing procedures; • Creep rates as a function of temperature; • Effect of temperature gradient on creep rates at different locations in EDZ	• Rock properties as a function of temperature and saturation; • Thermal input rate from the waste; • Seal designs	• Rock strength, creep and suction properties as a function of humidity or saturation and temperature
Some Technical Factors	• Interplay between thermal compression and resaturation in EDZ • Effect of back-pressure from bentonite buffer and backfill on EDZ	• Backfill provides confining pressure. • Heating induces compressive stress in most areas, while increases differential stress in others.	• Resaturation weakens rock, enhances creep, and induces closure, especially normal to bedding planes • Heating may cause near-field drying in EDZ, working opposite resaturation from the rock, resulting in varying wetting and pore pressures in EDZ • Sealing bentonite provides loading on EDZ and reduces k	• Resaturation causes suction to disappear. • Open fractures start to heal. • After a few weeks, piezometric measurement indicates open fractures do not extend beyond 1 m of wall. • Temperature assists in healing; may also cause gas pressure buildup in pores. • Temperature with saturated conditions increases creep.

EDZ = excavation disturbed/damaged zone; ΔK = increase in permeability K; $\Delta\phi$ = change in porosity; om = orders of magnitude.

Table 2.8 Late closure stage (cooling; support degradation; self-sealing)

	Crystalline rocks	Salt rock	Argillaceous rock	Plastic clays
Processes and Events	• Rock, EDZ and bentonite/backfill fully saturated • Decreasing temperature with smaller gradients as compared with heating stage • Chemical processes (with probably smaller effects), bacterial activities may be significant; potential decrease in permeability due to clogging of transported particles • Degradation of rock support system (rock bolts, drift lining) with its physio-chemical effects: (a) chemical reactions with backfill and bentonite; (b) high permeability paths along locations of degraded rock bolts and drift linings	• Stress redistribution; process of self healing and reduction of permeability • Stress conditions decrease below dilatancy criterion; no micro cracking • Perhaps occurrence of cooling fractures • Healing (field data at 700 m show change in k in EDZ from E-18 m^2 back to E-16 m^2 in 90 years; not full healing yet to intact rock value of ~E-21 to E-20 m^2)	• Cooling and fluid pressure restoring in the rock mass • Potential self-sealing due to mechanical fracture closure of precipitation of infill minerals; • Effect of fluid pressure in drift (gallery), • Lining degradation with corresponding mechanical (support) and chemical changes (e.g., high pH plume); • Gas pressure buildup generated by corrosion and waste; • Changes in bentonite and cross-cut seals	• Generally, processes are much slower. • Potentially sealing on cooling • Potential carbonate transport and precipitation in the EDZ with possibility of changing rock strength or cohesion • Healing is slow; 2 kinds of processes: – closing up fractures during consolidation due to swelling and creep (data show visible fractures after 15 years, but k is at intact rock value) – volume increases around large open fractures or gaps; increasing plasticity index

Table 2.8 (continued)

	Crystalline rocks	Salt rock	Argillaceous rock	Plastic clays
Property and Design Parameters	• Properties and parameters related to bacterial activities • Physico-chemical properties of rock bolts and drift lining	• Parameters controlling creeping and self-healing process. • Compaction • Temperature changes • Moisture content changes	• Liner degradation properties; • Parameters controlling gas release and buildup from corrosion and waste. • Properties of bentonite seals and cross-cut seals	• Water movements • Pore pressure changes • Clay swelling (smectite group)
Some Technical Factors	No self-sealing; however, long-term heating with thermohydrologic effects over 2000 year time frame may lead to clay minerals in fractures, thereby clogging them.	• Effect of impurities, e.g. anhydrites • Δk (sealing) depends on time, effective stress, deviatoric stress • Healing can be by (a) viscoplastic deformation or (b) recrystallization in the presence of brine • Under confining pressure the effect of healing is immediate and then progresses with time	• Sealing is time-dependent process – Stress state for contractancy – Dilatancy – Swelling – Newly formed minerals • With saturation, ΔK of single fracture is reduced by 1 om in 1 year and Δk of fracture network reduces by factor of 50 in 110 days. • With gas injection, fracture opens. When gas depressurizes, fracture remains open and closes slowly over 5 months. • Sealing needs more studies to confirm.	• Observed: k in EDZ restored in several years: (a) clay closes spontaneously against borehole casings; (b) open boreholes closed completely, and (c) clay flows into open boreholes • Fractures closed by increase in stress; swelling and creep • Discontinuities are still present, but k restored

EDZ = excavation disturbed/damaged zone; ΔK = increase in permeability K; $\Delta \phi$ = change in porosity; om = orders of magnitude.

Table 2.9 State of science and technology

	Crystalline rocks	Salt rock	Argillaceous rock	Plastic clays
State of Field Information (Including Natural Analogues)	• Stripa: Buffer Mass Test • Pinawa URL, Canada • Aspo: Zedex Drift • Febex Experiment (including 6-year heating and cooling) • Kamaishi Mines, Japan	• Tests in Asse Mines (ALOHA and BAMBUS II projects)	• Tests in Mont Terri (Switzerland) and Tournemire (France) underground laboratories; new tests in expected URL at Meuse/Haute Marne • Studied effects of ventilation and dehydration, heating, desaturation and resaturation • Studied self-sealing for excavation-induced fracture and gas pressure-induced fracture opening	• URF at Mol: boom clay. First shaft, 1980–1982; Test drift, 1987; Second shaft, 1997–1999; Connecting drift, 2001–2002. • HADES Project: 20 years of data on hydromechanical effect and fracture patterns around drift • CLIPEX instrument program to study effects of advancing Connecting Drift • Analog: London Underground • Analog: St. Petersburg blue clay
State of Modeling	• Much efforts on constitutive relationships; shear stress permeability relationship still open • A large number of simulators have been developed and applied, including THAMES, ROCMAS, TOUGHFLAC codes	• Constitutive models developed for damage, dilatancy – Gens-Olivella, Hou-Lux models • First simulations performed by various codes (e.g., BRIGHT, MISES 3, JIFE, percolation model) that includes creep and viscoplasticity model • Models are for isotropic system – anisotropy to be included later.	• Palmer and Rice's model of stress localization, fracture initiation and propagation • Elastic and isotropic elasto-plastic model into codes BRIGHT, FLAC3D, MHERLIN, PFC • Need a threshold for onset of creep • Can model main features of EDZ development	• Model with modified CAMCLAY and Mohr-Coulomb model • Predicts well wall displacement and equilibrium pressure on lining • Cannot explain large extent of pore-pressure response (60 m) • Model of strain localization, fracturing and self-sealing is being developed.

Table 2.9 (continued)

	Crystalline rocks	Salt rock	Argillaceous rock	Plastic clays
Open Technical Questions	• Understand and reconcile data and results from different results from various field and laboratory experiments; correlate results with local conditions and drift sizes and shapes	• Need to confirm healing and study behavior for full healing • Need to test constitutive models to confirm reversibility of EDZ evolution by comparing results against excavation of different ages and around plugs installed long ago.	• Differences exist between laboratory and field results (possibly due to sampling and coring problems) • Interactions among anisotropy of intact rock, anisotropy of EDZ stress changes and anisotropy of ΔK interaction to be studied • Lack longer term experiments – self-sealing effect still not fully solved	• Explanation for 4 m communicating fractures and 60 m pore pressure changes ahead of the excavation face. • Property parameters as a function of humidity and temperature conditions • Anisotropic modeling
Long-Term Safety Impact issues	• Potential connected fast flow paths in the EDZ; effect of EDZ on resaturation rate. • Water flow between EDZ and fracture zones intersecting drift • Assessment of THC and THB impacts	• Role of contact between seal and EDZ and host rock • Possibly EDZ only important in the first decades because of self sealing effects; need more field data to confirm model and to test its capability for long-term prediction of self sealing (e.g., with respect to temperature and moisture changes) • Need to model anisotropic behaviors • Gas transport needs study	• On one hand, EDZ possibly a preferential pathway for gas release and reactivation of closed discontinuity, need further study • On the other hand, EDZ probably no significant effects, limited by availability of water from the low K formation, but – Confidence building – Study of sealing methods – Need bounding and scoping calculation	• Potential EDZ pathways; however, even if EDZ has high k, flow along EDZ may be limited by availability of water from the low k formation • Transport probably mainly by diffusion, need account for property changes, such as ϕ, K_d, D_l and solubility limits, • Interactions with other repository components to be simulated

Table 2.9 (continued)

	Crystalline rocks	Salt rock	Argillaceous rock	Plastic clays
Issues in Repository Design	• Improved tunnel construction and support method to reduce EDZ; • Optimal depth and orientation of tunnels relative to in situ stress condition and fracture/fault directions; • Design of methods to cutoff/interrupt EDZ; • Methods to "improve" rock-seal interface	• Possibility of installing stiff liner during or immediately after excavation • How to position seal in the repository; how to remove EDZ immediately before seal emplacement • Because of creep, special measures to reduce or cutoff EDZ may not be necessary	• Optimal design of drift (gallery) geometry, • Excavation method, lining properties (stiffness, settling time), • Orientation of drift with respect to bedding planes, potential discontinuities and in situ stress direction; • Seals locations. Design of EDZ cutoffs.	• Consideration of new excavation procedures to limit EDZ formation; • Avoid or control rock desaturation during heating • Development of methods to reduce or cutoff EDZ. • New excavation methods to minimize over-excavation to reduce convergence; use of shield for drifts; put in lining ASAP
"Bottleneck" Issues	• Need for establishing acceptable methodologies for complete PA scoping, bounding and sensitivity studies on effects of EDZ (including presence of fracture zones closed to drift) • Need to establish and confirm EDZ cutoff strategies and effectiveness of seals	• Understanding of processes in seal-rock interface and effectiveness of drift seals and cutoffs • Need for establishing acceptable methodologies for complete PA scoping, bounding and sensitivity studies on effects of EDZ (including sensitivity to degrees of self sealing)	• Understanding of processes in seal-rock interface and effectiveness of drift seals and cutoffs • Need for establishing acceptable methodologies for complete PA scoping, bounding and sensitivity studies on effects of EDZ (including sensitivity to degrees of self sealing)	• Understanding of processes in seal-rock interface and effectiveness of drift seals and cutoffs • Need for establishing acceptable methodologies for complete PA scoping, bounding and sensitivity studies on effects of EDZ (including sensitivity to degrees of self sealing)

EDZ = excavation disturbed/damaged zone; ΔK = increase in permeability K; $\Delta\phi$ = change in porosity; om = orders of magnitude.

References

1. Pusch R, 1997. Discontinuities in granite rocks. Suggested categorization for general description and rock mechanical calculations. SKB Report IPM D-97-03. SKB, Stockholm.
2. Pusch R, 1995. Rock mechanics on a geological base. Developments in Engineering Geology, 77. Elsevier Publ. Co (ISBN:0-444-89613-9).
3. Hoek E, Brown ET, 1980. Underground excavations in rock. The Institution of Mining and Metallurgy, London.
4. Pusch R, 1997. Boring-induced disturbance of granitic rock. SKB Report AR D-97-05. SKB, Stockholm.
5. Pusch R, 1994. Waste disposal in rock. Developments in Geotechnical Engineering, 76. Elsevier Publ. Co. ISBN:0-444-89449-7.
6. Svemar Ch, 2005. Cluster Repository Project (CROP). Final Report of European Commission Contract FIR1-CT-2000-2003, Brussels, Belgium.
7. Pusch R, 1982. Mineral-water interactions and their influence on the physical behavior of highly compacted Na bentonite. Canadian Geotechnical Journal. Vol. 19, No. 3 (pp. 381–387).
8. Pusch R, Adey RA, 1998. Accurate computation of stress and strain od rock with discontinuities. Advances in Computational Structural Mechanics. Civil-Comp Ltd, Edinburgh, Scotland (pp.233–236).
9. BEASY User Guide, Computational Mechanics BEASY Ltd, Southampton, UK. 1996.
10. Munier R, Follin S, Rhén I, Gustafson G, Pusch R, 2001. Project JADE. Geovetenskapliga studier, Report R-01-32. SKB, Stockholm.
11. Gale J, Roleau A, 1986. Hydrogeological characterization of the Ventilation Drift (Buffer Mass Test) area, Stripa. Stripa Project Technical Report 86-02. SKB, Stockholm.
12. Gray, MN, 1993. OECD/NEA International Stripa Project Overview Volume III: Engineered Barriers, Published by Swedish Nuclear Fuel and Waste Management Co. (SKB), ISBN 91-971906-4-0, Stockholm.
13. Gale JE, Witherspoon PA, Wilson CR, Rouleau A, 1983. Hydrological characterization of the Stripa site. Proc. Workshop on geological disposal of radioactive waste – In situ experiments in granite, Stockholm. OECD/NEA, Paris.
14. Pusch R, Börgesson L, Ramqvist G, 2003. Hydraulic characterization of EDZ in a blasted tunnel in crystalline rock – measurements and evaluation. Proc. European Commission Cluster conference and workshop, Luxembourg, 3 to 5 November 2003.
15. Börgesson L, Pusch R, Fredriksson A, Hökmark H, Karnland O, Sandén T, 1992. Final Report of the Rock Sealing Project – sealing of Zones Disturbed by Blasting and Stress Release. Stripa Project Technical Report 92-21. SKB, Stockholm.
16. Bauer C, Homand F, Ben Slimane K, Hinzen KG, Reamer SK, 1996. Damage zone in the near field in the Swedish ZEDEX tunnel using in situ and laboratory measurements. EUROCK Congress in Turin, September 1996.
17. Liedtke L, Shao H, Alheid HJ, Sönnke J, 1999. Material transport in fractured rock – rock characterisation in the proximal tunnel zone. Federal Institute for Geosciences and Natural Resources, Hannover, Germany.
18. Chandler N, Cournut A, Dixon D, Fairhurst C, Hansen F, Gray M, Hara K, Ishijima Y, Kozak E, Martino J, Masumoto K, McCrank G, Sugita Y, Thompson P, Tillerson J, Vignal B, 2002. The five year report of the Tunnel Sealing Experiment: an international project of AECL, JNC, ANDRA and WIPP. Atomic Energy of Canada Limited Report AECL-12127.

Chapter 3
Engineered Barriers and Their Interaction with Rock

3.1 Engineered Barriers

The multibarrier principle has been used by all the Organizations in designing their HLW concepts since the redundance gives an apparent and partially true back-up of the safety, which is basically the ratio of the allowed and predicted concentrations of major long-lived radionuclides in wells with drinking water. The barriers are:

1. The waste itself ("HLW")
2. Containers of metal in which the waste is confined ("canisters")
3. Low-permeable clay embedding the containers ("buffer")
4. Seals and fills in holes and tunnels ("plugs and backfills")

The waste has a very low solubility and hence offers resistance to release of radionuclides. The canisters can be made of chemically very stable material and be made tight by appropriate closure. The low-permeable clay minimizes percolation of groundwater and offers ductile environment of the canisters and hence moderation of the impact of external forces. The plugs serve to separate permeable and tight backfills in holes and tunnels.

We will examine the four barriers here, focusing on the three firstmentioned and their mutual interaction and leaving the last one to be treated in greater detail in the next chapter.

3.2 HLW

The international Atomic Energy Agency (IAEA) has defined five categories with respect to final disposal as given by Table 3.1. The international committee for radiation protection (ICRP), which introduced the ALARA principle implying that all radiation doses shall be kept as "low as reasonably achievable" with respect to economic considerations and to society, has defined a number of criteria concerning exposure of the whole body or part of it to radiation. National radiation protection

R. Pusch, *Geological Storage of Highly Radioactive Waste*,
DOI: 10.1007/978-3-540-77333-7_4, © Springer-Verlag Berlin Heidelberg 2008

Table 3.1 Definition of radioactive waste for final disposal

Waste classes	Typical characteristics	Disposal options
1. Exempt waste (EW)	Activitiy levels at or below clearance levels representing an annual dose of members of the public of less than 0.01 mSv	No radiological restrictions
2. Low and intermediate waste (LILW)	Activity levels above clearance levels as for 1) and thermal power below about 2 kW/m^3	No radiological restrictions
2.1 Short-lived waste (LILW-SL)	Restricted long-lived radionuclide concentrations	Near surface or geological disposal facility
2.2 Long-lived waste (LILW-LL)	Long-lived radionuclide concentrations exceeding limitations for short-lived waste	Geological disposal facility
3. High-level waste	Thermal power above about 2 kW/m^3 and long-lived radionuclide concentrations exceeding limitations for short-lived waste	Geological disposal facility

IAEA Safety Guide, Classification of radioactive Waste, Safety series No. 111-G-1.1, Vienna (1994).

requirements are established for purposes broader than radioactive waste management. In the establishment of acceptable levels of protection, account is typically taken of the recommendations of the International Commission on Radiological Protection (ICRP) and the IAEA and specifically the concepts of justification, optimization and dose limitation. The relevance of these concepts depends on the type of radioactive waste management activities. A recent classification system is presented in a Safety Guide issued by IAEA providing the waste classes in Table 3.1. The classification and nomenclature are currently reconsidered.

For ordinary reactors nuclear fuel usually consists of cylindrical pellets of uranium dioxide stacked in cladding tubes of a durable zirconium alloy (Fig. 3.1). The tubes are bundled to form fuel assemblies of about 4 m length in SKB/POSIVA's planned canister design. In the reactors radionuclides are formed by nuclear fission of uranium-235 and plutonium-239 and by neutron capture by nuclei in the fuel elements. Most of them become embedded in the uranium dioxide fuel matrix but some actinides can migrate.

There are two options for treating the spent fuel from nuclear reactors, direct insertion of the zirconium claddings with fuel rods in canisters to be placed in deposition holes or tunnels, or vitrification of the rest products of processing the material for separating plutonium to be used as nuclear fuel. Either way, the uneconomic disposal of used fuel or the more economic way of using the spent fuel ending with

Fig. 3.1 Cylindrical fuel pellets in cladding tube of Zircaloy [1]

burial of processed waste is possible. At present, most countries tend to apply the firstmentioned technique, which will be in focus in this book.

The radioactive decay of spent fuel depends on the composition of the fuel, which is the so-called the "inventory". Taking the fuel used for the Swedish reactors, i.e. BWR fuel from "boiling water reactors" as an example, the predicted drop in radioactive toxidity, is as shown by Fig. 3.2. It indicates that the required time for isolation from the biosphere is at least 100 000 years. In this period the radiotoxicity of the fuel declines to about 0.05 percent of its initial value and then remains on the same level as uranium ore exploited for producing the fuel [2]. By intermediate storage of the HLW the radioactive decay and heat production will drop to an acceptable level for placing it in the repository. Such storage can be made by keeping the waste under water for a suitable period of time, which is about 40 years

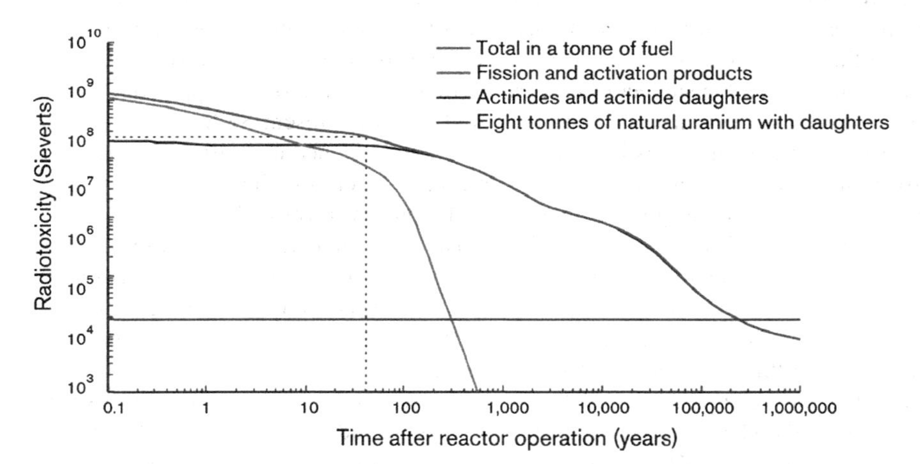

Fig. 3.2 Example of the activity content of spent fuel as a function of time. The graph represents waste to be encapsulated in SKB canisters [1]

in the Swedish Center for Intermediate Storage. The toxicity will thereby drop to a low value but even more important is that the heat production will be sufficiently low to avoid problems with overheating the canisters and embedding clay. It is of course possible to extend the intermediate storage for much longer time to reduce the heat production significantly but such prolongation would have to be for centuries. The responsibility and practical conditions for long-term intermediate storage would certainly be matters of dispute.

3.3 Canisters

3.3.1 Design and Material

The function of the canisters is of fundamental importance. As long as they are tight, radionuclides can not escape and cause contamination. If, or rather when, the canisters leak, radionuclides will migrate into the surrounding clay and rock and soon contaminate the biosphere.

Internationally, the plan of treating HLW is commonly to vitrify it for containment in relatively thin carbon steel or iron containers providing isolation for down to 1000 years. For spent fuel several canister types have been proposed and prototypes manufactured and tested by SKB with the intention of providing tight confinement of the hazardous waste for much longer time. Early versions had the fuel bundles placed in hollow canisters of highly compacted corundum or cast in concrete lined with titanium, while later types had copper as outer shield and the fuel bundles mounted inside or cast in lead. A special concept implied hot isostatic compression of copper powder in a copper tube with the fuel bundles submerged in the powder, a solution that we will come back to. Canisters of copper or lined with copper are believed to be chemically very stable for long periods of time and appear to become of interest also to other Organizations. SKB's present version is shown in Fig. 3.3.

In the present chapter we will focus on long-lived canisters and examine them after touching on the subject of how canisters made of iron or steel perform, which is of interest because iron is a component also of the present SKB canisters. Historically, the Organizations in Sweden and Finland planned to confine unprocessed spent fuel in very corrosion-resistant canisters of copper tubes but realizing that they would be weak the design was changed to the present form with an iron insert for the fuel bundles. The copper tube with 1.05 m outer diameter has 50 mm thickness and a length of about 4.8 m surrounds the insert of cast iron that has channels for 12 fuel bundles of BWR fuel. The canister insert is sealed in an environment of 90% noble gas which limits the water content per canister to 600 g. There is a gap of a couple of millimeters between the iron and the copper lining that can and will be closed by the water pressure and the pressure exerted by the surrounding clay except near the ends where the lids and bottoms prevent radial compression. The water pressure will be 4–5 MPa and the clay pressure can be even higher. The weight of the canister is 24 600 kg, of which the fuel represents 3600 kg, the copper 7400 kg, and the iron insert 13 600 kg. Naturally, these heavyweighters present difficulties in

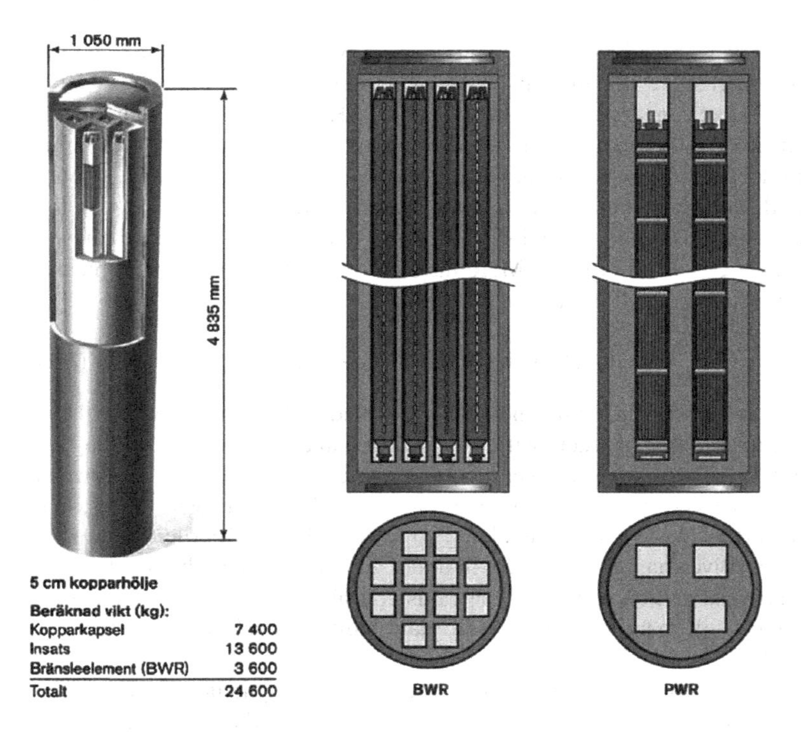

Fig. 3.3 Presently proposed SKB canisters. *Upper*: Schematic picture of the design. *Lower*: Iron columns with channels for 12 bunches of spent fuel to the left, and copper liner to the right [1,2]

placing them in deposition holes and tunnels especially since the radiation that they give off requires effective shielding and remote handling.

For iron and steel canisters and the iron insert of SKB's presently favoured canister concept, it is commonly stated that anaerobic conditions prevail soon after placement in deep repositories and that this delays corrosion very much. Magnetite will be formed early by oxidation through the oxygen content that is enclosed in the voids of the clay embedding iron and steel canisters and this compound is believed to be a dominant corrosion product. More magnetite will be formed anaerobically following the reaction equation Eq. (3.1):

$$3Fe + 4H_2O \text{ yielding } Fe_3O_4 + 4\,H_2 \tag{3.1}$$

The Fe^{2+} concentration in the magnetite is controlled by its non-congruent dissolution of the mineral, which reacts with the hydrogen and yields hematite (Fe_2O_3) and (FeOOH), [3]. There are also several other expected corrosion products formed in the absence of oxygen like ferrous hydroxide, $Fe(OH)_2$, and ferric hydroxide, $Fe(OH)_3$, the latter having a strong tendency to form colloids of particles that normally carry a positive charge. Iron ions will under all circumstances be free to react with other materials and be sorbed by the canister-embedding clay through cation exchange, which will make the buffer stiffer and reduce its self-sealing ability in case of large strain.

A phenomenon of great practical importance is that magnetite has a specific density that is only about 50% of that of iron, which means that it would occupy twice the space. This will compress the buffer and raise the pressure on the walls of the deposition holes to a level that can be critical. The production of hydrogen gas acts similarly: if the gas production rate is slow the gas can form a successively expanding bubble that compresses the buffer and makes it so tight that the pressure can cause breakage of the surrounding rock. If gas can penetrate the buffer clay and reach the boring-generated EDZ (Fig. 3.4) it will escape through permeable discontinuities that intersect deposition holes.

3.3.2 Physical Performance of the Presently Proposed SKB Canister

The presently proposed candidate canister has two components, an outer cylindrical copper tube as corrosion resistant liner and a supporting iron insert with isolated cells for fuel bundles as described in Fig. 3.2. Transport to the place where the canisters shall be inserted in the rock will be made such that they will not be exposed to external forces. As soon as they are on site a number of processes begin that can affect their physical state like:

- The temperature of the surface and thereby in the interior will start rising since the confinement serves as thermal isolation.

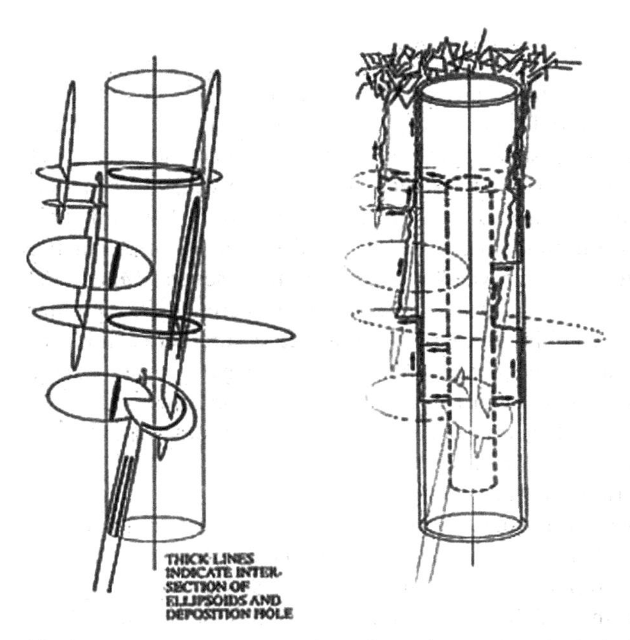

Fig. 3.4 Major transport paths for radionuclides and gas in the near field of deposition hole with canister embedded in buffer clay. *Left* drawing shows generalized fractures intersecting the hole or being parts of a fracture network in the rock [4]

- Groundwater will seep in through the confining, not water saturated buffer, and come in contact with the hot copper surface. Boiling will start and the clay adjacent to the copper liner will begin to desiccate.
- When the buffer has taken up water from the surrounding rock and expanded it will come in physical contact with the copper and a number of chemical processes begin.
- Shear displacement in the surrounding rock will induce deformation of the buffer, which transfers the forces to the canisters.

We will look at the various impacts for understanding the basis of the design of the various engineered barriers, starting with the thermal issues.

3.3.2.1 Evolution of Temperature

The heat produced in conjunction with the radioactive decay drops from a value determined by the power 1800 W per reference canister to 85% of this value after 10 years, to 30% after 100 years and further to 7% in 1000 years. The maximum surface temperature of the canister is estimated at 100°C after 10 years under ideal conditions, i.e. successive increase in thermal conductivity of the embedding clay, and drops to ambient after a few thousand years.

3.3.2.2 Internal Pressurization

One has to consider that water under the prevailing pressure of 4–5 MPa in the repository can enter a hole – a defect from the manufacturing or emerging from thermally or tectonically induced strain – in the copper liner and the iron insert and that this adds to the initial amount of water – and hence give a much higher water content in the insert of the canisters than the presently assumed 600 ml. This water will be heated to several hundred degrees in the first century after placing the canisters in the repository and the vapour pressure will be several MPa. This pressure will act on the buffer clay and compress it, which increases its swelling pressure. The two processes combine to raise the contact pressure between buffer and rock so much that it may break provided that the vapour gas can not escape through the clay (Fig. 3.5). This may not be possible if the buffer consists of very tight smectite-rich clay but is believed to occur if the clay is more permeable like clays containing mixed-layer clay minerals with an expanding component. A clay of this sort is the German Friedland Ton.

The example is certainly relevant since the swelling pressure of expanding buffer clay may well reach over 20 MPa as we will see in the next subchapter and the vapour pressure at 200°C is over 15 MPa. Ordinary granite has a tensile strength of 10–20 MPa, which would withstand the tension stress but weaknesses of 4th and 5th orders that are parallel to the axis of the hole may open and form strongly permeable channels into which buffer clay may enter and migrate.

3.3.2.3 External Pressurization

One can identify at least three types of mechanical impact on the canisters:

- Shearing of the rock along a fracture that intersects the deposition holes.
- Swelling pressure exerted on the canisters by the surrounding clay.
- Water pressure.

We will consider the shearing case first, focusing on the two possibilities that are illustrated in Fig. 3.6, i.e. one representing the relatively ductile SKB canister, and one represented by a very rigid canister in rather weak clay embedment. The problem with the firstmentioned is that the canister can fail by undergoing critically large strain, and the other that the canister, staying intact, can make the clay heterogeneous. The case with deformable canisters has been investigated in detail by finite element analyses using material data of smectite-rich clay of high density (2000–2050 kg/m^3).

The space between the copper liner and iron insert in the SKB canister is about 2 mm and most of it will be closed by the water pressure and the swelling pressure exerted by the surrounding clay except near the ends. However, at the junction of the straight part of the tube and the lid connected to it, a circumferential void will remain and fractures may grow from it under the very high stresses that prevail here. The same problem can appear if the tube is also exposed to axial tension. This would create critical three-dimensional stress concentrations near the ends of the canister

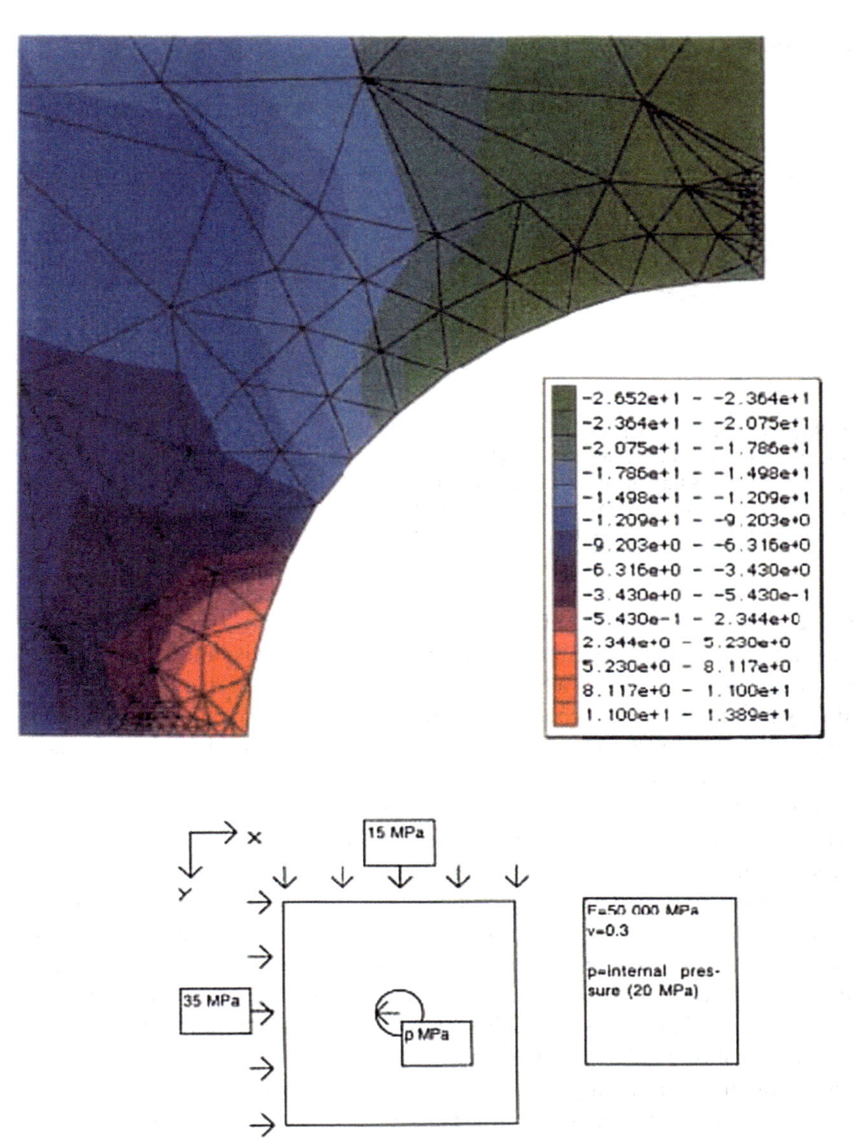

Fig. 3.5 Net hoop stress in rock around a 1.6 m diameter deposition hole exposed to an internal pressure of 20 MPa. The primary stresses in the rock were assumed to be 35 and 15 MPa, respectively. Tension stresses appear in the red parts reaching a maximum value of about 11 MPa (PLASTFEM code), [5]

and imply risk of brittle failure. Tension can be caused by upward expansion of the surrounding clay in a vertical deposition hole: the clay moves up while the canister is fixed in the clay at its lower end (Fig. 3.7).

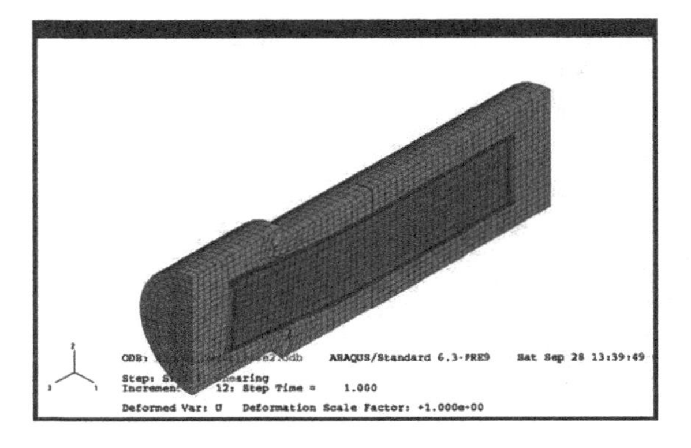

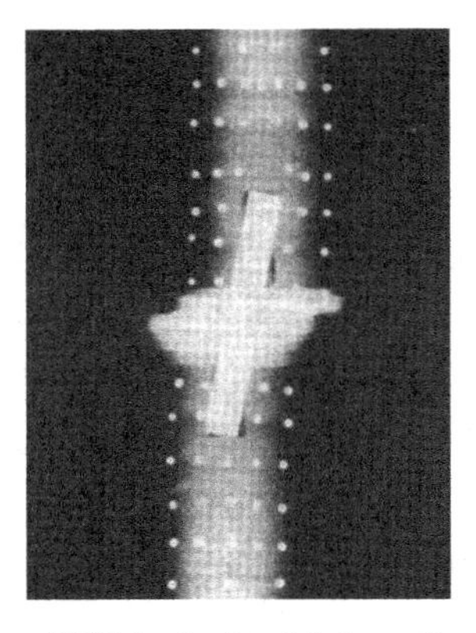

Fig. 3.6 Strain and displacement of SKB canister. *Left*: Illustration of FEM-calculated deformation of ductile clay-embedded (*yellow*) canister (*blue*) by 0.1 m tectonically induced instant shearing along a fracture intersecting the deposition hole normally to it. *Right*: Formation of heterogenenities in the clay (*dark wedges*) by shearing of a stiff canister. X-ray image of model test with leadshots in the clay to show the strain pattern. The bright center is the shear box arrangement [4]

Only recently it has been found that creep strain can be up to 5%, yielding cavern corrosion and fissures [6]. Specimens of different copper metals failed in tests under higher stresses than expected in a SKB-3 V repository but still of relevant magnitude considering possible stress concentrations originating from the conditions in Fig. 3.7 and from conversion of the inner iron component to magnetite. Failure associated with cavitation occurred even down to 75°C, indicating the sensitivity of copper/iron canisters of SKB's type. Fissures and holes, illustrated in Fig. 3.8, represent permeable features through which migration of water in liquid and vapour form into the canisters can take place.

A most important prerequisite for acceptable performance of the SKB canister is that the copper lids are tightly connected to the tube, for which electronic welding and "friction" – type jointing methods have been developed. Still, there is a risk that the joints can have defects or that mechanical stresses may yield corrosion and growth of defects of the copper tube. This has led to the assumption that a certain fraction of the canisters placed in the repository are defect from start, meaning that porewater from the surrounding clay material can migrate into the interior and come in contact with the iron core and fuel. Since this may be a general phenomenon water migration into the canisters and development of very high pressures on the rock may happen to all SKB canisters deposited in a repository.

The buffer may be exposed not only to high stresses but also to chemical impact of the following types:

- Dissolution of the iron core and migration of iron in elemental form and as iron complexes through the hole into the surrounding buffer clay.
- Dissolution of the uranium dioxide causing release and migration of radionuclides into the surrounding buffer clay.

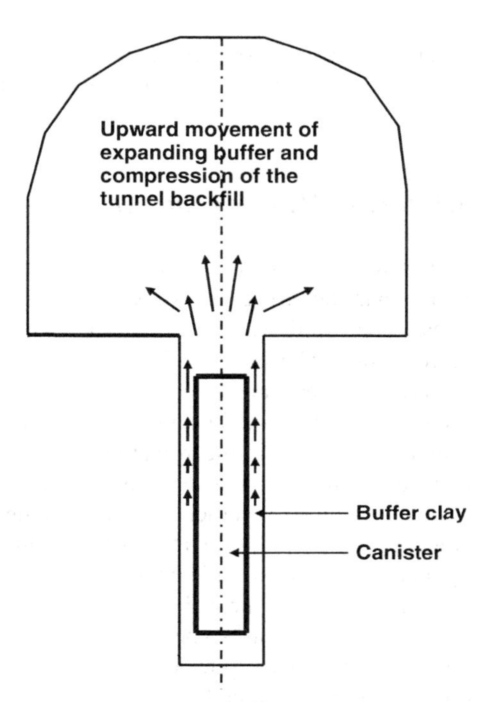

Fig. 3.7 Axial tension of canister by upward expansion of the buffer clay at its upper part in conjunction with compression of the tunnel backfill. The lower and central part of the canister is firmly held by the clay

- Vapourization of the water entering the canisters causing high pressure and migration of very hot vapour into and through the buffer clay causing dissolution of the mineral constituents. This would reduce the swelling pressure and hence the stresses in the rock.

For assessing this we need to consider the performance of the clay, which is one of the most important objectives of the book. However, before doing so we will

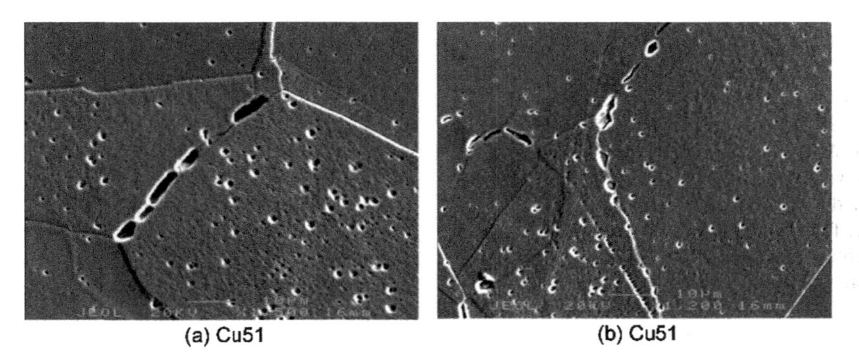

Fig. 3.8 Examples of intergranular creep cavities and cracks identified by scanning electron microscopy (SEM). Width of micrographs about 75 μm [6]

describe an alternative canister type that may provide a solution of the problem of safe disposal of HLW, i.e. the HIPOW canister.

3.3.3 The HIPOW Canister

The weaknesses of the presently proposed SKB canister are that defects caused by the manufacturing can not be avoided and that canisters will be exposed to high stress concentrations that can cause local breakage and exposure of the interior to the surroundings. Corrosion will weaken the canisters significantly after a few tens of thousands of years and through-corrosion by pitting may take place in less than that. An ideal canister would be a solid copper body homogeneously encapsulating spent fuel since this would mean that even very deep pitting corrosion would not cause exposure of more than a very small fraction of the fuel to water, namely where corrosion happens to hit a fuel element. A technique for producing canisters of this type was suggested in 1976 by the Swedish company ASEA Atom AB, which proposed application of hot isostatic compression, termed HIPOW [7]. The idea was to place the fuel bundles in their zirconium claddings in a copper tube in the compression chamber of a Quintus^R-type press with the fuel kept in constant position in the tube while pouring copper powder into it, followed by heating the object to 600°C while compacting it isostatically at 100 MPa (Fig. 3.9).

Half-scale tests by ASEA in the early eighties demonstrated the feasibility of the HIPOW process for encapsulation of spent nuclear fuel as illustrated by Fig. 3.10. The porosity of the copper was as low or lower than that of cast copper. The copper column was only 1.6 m long and with a diameter of about 1 m because of the

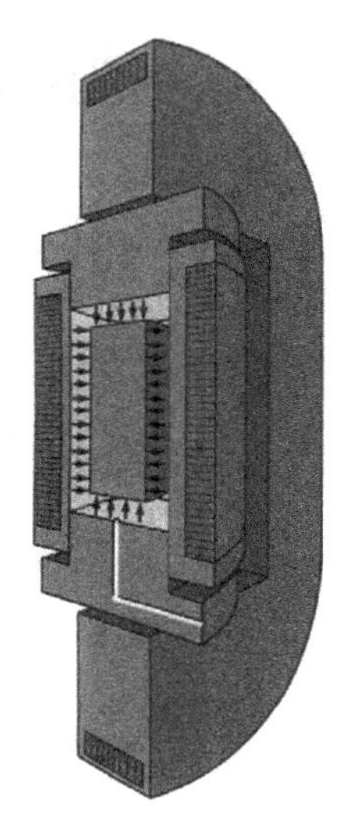

Fig. 3.9 ASEA's Quintus^R device for high isostatic ("HIPOW") compression. The powder material to be densified is enclosed in a flexible form which is submerged in a liquid pressure medium (oil, water) for cold isostatic compression. For hot isostatic compression, an inert high pressure gas is used, usually argon [7]

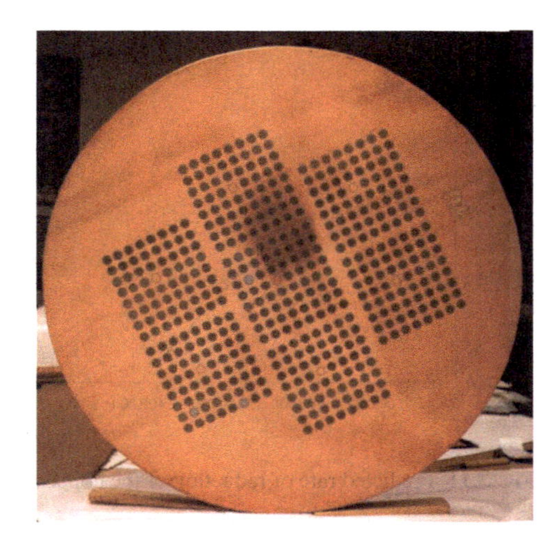

Fig. 3.10 Disc sawn from a 1.6 m long column of HIPOW-compressed copper powder in a copper tube. The spent fuel, simulated by steel pellets in the experiment, are the black spots in the copper mass

limited size of the press but today the length can be the same as of ordinary KBS-3 canisters [7].

For getting good metallic bonding between the copper powder grains, which is needed for achieving a non-brittle, high-strength product, surface oxide coatings of the copper grains must be removed. This can be made by reduction of the powder in hydrogen or some other gas at 250°C. Alternatively, the whole process of preparing a canister by compacting copper powder can be made in inert gas environment although this has not yet been demonstrated. The tensile strength of HIPOW canisters is at least 100 MPa.

In addition to the practically eliminated risk of exposure of the confined spent fuel to water for hundreds of thousands of years or more, the copper columns have the advantage of being stronger than SKB's presently proposed canisters but still relatively ductile. Fractures generated by extreme deformation would not expose more than an insignificant fraction of the waste to the surroundings. A further advantage that has to do with the requirement of straightness for bringing the cansisters down in deposition holes with good fitting is that perfect straightness would be easily achieved by minimum machining of HIPOW canisters, while SKB canisters are not straight and cannot be straightened.

Admittedly, there are practical difficulties in manufacturing HIPOW canisters because of the severe radioactivity conditions but reasonably extensive R&D is expected to overcome them [8].

3.3.4 Chemical Integrity of Copper Canisters

The most important criterion stated by regulators is that the canisters must guarantee tight confinement of the waste for 0.1 to 1 million years. A primary requirement is that through-corrosion will not take place in this period of time, which makes it necessary to estimate how corrosion takes place and what the rate of degradation really is. Copper is thermodynamically perfectly stable in pure water but undergoes

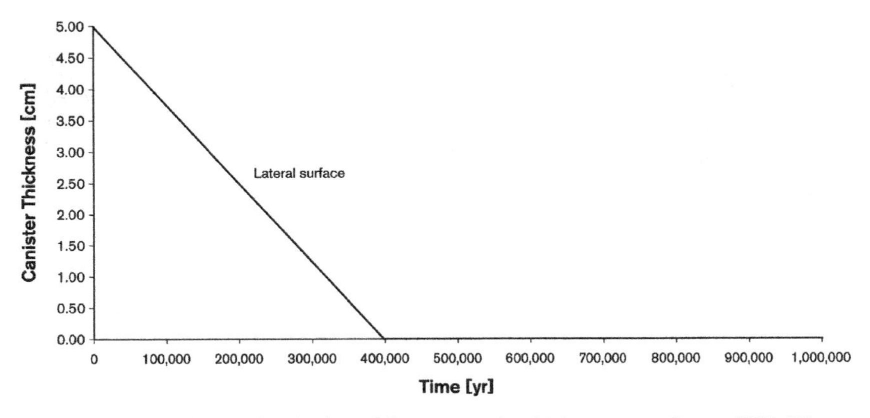

Fig. 3.11 Predicted rate of reduction of the copper tube thickness according to SKB [9]

corrosion in groundwater at a rate that is low where the concentrations of sulphur and oxygen, which are the most corrosive agents, are low. Since oxygen in the groundwater will be used up by various oxidation processes very soon after disposal and by microbes, elemental sulphur in the groundwater is the major threat to copper stored at depth. Corrosion of the initially 50 mm thick copper tube is predicted to be about 15 mm in a hundred thousand years according to experiments and theoretical models and it is estimated to be totally corroded in 400 000 years (Fig. 3.11). Pitting corrosion can speed up the degradation by an order of magnitude.

Applying the same rate history to the HIPOW canister it is clear that it offers total confinement of the waste for the stipulated 1 million year period and beyond that. Chemical impact of canisters on the surrounding buffer clay is another issue that we will look at later in this chapter.

3.4 Clay

3.4.1 The Role of Clays in a Repository

Canisters need to be embedded by a medium with very high tightness for minimizing percolation of groundwater, and with sufficient bearing capacity for minimizing movement of the heavy canisters but being sufficiently soft to transfer only a fraction of the stresses generated by tectonic movements to the canisters. A necessary requirement of the embedment is that it must have a potential for expansion in order to support the contacting rock and to heal voids that may arise from strain imposed by external forces or internal processes, like temporary shrinkage due to desiccation. Further desired properties are that it must be chemically compatible with the canister material and surrounding rock with its groundwater chemistry, and that it has to be largely inorganic for eliminating the risk of producing organic colloids that can carry radionuclides, or of creating new forms of life.[1] All these properties should be preserved for a period of time that is required by the regulators, commonly representing more than tens to hundreds of thousands of years. The only material that can provide all this is smectitic clay, which is a typical component of clays formed from volcanic glass in nature (bentonites), or resulted from weathering of granitic and amphibolitic rock. The important role played by clays in repositories justifies the detailed treatment of their constitution and properties given in this and other parts of the book.

3.4.2 Smectite Minerals

Clay has been taken as an effective barrier because of its low hydraulic conductivity and ion diffusion capacity and since it provides a ductile surrounding of the canisters. Smectites, like montmorillonite, are "expanding" and have been proposed as most suitable embedment, "buffer", of the canisters since they are tighter than other clay types [4,9,10]. The Organizations have paid much attention to the physical properties of commercially available clays and this had led to data bases that provide information on the hydraulic conductivity, expandability (swelling pressure), and ion diffusion, as functions of the density.[2] The buffers are often called bentonites, a term that should be reserved for clays formed from volcanic ash sedimented in the sea; many smectites originate in fact from weathered rock.

The smectites particles, which consist of very thin lamellae with a length of less than thousands of a millimetre, have a chemical composition that can be generalized as: $(Na,Ca,Mg,H_2O)(Al,Fe,Mg)_2(Si,Al)_4O_{10}(OH)_2$. The bonds between the lamellae, which appear as stacks of 3 to 30 units, are weak and water and other molecules can occupy the space between them. The surface area of a gram of smectite can be $1000\,m^2$, which makes the electrically charged stacks very active in forming stable gels even at very low densities. This extreme surface area also means that ion exchange processes take place fast and that the ion exchange capacity is high. At high densities such clays have an extremely low hydraulic conductivity because the dominant part of the porewater is nearly immobile. If the charge of the crystal lattices is neutralized by uptake of certain cations, like potassium, in the interlamellar space this fraction is reduced and the hydraulic conductivity raised, while the mechanical strength is reduced if the bulk density is unchanged. This process, which is generated by heating and uptake of potassium ions, is termed illitization since the smectite minerals are converted into illite minerals, or more correctly hydrous mica. It is believed to take place by successive transition from true smectite via mixed-layer smectite/illite (S/I), but may also result from neoformation of illite particles

---

[1] Jokes about little green men being created do not have to be taken seriously but it is clear that radiation can cause a special environment favouring changes in organic life forms.

[2] The density is often expressed as "dry density" i.e. the ratio of the solid mineral content and the total volume including voids but with no water content.

in the porewater on the expence of the smectite. In nature, rotation and shifting of structural layers and units can lead to polytypes that can differ in morphology and result in different physical properties.

The main smectite mineral species are:

Montmorillonite, $(M_{0.33}$ x $nH_2O)(Al_{1.67},Mg_{0.33})$ Si $_4O_{10}(OH)_2$
Beidellite, $(M_{0.33}$ x $nH_2O)Al_2(Si_{3.67}, Al_{0.33})$ $O_{10}(OH)_2$
Nontronite, $(M_{0.33}$ x $nH_2O)Fe_2{}^{3+}(Si_{3.67}, Al_{0.33})$ $O_{10}(OH)_2$
Saponite, $(M_{0.33}$ x $nH_2O)($ $Si_{7.34},All_{0.67})O_{20}(OH)_4$

"M" is a mono-valent cation or molecule: Li^+, Na^+, K^+, $NH_4{}^+$ or in some cases: $[Al(OH)_2]^+$ or organic molecules $RNH_3{}^+$. The exchangeable ions can be substituted for divalent cations: Ca^{2+}, Mg^{2+} etc. Electron diffraction patterns often indicate low crystallinity due to superposition of sheets in a completely random fashion. The XRD diffraction patterns generated by common materials typically reveal a number of broad peaks. Isomorphous replacement of lattice cations and the amount of interlamellar water molecules are related to the spacing of the lamellae, which give diagnostic values for the reflection peaks of XRD spectra.

The minerals formed in the illitization process are primarily illite and chlorite, which are largely non-expanding. Also glauconite, an iron-bearing equivalent of illite, can be formed. They have the following compositions:

Illite group – chemical formula: $K(Al,Fe,Mg)(Si,Al)_4O_{10}(OH)_2$

The members of this group incorporate a small quantity of water molecules and their diffraction patterns are almost the same as that of the mica mineral muscovite.

Glauconite group – chemical formula: $K(Fe,Al,Mg)(Si,Al)_4O_{10}(OH)_2$
Chlorite group – chemical formula: $(Fe,Mg)_6(Si,Al)_8O_{20}(OH)_4$

The particles typically aggregate and form stacks of 3–30 lamellae. Figure 3.12 shows two coupled lamellae of the commonly favoured model of montmorillonite. In fact, the true crystal lattice constitution of montmorillonite is not known with great certainty, it may well be complex with a certain fraction of the silica/oxygen tetrahedrons inverted depending on the inerlamellar cation and temperature [10].

A most important matter is that the interlamellar space can take up and give off water molecules depending on the temperature and the confining, effective, pressure. Most designers of waste disposal systems neglect the different in hydration potential of the various smectite species and can therefore not justify their preference of one of them. In fact there is considerable difference in hydrophilic performance as illustrated by Table 3.2.

Three hydrate layers can only be formed in some of the smectite species, and normally only when sodium, lithium or magnesium are in interlamellar positions. One finds from this table that montmorillonite is the only smectite mineral of the

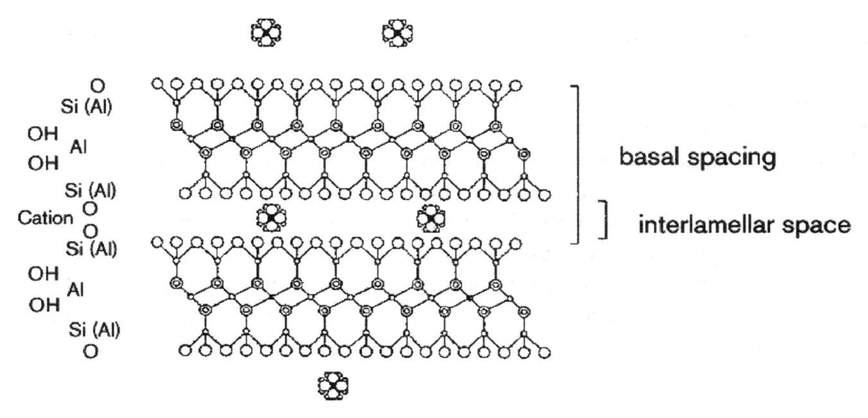

Fig. 3.12 Commonly assumed constitution of a stack of two montmorillonite lamellae with 10 Å thickness in dehydrated form. Cations and water molecules, which can form up to 3 hydrates are located in the inerlamellar space

three mentioned that can expand by forming 3 interlamellar hydrates. This is the reason why montmorillonite clay material expands more than the other smectites and why it has the strongest gel-forming capacity, and why Na or Mg montmorillonite should be selected as clay buffer material if expandability and self-sealing are preferred. However, the conditions are different concerning the chemical stability at increased temperature and pressure. Thus, Mg^{2+} appears to promote better stability of smectites then Na^{+} and counteracts formation of hydrous mica [13], and improved chemical stability is also offered by certain mixed-layer minerals, to which will return in Chap. 4. We will consider these matters in a detailed examination of the long-term performance of the buffer clay over longer period of time in a repository. In that context we will also examine some other clay types that may serve equally well or better than montmorillonite.

Table 3.2 Number and thickness of interlamellar hydrates in Å [11,12]

Smectite clay	M[1]	1st hydrate	2nd hydrate	3rd hydrate
Montmorillonite	Mg	3.00	3.03	3.05
	Ca	3.89	2.75	–
	Na	3.03	3.23	3.48
	K	2.42	3.73	–
Beidellite	Mg	2.69	2.69	–
	Ca	2.30	2.30	–
	Na	2.15	2.15	–
	K	2.54	–	–
Nontronite	Mg	2.92	3.00	–
	Ca	3.05	3.37	–
	Na	2.70	2.79	–
	K	2.60	–	–

[1]Cation

3.4.3 Microstructural Constitution of Smectite Clays – A Key Issue

The microstructure of smectite clays determines their physical properties. This is illustrated by Fig. 3.13, which shows the build-up of networks of particle aggregates and demonstrates that more water is in interlamellar positions in smectites than clays containing illites (hydrous mica) and kaolinite [4,10]. The huge specific surface area of the smectite lamellae system that binds much of the porewater, and the fineness of the particle aggregates that makes the void size very small, combine to cause the extremely low hydraulic conductivity and the significant expandability that are the main reasons for using smectitic clays, like bentonites, as engineered barriers in HLW repositories.

3.5 Buffer

3.5.1 Function

The canisters confine the HLW and need protection against external impact like tectonically induced shearing of discontinuities of 4th order that intersect the holes they are placed in, and chemical attack by dissolved elements and microbes. This requires embedment in a ductile, very tight medium – "buffer" – with sufficient bearing capacity to keep the canisters in position and sufficient softness to absorb strain, and also with an ability to maintain tight contact with the canisters and the surrounding rock. Very dense smectite clay is suitable for this purpose if the dry density – expressed as the ratio of the mass of solid minerals within a defined volume

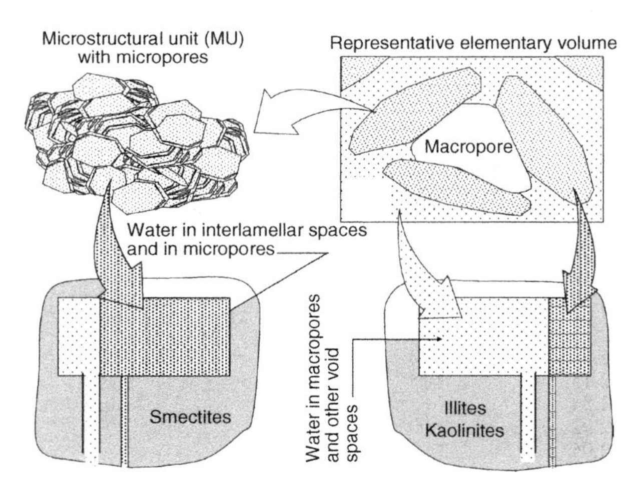

Fig. 3.13 Microstructural units in clay of smectite type (*left*) and clays with illite or kaolinites (*right*) as major mineral constituents [10]

and the volume including all sorts of voids – is in the range of 1450 to 1900 kg/m^3, which corresponds to 1900 to 2200 kg/m^3 in fully water saturated form.

The reason for the desired high density of the buffer is that transport of dissolved species and water takes place predominantly through slow diffusion and not flow, while for lower densities they can migrate quicker because flow contributes to the transport. The swelling pressure and hydraulic conductivity are determined in the laboratory by the type of rigid "swelling pressure oedometer" illustrated in Fig. 3.14. The evolution of the swelling pressure is rather slow and dependent on the rate with which water is sorbed in the wetting process. Montmorillonite is tighter than the other smectites because of the higher amount of interlamellar water, which is largely immobile (Table 3.3). As indicated in Fig. 3.15, the porewater salinity controls the conductivity because the softest parts of the microstructural network coagulate in salt water by which the voids increase and become more interactive.

In performance analysis and assessment one needs to consider not only the hydraulic properties, but also the following issues referred to as Thermal (T), Hydraulic (H), Mechanical (M), Chemical (C), Biological (B) and Radiological (R) processes and properties. In the present subchapter we will consider mainly T, H, and M but we will see later how the others can affect the buffer material.

It is obvious that if the buffer clay has a density of about 2000 kg/m^3 its hydraulic conductivity is lower than that of the crystal matrix of the rock. The rate of chemical processes in the buffer involving transport of ions into and out from the buffer in a

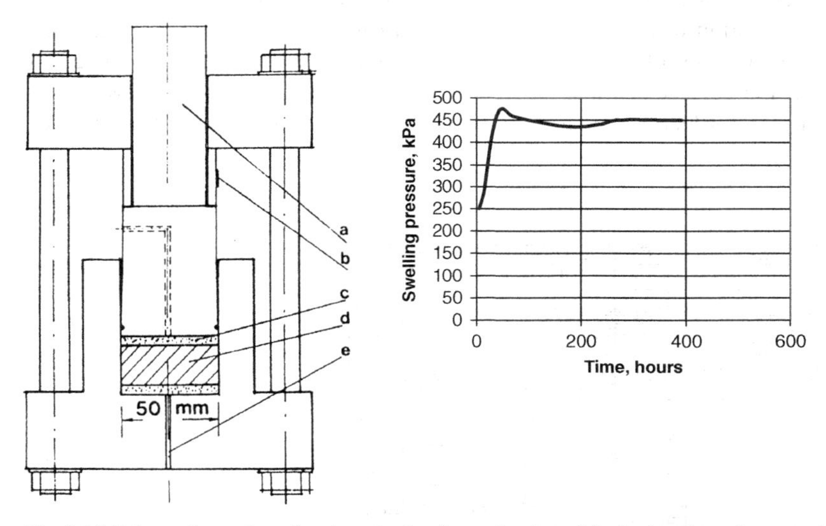

Fig. 3.14 Schematic section of oedometer for determination of the hydraulic and gas conductivities and the swelling pressure. The common procedure is to precompact the air-dry clay grains to a desired density and then let the material take up water or solution under confined conditions while recording the swelling pressure (right picture showing montmorillonite-rich clay with 1700 kg/m^3 density at saturation with distilled water). When equilibrium is reached, a hydraulic gradient is applied across the sample and the percolation rate measured as a function of time. a) is piston, b) strain gauge, c) fine filter, d) clay sample and e) inlet of fluid

Table 3.3 Generalized hydraulic conductivity (*K*) of nearly pure clays with different clay minerals

Dominant clay mineral	Density at water saturation, kg/m^3	*K* m/s percolated by distilled water	*K* m/s percolated by 3.5% $CaCl_2$ solution	Fraction of the porewater in interlamellar positions, Nasaturated	Fraction of the porewater in interlamellar positions, Casaturated
Montmorillonite	1500	E-13	5E-13	40	20
Montmorillonite	1750	3E-14	E-13	80	40
Montmorillonite	2000	2E-14	5E-14	90	85
Illite	1500	2E-11	E-10	<5	>5
Illite	1750	5E-12	E-11	5-10[1]	5-10[1]
Illite	2000	E-12	3E-12	5-30[1]	5-30[1]
Kaolinite	1500	>E-8	>E-7	0	0
Kaolinite	1750	E-9	E-8	0	0
Kaolinite	2000	E-10	E-10	0	0

[1] Theoretically zero but of the indicated order for commonly somewhat weathered species

deposition hole is controlled both by the rate of water flow in rock fractures interacting with the clay in its hole and by the rate of ion diffusion in the rock crystal matrix surrounding the hole, and in the clay. The clay has a very low diffusion capacity for anionic species because the large majority of the clay particle surfaces are negatively charged. The cation diffusivity is higher but still low as illustrated by Fig. 3.16. It is of particular importance that the migration rate of anions can be reduced to nearly nil if the density at water saturation can be raised to 2100 kg/m^3 (dry density 1750 kg/m^3). Electrical equilibrium can of course be maintained irrespective

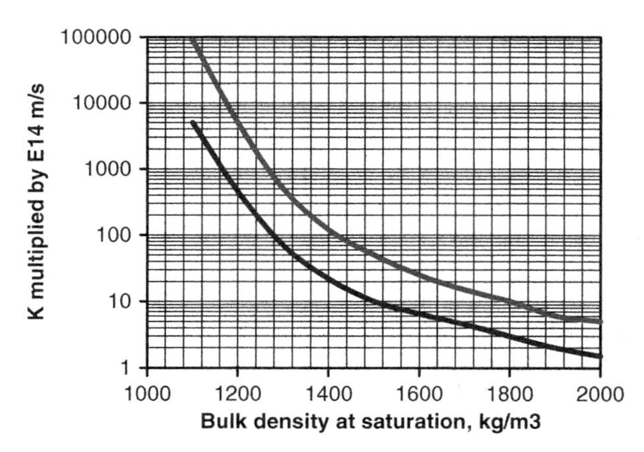

Fig. 3.15 Hydraulic conductivity of smectite-rich clay with Na as major adsorbed cation. The lower curve in the diagram represents low-electrolyte water and the upper ocean salinity. For the same clay saturated with Ca the conductivity is about 10 times higher than for the Na form

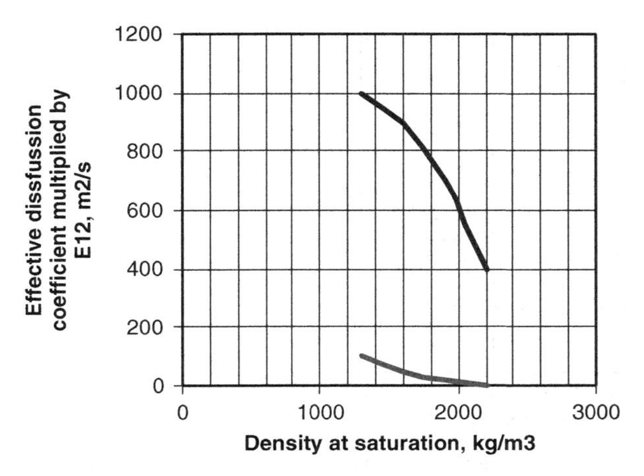

Fig. 3.16 Measured ion effective diffusivities for smectite clay [10]. *Upper*: Monovalent cations. *Lower*: Monovalent anions like chlorine

of the rate and type of cations if their migration is controlled by place exchange mechanisms.

The diagram in the upper right of Fig. 3.14 illustrates a most important property of smectite-rich clays, i.e. to exert a swelling pressure on the confinement. This ability depends on the nature of the cations that are adsorbed to balance the negative charge of the crystal lattice of the individual lamellae. At higher densities the pressure is due to the potential of "steric", interlamellar water establishing the maximum number of hydrate layers that are indicated in Table 3.2, while at lower densities it has the character of osmotic pressure, controlled by the force fields between adjacent stacks of lamellae as concluded from numerous scientific studies [10]. The diagram shows the typical evolution of the pressure at unlimited access to water, like in the oedometer, i.e a quick rise to a peak value and then a slight drop followed by a period with small changes but ending up in a constant pressure. In practice, the evolution of the swelling pressure is controlled by the ability of the surrounding rock to give off water and this can be a threat to acceptable performance of the buffer as we will show in this chapter.

The diagram in Fig. 3.17 illustrates the relationship between swelling pressure and density at complete saturation of smectite-rich clays with Na and Ca as major adsorbed cations. Thus, while Na-clays show some minor swelling pressure at densities even down to 1050 kg/m^3, Ca-clays need to have a density of at least 1600 kg/m^3 to exert a measurable swelling pressure. The reason is the strong coagulation of Ca-smectite, which makes it impossible to prepare a coherent sample of Ca-clay of lower density than his value; it would simply lead to separation of clay and water. For the high density required for the clay surrounding HLW canisters, changes in porewater composition and concentration (TDS) have only a rather small impact on conductivity and swelling pressure. This is because the frequency of microstructural elements with sufficiently low density to react on such changes is very low. For clays

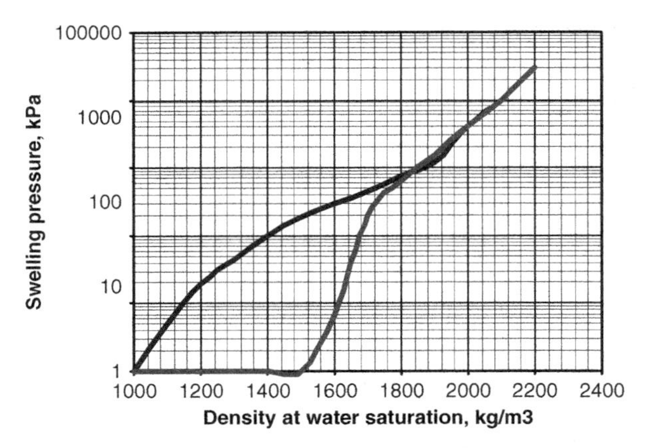

Fig. 3.17 Typical relationship between swelling pressure and density of smectite-rich clay. The curve starting at origo represents Na montmorillonite (MX-80) saturated with low-electrolyte water and the other Ca montmorillonite saturated with low-electrolyte water

with bulk densities lower than about 1800 kg/m^3, this frequency is higher and the influence of porewater chemistry significant.

It should be noticed that for densities at saturation higher than about 2150 kg/m^3 the swelling pressure increases beyond 20 MPa.

3.5.2 Evolution of the Buffer

Realizing the fundamental importance of canister tightness, it is clear that the embedding clay must serve to protect the canisters from mechanical impact and be so tight that practically no permeation by groundwater can take place. These are the main objectives of the buffer, while retardation of the migration of escaping radionuclides is a secondary and less important task. Hence, the primary requirement is that the buffer should maintain its tightness and ductility and not exert shearing or bending of hollow copper/iron canisters. We will examine these issues and start by considering the first phase of the maturation of the buffer, taking the SKB concept as a basis. It is termed KBS-3 V and shown in Fig. 3.18. The design principle was proposed back in the late seventies and has later been taken as candidate concept also by other representatives of the Organization, like Canada, South Korea, South Africa and Japan. It has the nice property of allowing for intermittent placement of the waste, offering much time for preparation of the deposition holes, emplacement of buffer and canisters, and for closing them. Backfilling of the tunnels does not need to be made in conjunction with the waste disposal but should be made relatively soon afterwards for making it possible for the piezometric pressure to rise and supply the buffer with water.

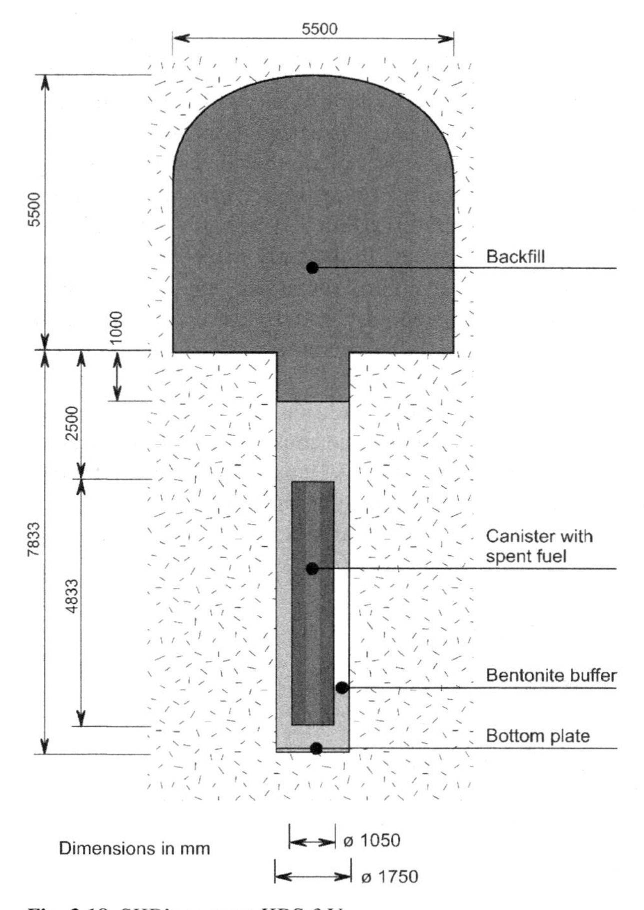

Fig. 3.18 SKB's concept KBS-3 V

3.5.3 Wetting Rate of the Buffer

The buffer must be manufactured and placed so that the homogeneity of the matured clay becomes high and its density so high that transport of ions will take place by diffusion and not by flow. This has led to the criterion that the minimum final density of the matured buffer should be 1950 kg/m^3. However, using soil mechanical THM models that are used for predicting the soil mechanical performance of the buffer [2], the ultimately formed density will not be uniform, which has also been verified by field experiments. This is because the internal friction prevents complete evening out of the initially very obvious density differences. Theoretically, at least, the density will be highest at mid distance between canister and rock and lowest near the rock.

The principle of using buffer consisting of blocks of highly compacted smectite clay as major component according to the Swedish and several other concepts, means that the blocks will expand by taking up water from the rock and thereby close the gap between blocks and rock that is required for placing the block. An often raised question is if the joints between the blocks serve as paths for quick inflow of water and thereby cause local expansion with irregular uplift and breakage of individual blocks. This seems to be the case and it is definitely proven that steep desiccation fractures are formed in the hottest part of the buffer before swelling by water uptake ultimately closes them. Another question is if the density in the ultimately matured buffer can be lower than the stipulated 1950 kg/m^3 or be higher than 2100 kg/m^3, which may damage the canisters in the early mentioned fashions. This risk can be minimized by careful material control, but a risk remains that water seeping into canisters of SKB's presently favoured canisters can cause such high vapour pressure that the buffer is compressed. Also, pressure of hydrogen gas given off from the corroding iron insert adds to the pressure.

Neglecting for the moment such extraordinary conditions we will see here how the normal evolution of the buffer proceeds and take the constellation of engineered barrier components of the KBS-3 V concept in Fig. 3.19 as a basis for predicting the maturation process. As shown in the figure the 5 cm wide space between clay blocks and rock will be filled with compacted smectite-rich pellets for reaching the

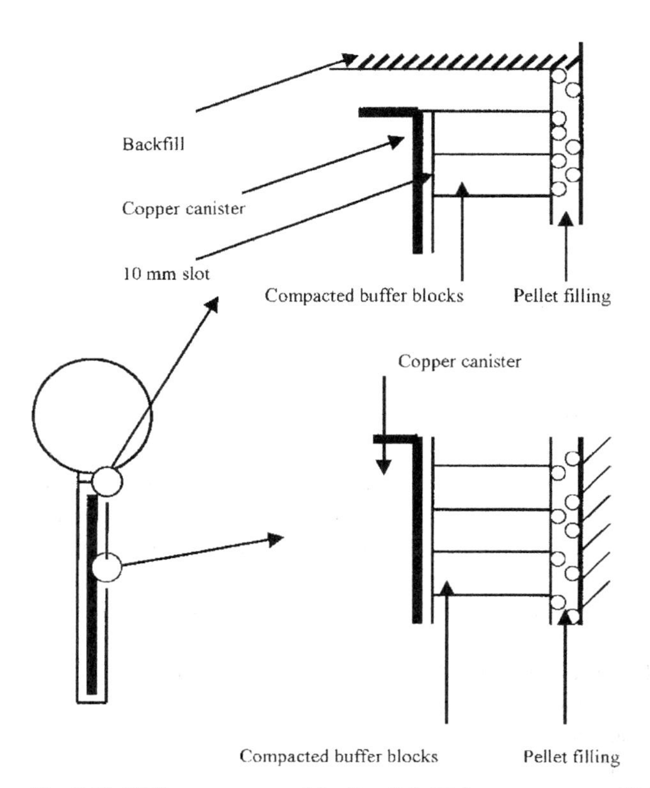

Fig. 3.19 EBS components of the Swedish EBS components, valid in principle also for the other concepts for crystalline and argillaceous rock, with or without pellet fills

required minimum density. Tests in SKB's underground laboratory have shown that placement of pellets is impractical and a better solution would be to change the design by letting the outer gap be unfilled but reduce it by 50% for maintaining the same average density of the buffer. This would, however, require precision boring for obtaining perfectly straight and vertical holes, requiring techniques that are presently not available.

An option is to fill the hole with smectitic mud so that its upper level currently coincides with the top of the stack of blocks when the buffer blocks are inserted. The mud should of course have a sufficient density for yielding the required net density after complete maturation of the buffer. This case gives quicker and more homogeneous wetting of the blocks but has not yet been tested.

Several coupled processes take place in the maturation of the buffer in Fig. 3.20. In a strongly simplified form they can be described as follows, assuming like the modellers in the CROP project, that the buffer blocks are homogeneous and completely occupy the hole:

- Porewater in the buffer migrates from the hot part of the buffer towards the colder walls of the deposition hole. The outer, colder part of the buffer expands and exerts a swelling pressure that tends to compress the warmer, less wetted buffer.
- Water enters the deposition holes, which causes additional wetting of the buffer from the rock/buffer contact and successive migration of water towards the hot canister. A swelling pressure is first built up in the porous pellet fill next to the blocks and in the coldest part of the dense clay blocks by which the blocks are pressed against the canister. Water is successively sorbed also in the hotter part of the blocks and they expand and start compressing the then fully water saturated pellet fill.
- The wetting involves transport of water in liquid form from the cold towards the hot part of the buffer where partial vapourization takes place followed by

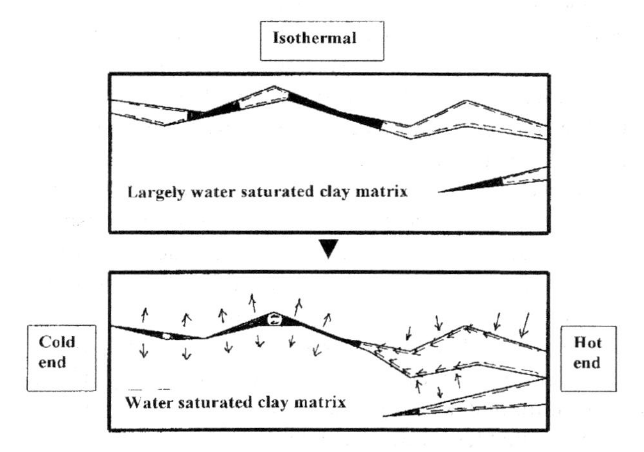

Fig. 3.20 Schematic picture of microstructural changes in the initial process of porewater redistribution under a thermal gradient

flow back in vapour form towards the colder part where condensation takes place followed by migration of water towards the hot side where vapourization takes place etc.

The true behaviour is much more complex and requires that the microstructural constitution is taken into consideration. The most important feature is the aggregated character of the buffer that is inherited from the initial state of clay powder compressed to form the blocks. The smectite powder grains have a water content that is related to the relative humidity of the air in which it is stored. For pure smectite stored at 50–70% relative humidity the water content of the grains is about 10%, implying that the interlamellar space holds 1–2 hydration layers. Compaction of the grains under 100 MPa pressure welds the grains together leaving only small voids between them. The voids have the form of channels of varying aperture, the tightest parts being completely filled with capillary water while wider ones contain air. This "isothermal" state is shown in the upper part of Fig. 3.20, which also indicates the changes in the microscopic channels caused by a thermal gradient yet with no water uptake at the cold side. At this stage the drying of the hottest buffer makes stacks of lamellae contract, causing widening of the channels voids and formation of steep, radial fractures to several centimetres distance from the hot canister. The accumulation of water in the cold part of the buffer increases the degree of water saturation and causes expansion.

Assuming that buffer blocks fully occupy the deposition hole and that the rock provides unlimited amounts of water for uptake of the buffer it does so by exerting a tremendous suction, up to 100 MPa, on the rock. The suction of dense smectite clay is known to be this high if the initial degree of water saturation of the clay is below 20–30%, while it drops to very low values when the clay is nearly saturated. The suction is the driving force of the wetting, which takes place by flow and diffusion [10].

Considering the pellet fill at the buffer/rock contact, its density is low and the suction hence low, which implies that water that initially enters the fill flows in from the rock under a low hydraulic gradient. The dense blocks suck water from the pellet fill, which becomes compressed. Still, its density will stay lower than that of the blocks and it will therefore make up the most permeable part of the buffer, disregarding, for the moment, from the buffer closest to the hot canister where physico/chemical processes will raise the conductivity. The denser the compressed fill, the lower its hydraulic conductivity, which therefore partly determines the rate of water transport from the rock via the fill to the blocks. With increasing degree of water saturation follows a drop in suction and in the latest saturation phase it is the water pressure in the rock that determines the rate of water uptake by the whole buffer.

One realizes from all this that the ability of the rock to give off water to the contacting buffer has a strong effect on its wetting rate: if the conductivity of the rock is sufficiently high, there is unlimited access to water for wetting of the buffer, while a low rock conductivity means that less water is available to the buffer and that this retards the water uptake and maturation of the buffer [10]. This is obvious from the numerical calculation of the saturation rate of a KBS-3 V buffer in Fig. 3.21.

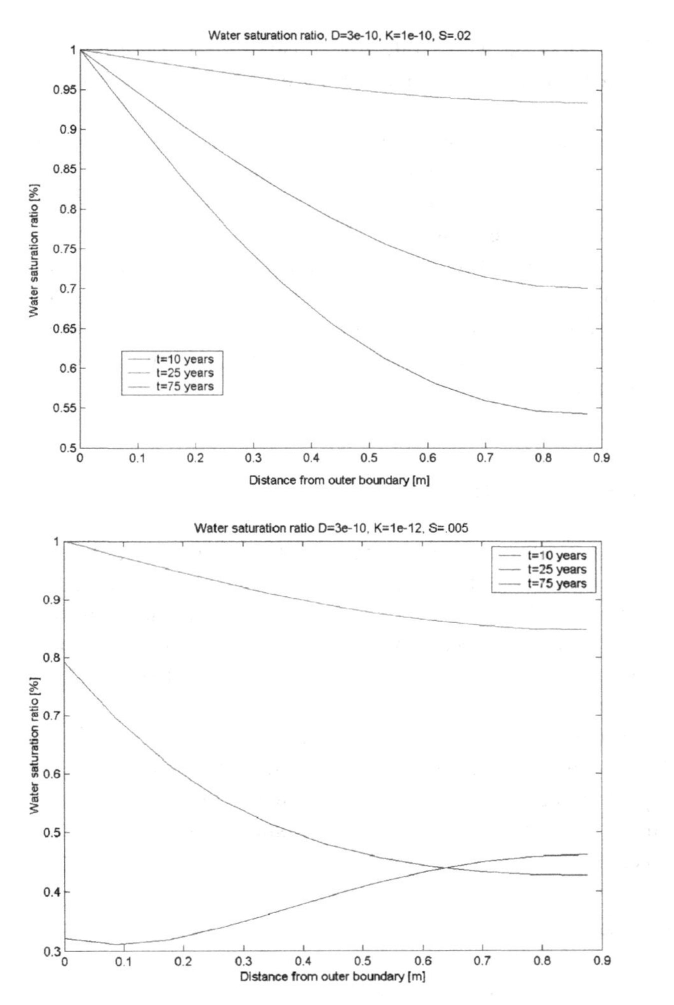

Fig. 3.21 Evolution of saturation of KBS-3 buffer with 1.8 m diameter for the hydraulic conductivity E-10 m/s (*upper*) and E-12 m/s of the rock (*lower*), [10]

In fact, for very tight rock, the buffer dries rather than becomes wetted in the first 10 years.

Water saturation does not mean that the buffer becomes homogeneous. For this redistribution of minerals and asoociated migation of water are required, which are both extremely slow processes because the driving forces for homogenization are negligibly small. Complete uniformity with respect to density and particle orientation may in fact never be reached.

It is interesting to see that the rate of saturation of buffer clay is not particularly affected by temperature and moderate temperature gradients. Thus, as we will see in a later sub-chapter dealing with prediction of buffer wetting rates, initially dry buffer becomes saturated at nearly the same rate under isothermal conditions as under a temperature gradient of about 1°C per centimetre, provided that boiling does not take place.

Very tight rock on macroscopic and megascopic scales, exemplified by one of the candidate sites in Sweden and by most types of argillaceous rock, can delay the water saturation so much that the hottest part of the buffer is damaged by permanent microstructural contraction and cementation. Occasionally, deposition holes are intersected by as much as three to four 4th order discontinuities, which will give off sufficient amounts of water via the boring-induced EDZ (cf. Chap. 2) to eliminate this risk. However, for the common case of fewer fractures intersecting the holes the wetting may be so slow that critical conditions prevail with respect to the chemical stability of the buffer.

3.5.4 Mechanical Processes in the Buffer

The mechanical and hydraulic interaction of the pellet fill and the dense blocks has been studied in laboratory experiments using the test arrangement in Fig. 3.22 with compacted MX-80 contacting MX-80 pellets. Distilled water was let in to the pellet-filled space, thereby simulating the true conditions in a deposition hole in rock with low-electrolyte groundwater. The water uptake and development of swelling pressure are shown in Fig. 3.23.

The hydration rate was represented by a straight line in a log-time diagram indicating diffusion-controlled saturation. The evolution of the swelling pressure is typical with an initial small peak and a second main peak a few weeks later, followed by a slight drop and finally a continuous slight rise in pressure caused by delayed expansion of the very dense pellets.

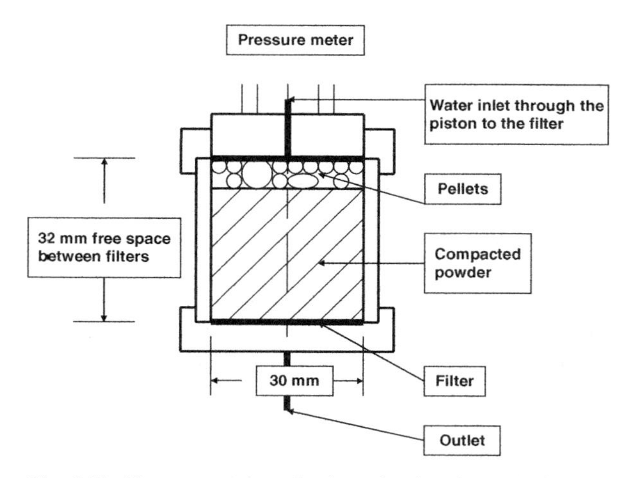

Fig. 3.22 Test arrangement for investigating the evolution of compacted buffer contacting soft pellet fill

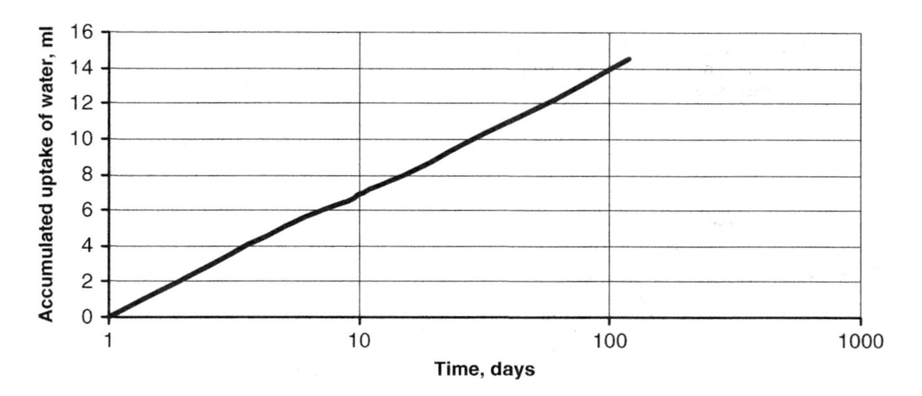

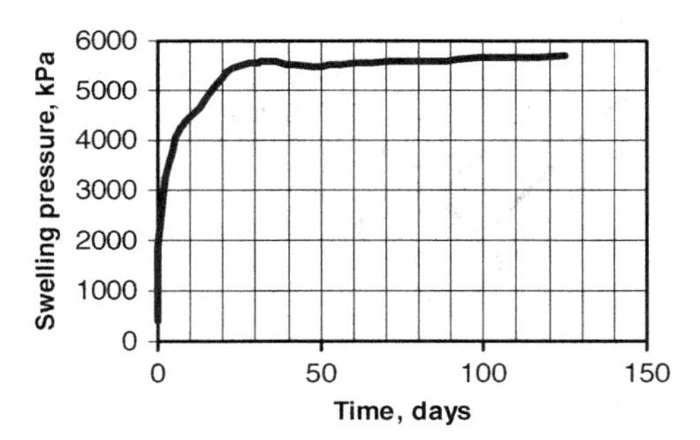

Fig. 3.23 Maturation of highly compacted MX-80 powder contacting pellets of MX-80 clay. *Upper*: Hydration rate. *Lower*: Development of swelling pressure

At termination of the test the sample was sectioned for determination of the density of diffferent parts. The clay was apparently homogeneous with an average density of 2050 kg/m^3.

3.5.5 Maturation of Buffer Submerged in Smectite Mud

Filling of pellets around the highly compacted clay blocks according to the present SKB concept KBS-3 V is not only impractical but can also lead to asymmetry if wetting of the fill and blocks is not symmetric. If one follows the idea of using super-containers, uniform and symmetric wetting of the buffer blocks would be achieved if they are surrounded by smectitic mud. The supercontainers, which can be made of navy bronze consisting of more than 90% copper and having a tensile strength approaching that of iron, can be prepared outside the deposition area by filling them with highly compacted smectite-rich blocks and centrally placed containers. Filling the holes with smectite mud to a suitable level and submerging the supercontainers in them would offer considerable advantages by providing the buffer with a large

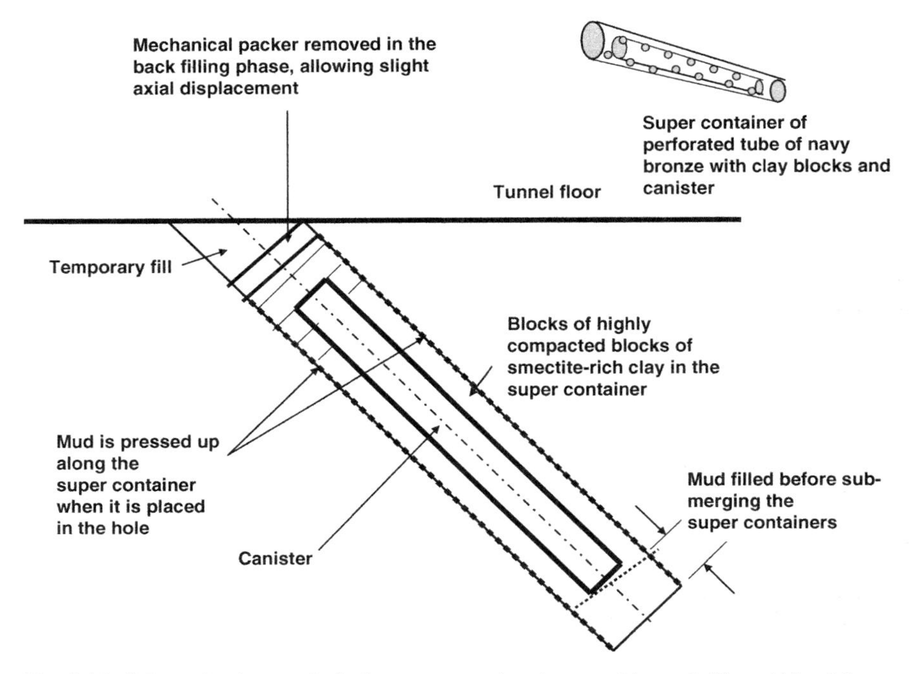

Fig. 3.24 Schematic picture of placing supercontainer in smectitic mud. The width of the gap between the supercontainer and the rock should be less than 3 cm

amount of water for its hydration, and by transferring heat to the rock. A practical technical solution would be that in Fig. 3.24, which illustrates a case with inclined deposition holes that we will consider further in Chap. 4. The technique is of course applicable also to vertical deposition holes.

The mechanical and hydraulic interaction of smectite mud with dense smectite blocks with a dry density 1675 kg/m^3 is exemplified by laboratory experiments under isothermal conditions, using the test arrangement in Fig. 3.25. It represents a technique proposed for borehole plugging and implies that highly compacted smectite blocks are fit in a perforated tube that is submerged in smectite mud. In the experiments, as well as in full-scale application of the supercontainer concept to canister deposition holes, the dry density of the dense clay is about 1700 kg/m^3, which yields an ultimate density after water saturation of 2070 kg/m^3.

In principle, suitable muds can be prepared by dispersing smectite clay in Na form, and possibly some quartz particles of silt size, to reduce the erodability, in low-electrolyte water to a dry density of a few 100 kg/m^3. They are low-viscous fluids at pumping but stiffen thixotropically when left in the hole (Fig. 3.26).

The sample extracted from the oedometer in Fig. 3.25 appeared as in Fig. 3.27. Clay specimens were cut from three parts of the sample: (1) the outer "skin" with 1–2 mm thickness, (2) the "plugs" extending from the perforated tube to the "skin",

Fig. 3.25 Growth of soft clay through the perforation of an 80 mm copper tube with dense MX-80 clay in an oedometer for simulating maturation of a borehole plug. After 8 hours the central part of the clay core is still unaffected. After a few days the soft gel is densified by being compressed by clay moving from the core through the perforation [14]

Fig. 3.26 Clay mud of MX-80 bentonite with the density 1100 kg/m^3 prepared by mixing clay powder with distilled water and left to rest for 2 minutes. The shear strength is sufficient to avoid "slope failure" of the clay gel

and (3) the clay between the "plugs", extracted from the tube and 2–3 mm outwards, i.e. to the "skin".

The clay "skin" surrounding the perforated tube, representing a model version of the supercontainer, had a rather uniform density after 24 hours, when it had increased from the initial value 1100 kg/m^3 of the mud to 1200 kg/m^3 and to about 1700 kg/m^3 after 14 days. The pressure increased to several tens of kPa in a few

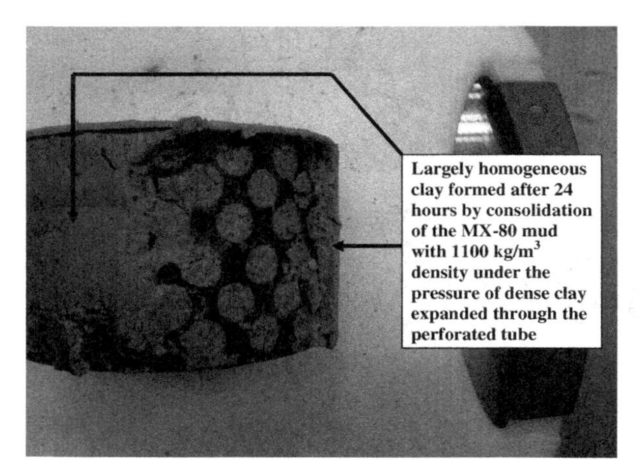

Fig. 3.27 Appearance of 24 hour old clay plug of "basic" type after cutting off clay specimens for investigation with respect to water content and density

days, which provides very good mechanical support to the supercontainer and the surrounding rock.

3.5.6 Modelling of Buffer Evolution – Conceptual Version

3.5.6.1 Migration of Liquid Water and Vapour

Prediction of the evolution of the buffer has to be verified by relevant field experiments to be thrustworthy. Comprehensive national and co-operative international research work have been made for testing proposed models of buffer evolution. We will examine here the basic principles on which the theoretical models are based and see how they fit with recorded data.

The processes that have been modelled are indicated in Fig. 3.28.

Considering first movement of liquid water under temperature gradients there are five possible reasons for water movement in the liquid phase in an unsaturated soil under a thermal gradient [10]:

- Surface tension gradients.
- Temperature influence on soil-water potential.
- Thermo-osmosis; flow of water from the less warm regions to the warmer regions because of differences in the respective heat contents between the liquid water layers sorbed onto the solid particle surfaces and the bulk liquid in the pore spaces.
- Water movement generated by the changes in the random kinetic energy of the hydrogen bonds.

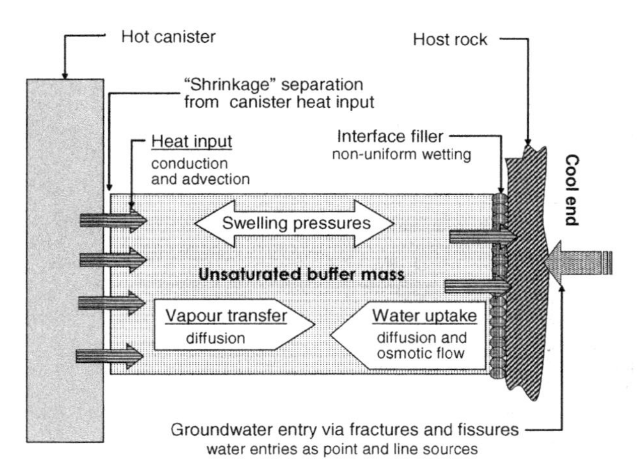

Fig. 3.28 Simple schematic showing basic elements of the canister-buffer problem to be modelled [10]

- Ludwig-Soret effect. This cross effect between temperature and concentration results in the phenomenon of thermal diffusion, as for example in the diffusion of solutes in the pore water from the warm regions to the less warm regions.

Vapour migration in the buffer mass is basically a diffusive process. In practice, assessment of the importance of vapour migration needs that the boundary conditions of the system are taken into consideration. Thus, vapourization of porewater takes place in the hottest part of the buffer and results in loss of water to the overlying backfill in the KBS-3 V case, meaning that the total amount of water can change significantly.

3.5.6.2 Swelling and Compression

Swelling of a buffer mass upon water entry and uptake produces compressive forces throughout the wetted portion, which will be transmitted to the unwetted portion of buffer resulting in progressive compression of this part (Fig. 3.29).

The compressive forces resulting from the constrained swelling performance of the buffer material depend on the amount of actual volume change occurring in the wetted portion, and on the rigidity and degree of uniformity of the unwetted portion, all related to the mechanical behaviour of the system (M, being one of the earlier defined parameters T,H,M,C,B). The latter includes compression of joints between clay blocks and of the gap between blocks and canister.

In the early maturation phase, chemistry (C) plays a major role and biological activities also play a role. One can identify two broad scenarios that include transport of dissolved solutes due to the presence of thermodynamic gradients provoked by temperature (T) and hydraulic gradients water (H), and impact on the clay material in the buffer mass due to chemical reactions and biological activities

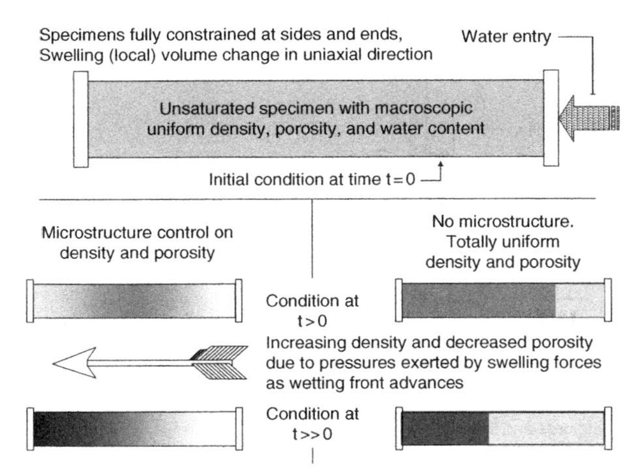

Fig. 3.29 Illustration of a laboratory experiment demonstrating the effect of swelling of buffer material in a constrained sample. Note the initial macroscopic uniform condition prior to water entry into the sample. Increased darkness in shading illustrates increasing density of sample [10]

(B). Transformations in the buffer material take place over long time periods. Early chemical reactions promote dissolution and precipitation, depending on the type of clay mineral, and they are greatest at higher temperatures and higher moisture contents.

Considering C-effects first, the most important impact in the early maturation phase, the water saturation stage, is the accumulation and precipitation of compounds with lower solubility at higher than at lower temperatures. The elements are Na, Ca and Mg, and the anions Cl and SO_4, yielding the compounds sodium or calcium chloride, or sodium, calcium and magnesium sulphates. They precipitate at the moving front of water saturation and serve as cement, changing the swelling capacity and hydraulic conductivity both as precipitates and by altering the microstructure through cation exchange from the initially dominant sodium. Later, when thermal gradients are evening out some of them can dissolve again and diffuse out of the buffer. Figure 3.30 illustrates this phenomenon as verified in laboratory experiments where a 7 cm long MX-80 clay column with a dry density of 1279 kg/m^3 and initial water content of 24% was confined in a tube with filter at one end that was held at room temperature and the opposite, closed end warmed to 100°C. At the colder end the clay could take up 3.5% NaCl solution for 45 days. Figure 3.31 shows scanning electron micrographs of precipitated sodium chloride and sodium sulphate.

One concludes from the diagram in Fig. 3.30 that the salt content increased considerably near the wetting front but that it was lower in the wetted clay, indicating that excess salt successively diffused back towards the water inlet. If the solution had been rich in calcium, comprehensive cation exchange from the initially sorbed sodium to calcium had taken place, causing significant changes in microstructure, primarily contraction of the stacks of lamellae and widening of voids. This would have caused a higher hydraulic conductivity of the water-saturated part of the

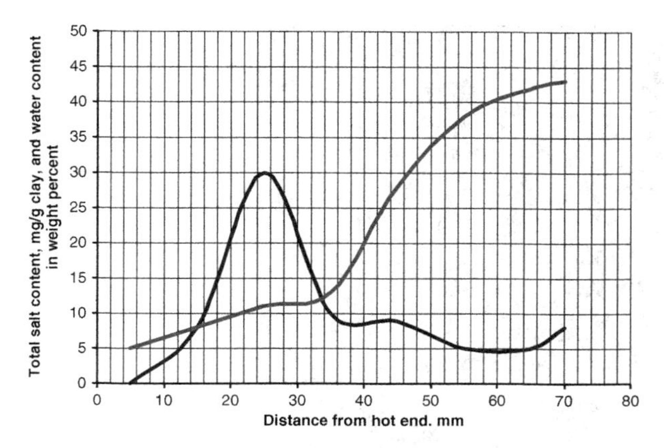

Fig. 3.30 Wetting of the clay from the cold end (steadily rising curve) and accumulation of salt at the wetting front (curve with peak) in laboratory experiment [4]

sample and quicker wetting of all parts of the sample, which is verfied by Table 3.4 giving data from a saturation experiment with a MX-80 sample with 50 mm length and diameter. The dry density was 1800 kg/m^3 and the initial water content 10%. The longitudinal temperature gradient was 10°C/cm. An external water pressure of 50 kPa was applied at the cold end. Figure 3.32 illustrates schematically the difference in the microstructural constitution of smectite clay saturated with low-electrolyte water and salt water. The rate of saturation is hence a function of the chemical composition of the water taken up by the clay.

3.5.7 Modelling of Buffer Evolution – The "Codes"

3.5.7.1 General

Several attempts have been made to express the conceptual model for the wetting/drying and expansion/compression of smetcite-rich buffer in terms of theoretical models, termed codes because of their use in numerical calculations. They have been developed by international research groups with the intent of exercising models of the behaviour of the buffer and backfill in relation to T,H,M,C and B. The codes, which have in fact not yet reached the level of sophistication to incorporate C and B, have been described in detail in international projects [14] to which the interested reader is referred. For the basic models attempt have been made to include C for coupling with a geochemical speciation model but it has still to reach the stage where it can directly be superposed onto the earlier developed THM platform. We will see that the assumed physical performance assumed as basis of the codes is certainly debatable.

The stochastic nature of the system of particles and porewater would cause great difficulties in modelling the maturation of the buffer on the microstructural scale and the modellers have instead, neglecting the actual physico/chemical processes

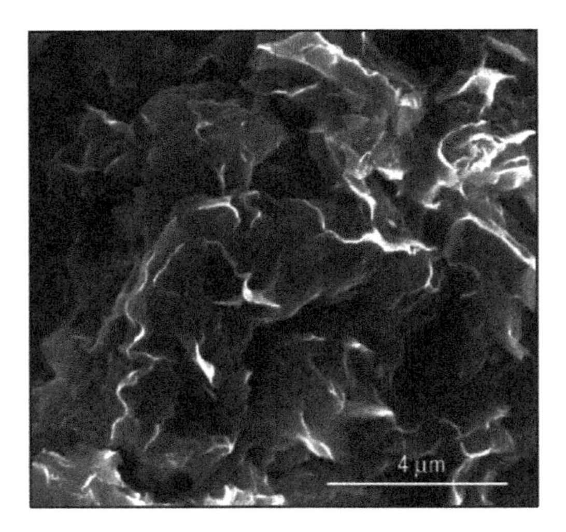

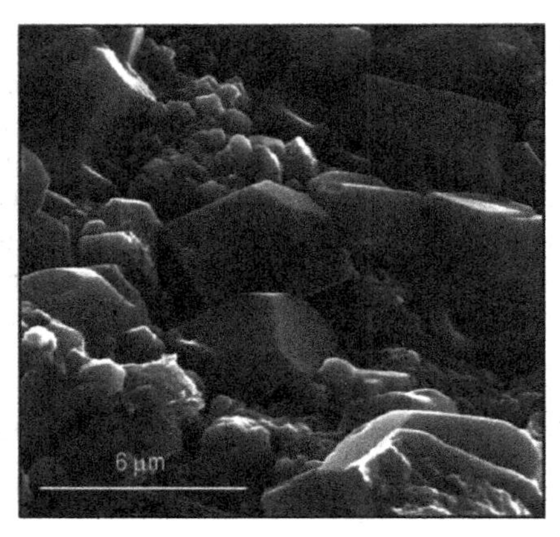

Fig. 3.31 Scanning electron micrographs of precipitated salt in hydrothermal experiment (Micrographs by Greifswald Univ., Geogr. And Geol. Dept.). *Upper*: Normal MX-80 clay with typical interwoven stacks of very thin montomorillonite lamellae. *Lower left*: Precipitated NaCl crystals in the clay matrix. *Lower right*: Precipitated $NaSO_4$ (gypsum) in the matrix

on the microstructural scale, considered the system to be a homogeneous, isotropic porous medium with bulk properties that can be determined by use of traditional laboratory techniques and to interpret and explain the results with respect to assumed processes on the microstructural scale. This has led to shortcomings and in some cases completely misleading results, which have forced the designers of the models to introduce repair coefficients and to derive parameters backward from true experiments. We will focus here on the physico/chemical background of the codes for judging their relevance.

For the models developed to predict performance of the engineered buffer system it is essential to find means to provide the basis for model validation, for which

Table 3.4 Wetting rate of confined sample of MX-80 clay using Na and Ca solutions

Distance from cold end, mm	Water content after 45 days for 3.5% NaCl solution	Water content after 45 days for 3.5% $CaCl_2$ solution
10	47	45
20	45	41
30	32	38
40	12	33
50	8	28

the EU-supported "Prototype Repository Project" has served [14]. This project is a full-scale test of the KBS-3 V concept at about 400 m depth in crystalline rock from which all pertinent parameters and data relating to buffer performance were derived. The data were specific to the type of buffer material used, i.e. the MX-80 bentonite, the specific initial and boundary conditions, and the specific test conditions.

None of the codes can take the true geometrical conditions into account, neither the gaps – unfilled, filled with pellets or with smectite mud - that are required to bring the buffer down in the deposition holes, nor the desiccation-induced fracturing and loss in coherence of the buffer block in the hottest part.

The ways in which the codes deal with the maturation of the buffer is obvious from their constitutions, which are exemplified here by the code "COMPASS" worked out by Hywel Thomas and his colleagues at Cardiff University in Wales, UK.

3.5.7.2 "COMPASS", a Code Used for Predicting the Evolution of Buffer in Deposition Holes

COMPASS is based on a mechanistic theoretical formulation, where the various aspects of soil behaviour under consideration, are included in an additive manner. In this way the approach adopted describes heat transfer, moisture migration, solute transport and air transfer in the material, coupled with stress/strain behaviour [15].

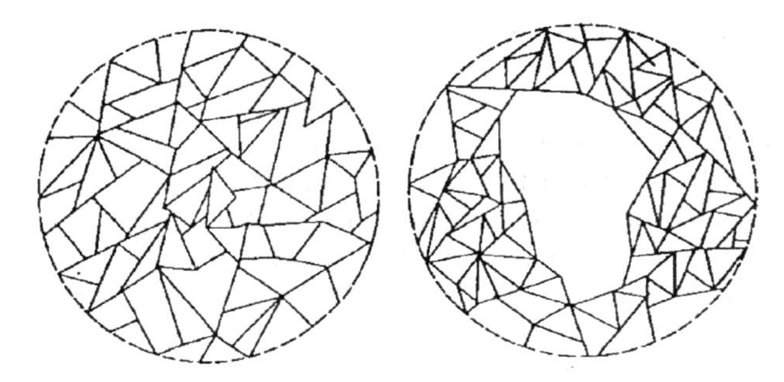

Fig. 3.32 Coagulation of smectite clay by increasing the salt content in the porewater. *Left*: uniformly distributed smectite particles in fresh water. *Right*: Coagulation in salt water yielding a number of larger voids

Partly saturated soil is considered as a three-phase porous medium consisting of solid, liquid and gas. The liquid phase is considered to be pore water containing multiple chemical solutes and the gas phase as air. A set of coupled governing differential equations are developed to describe the flow and deformation behaviour of the soil.

The main features of the formulation are as follows:

- Moisture flow includes both liquid and vapour flow. Liquid flow is assumed to be described by a generalised Darcy's Law, whereas vapour transfer is represented by a modified Philip-de Vries approach.
- Heat transfer includes conduction, convection and latent heat of vaporisation transfer in the vapour phase.
- Flow of dry air arising from an air pressure gradient and dissolved air in the liquid phase are considered. The bulk flow of air is again represented by the use of a generalised Darcy's Law. Henry's Law is employed to calculate the quantity of dissolved air and its flow is coupled to the flow of pore liquid.
- Deformation effects are included via either a non-linear elastic state surface approach or an elasto-plastic formulation. In both cases deformation is taken to be dependent on suction, stress and temperature changes.
- Chemical solute transport for multi-chemical species includes diffusion dispersion and accumulation from reactions due to the sorption process.

A large number of parameters are used of which some can be determined through careful experiments under controlled conditions, while others emanate from classic or applied physics.

The governing equation for moisture transfer in partly saturated clay is, as for all the proposed THM models, expressed as a function of a generalised Darcy's law and we see that this is not really valid since water transport is in fact a matter of diffusion except for a very high degree of water saturation. The modellers' inclusion of an osmotic flow term in the liquid velocity is a totally artificial measure that can be deleted by considering the actual surface diffusion transport mechanism.

Where a chemical solute is considered non-reactive and with no sorption the governing equation for conservation of chemical solute includes another debatable component, a hydrodynamic dispersion coefficient comprising both molecular diffusion and mechanical dispersion. This way of taking dissolved elements into consideriation totally ignores the most important process in the processes of water saturation and subsequent percolation of smectite buffers, i.e. cation exchange. One understands from Fig. 3.30 that the strongly raised salt content at the wetting front has a significant impact on the microstructural build-up there and that it largely controls the migration of liquid water. Also vapour transport is controlled by this phenomenon since the voids, widened by the cation exchange, enhance vapour flow as well.

Considering, finally, stress-strain relationships, the models define the total strain, ϵ, as consisting of components due to suction, temperature, chemical and stress changes. This can be given in an incremental form as:

$$d\epsilon = d\epsilon_\sigma + d\epsilon_{c_s} + d\epsilon_s + d\epsilon_T \tag{3.2}$$

where the subscripts σ, c_s, T and s refer to total stress, chemical, temperature and suction contributions.

The problem with this approach is that the relationship between stress and strain is not only non-linear but also changing with the salt content. With appropriate initial and boundary conditions the derived set of non-linear coupled governing differential equations can be solved and the respective unknowns, i.e. temperature, swelling pressure, porewater pressure and strain calculated.

Coming finally to the practical use of all this, one needs to arrive at a numerical solution of the governing differential equations by combination of the finite element method for the spatial discretisation and a finite difference time stepping scheme for temporal discretisation. Considering all uncertainties, neglection of scale-dependencies and application of incorrect submodels for migration of liquid water and vapour, and adding also the inability of this and similar codes to take different boundary conditions into consideration, they remain useless.

3.5.8 Can One Predict with Accuracy Even the Simpliest Process, i.e. Water Saturation, by Using the Proposed Codes?

3.5.8.1 Comparison of Predictions and Recordings

One of the first questions that was raised in the early stage of the CROP project was to see how accurately one can predict the temperature rise and saturation process under well defined test conditions. As to thermal processes simple estimates are usually sufficient in practice and they are easily made with reasonable accuracy. For the wetting it is different because it has an enormous impact on the physical performance of smectite clay. For testing how accurately one can predict it modellers were asked to consider a tube filled with macroscopically homogeneous MX-80 clay with 50% water saturation and a dry density of 1600 kg/m^3 the tube being exposed to a thermal gradient of 10°C/cm (max temp. 80°C). The initial water content was 10% corresponding to 50% degree of saturation. A laboratory experiment of the same type was conducted and several samples were taken for determination of the water content at different periods of time. Two such predictions were made and the results are shown in Figs. 3.33 and 3.34.

One finds that the measured values were consistently lower than the predicted ones, meaning that the saturation process was in fact significantly slower than predicted. The expected S-shape of the curves was not found in the experiments; instead the actual curves were of diffusion type.

A second attempt with the same test arrangement and use of another code, "ABAQUS", developed for SKB, gave the results shown in Fig. 3.34. Like the Barcelona prediction the one based on ABAQUS gave significantly quicker saturation for the last two weeks and within 15 mm distance throughout the test period. In the last three weeks the actual saturation degree at the distant end was 70% and the predicted saturation 80%.

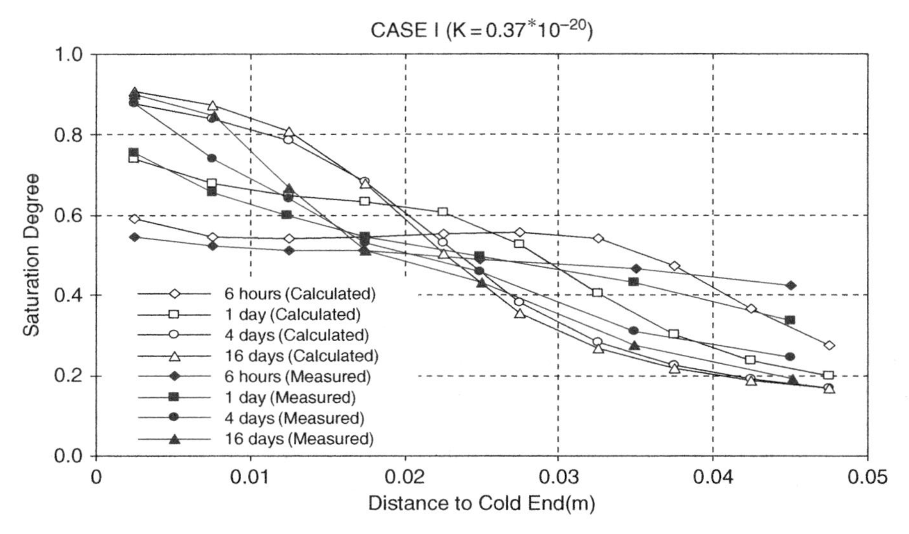

Fig. 3.33 Comparison of recorded and predicted ("Barcelona basic model") wetting rate of a 5 cm long sample (After Ledesma et al.), [2,15]

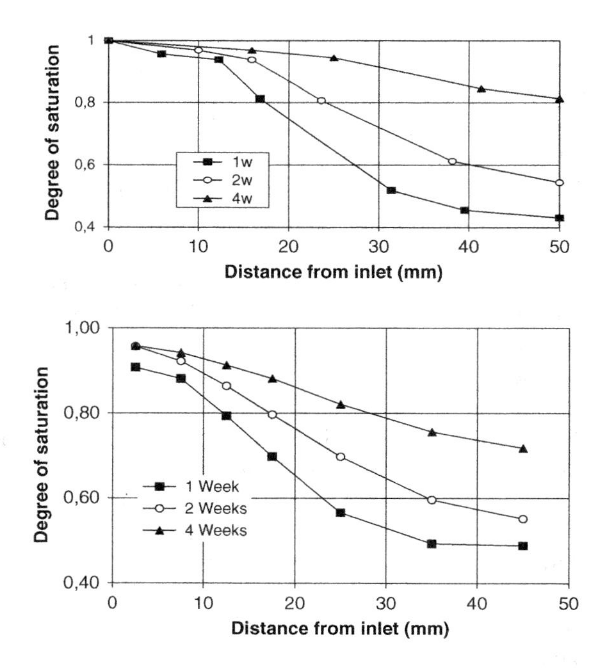

Fig. 3.34 Comparison of recorded and predicted wetting rate (ABAQUS code). Data refer to 1, 2, and 4 weeks. *Upper*: predicted, *Lower*: recorded [2,15]

One concludes from these comparisons that the predicted wetting rates agree poorly with the actual ones considering that the starting level was 50%. Any other prediction, like the development of swelling pressure, would be expected to show similar or poorer agreement, and this is also the case as demonstrated by the so-called retrieval test in SKB's underground laboratory where the KBS-3 V concept has been tested on full scale. We will consider a few of them in detail after outlining the true mechanisms that control the water saturation process.

3.5.8.2 Migration of Liquid Water and Vapour

Water uptake into smectite particles and particles aggregates is initially via hydration forces and subsequently via diffuse double-layer (DDL) forces [10]. The energy characteristics defined by the soil-water potentials determine how strongly water is held, a basic condition being that, at final equilibrium, the osmotic or solute potential ψ_π of the microstructural unit equals the matric suction potential ψ_m measured as a macro property that is easily measured in the laboratory. The latter is developed by the microcapillaries, van der Waals forces, weak osmotic phenomena, and other surface active forces existent in the macropores. Figure 3.35 schematically depicts the performance of microstructural units i.e. aggregates of stacks of smectite lamellae in the wetting phase.

The sequence of water uptake in the buffer depends on the nature of the exchangeable cations in the interlayer spaces and on its the initial water content. Initial exposure to liquid water or vapour causes hydration by formation of interlamellar hydrates as a first step. Then, continued hydration takes place due to double-layer forces [10], leading to two interlamellar hydrates and, in some smectites, to 3 hydrate layers when sodium is the dominant sorbed cation.

Turning back to Fig. 3.20 for a moment we see that the possibility for water to enter the initially only partly water saturated clay matrix is not by flow through the dense and tightly spaced aggregates of stacks of lamellae, but through interconnected voids or channels with less dense clay gels (Fig. 3.36). Practically all water migration takes place within the three-dimensional network of gel-filled channels

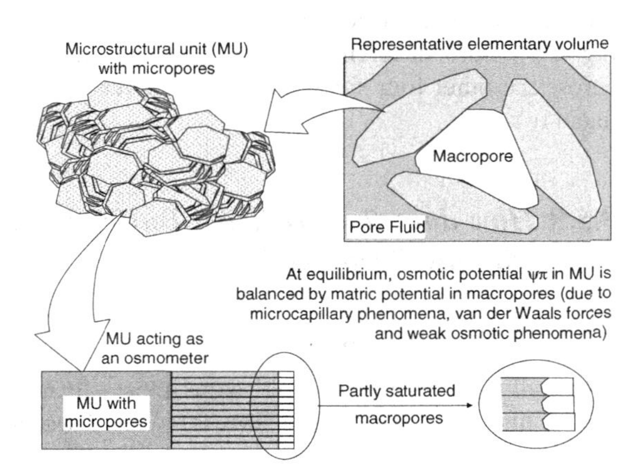

Fig. 3.35 Schematic diagramme showing the action of microstructural units in a representative unit volume. Uptake of water by a typical microstructural unit is initially via hydration forces, and later on by diffuse double-layer forces [10]

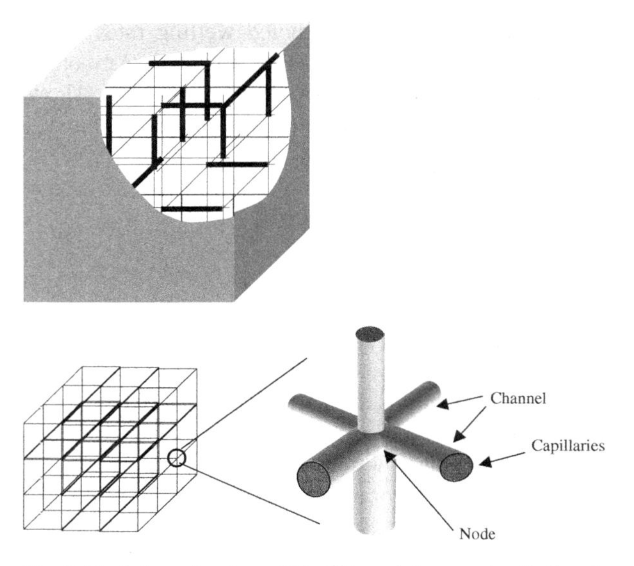

Fig. 3.36 Schematic view of the 3D model concept with the channel network mapped as a cubic grid with channels intersecting at a node in the grid

with stochastic properties. The channels are characterised by their lengths, widths, apertures and transmissivities. The clay matrix is assumed to be porous but impermeable. The basis of the development of the presently used model is the code 3Dchan [16].

The number of channels, which are assumed to have the length L, contain bundles of N capillaries with a diameter (d) that is proportional to the channel width is chosen to match the total porosity of the clay. After complete grain expansion the voids filled with homogeneous clay gels are assumed to have a normal size distribution with the same intervals as in the 2D model, i.e. 1–5 μm for the clay with 2130 kg/m^3 density, 1–20 μm for the clay with 1850 kg/m^3 density, and 1–50 μm for the clay with 1570 kg/m^3 density. The code generates a certain number of channels for a given volume. Using the Hagen-Poiseuille law the flow rate through the channel network is calculated for given boundary conditions assuming a pressure difference on the opposite sides of the cubic grid and no flow across the other four sides. The model agrees very well with recorded percolation data [16].

3.5.9 How does the Predicted Buffer Performance Agree with the Recorded?

3.5.9.1 Reference Case

Naturally, representative field experiments in underground laboratories (URLs) are required for assessing THMCB codes. Two SKB-operatated test sites with

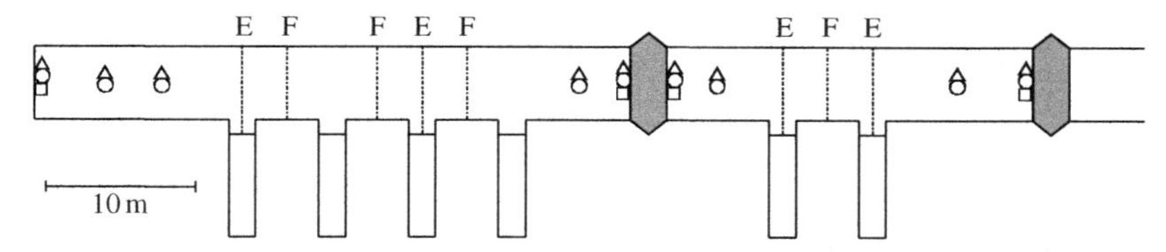

Fig. 3.37 Schematic section of the Prototype Repository Project at Äspö. Six deposition holes were core-drilled to about 8 m depth with 1.8 m diameter. Backfilling of the tunnel by mixture of 30% MX-80 bentonite and 70% crushed rock. The whole arrangement is a full-scale version of a KBS-3 V repository drift. The inner part is planned to be in operation for up to 20 years [10]

international R&D staffs involved in national and international projects, the Stripa URL established in the early eighties in an extended part of the old Stripa mine, and the Äspö URL constructed some tens years later. Both are located in granitic rock and have rooms down to more than 400 m. At Äspö a tunnel of KBS-3 V type was excavated for conducting an up to 20 year long experiment for investigating the maturation and performance of SKB's reference smectite-rich clay MX-80, which has Na as dominant adsorbed cation. The tunnel with deposition holes and plugs for separating tunnel segments is shown in Fig. 3.37. The tunnel was backfilled with a mixture of crushed rock and MX-80 clay that was instrumented for recording the wetting process and measuring temperature and swelling pressure. The deposition holes with copper/iron canisters surrounded by strongly compacted blocks of granulated MX-80 clay powder, and pellets were richly equipped with thermocouples, moisture sensors and pressure cells, which has made it possible to get a nearly complete picture of the early maturation phase of the buffer in rock with largely varying inflow of water in the deposition holes. The canisters had electrical heaters with a total maximum power of 1800 W.

As indicated earlier rock with few or no fractures of 4th order type intersecting deposition holes may delay the buffer wetting so much that it undergoes permanent miscrostructural changes. Two to four discontinuities of this type, crossing the deposition holes, should provide enough water to represent "wet" boundaries and no risk of permanent degradation. The Prototype Repository Project offered a possibility to investigate both extremes and intermediate cases. Inflow tests in the six holes gave the data in Table 3.5, which have been used for calculating the equivalent average hydraulic conductivity of the surrounding rock. The geometric mean of these values is slightly higher than E-11 m/s.

The evolution of heat in the wettest hole is shown in Fig. 3.38. The temperature still rose after more than 4 years and reached about 72°C in the clay adjacent to

Table 3.5 Water inflow in the six deposition holes of the Prototype Repository Project

Hole number	Rock hydraulic conductivity, m/s (After Rhen et al.)	Inflow of water, l/min
1	E-10	0.0800
2	E-11	0.0020
3	E-11	0.0030
4	E-12	0.0007
5	E-11	0.0027
6	E-11	0.0030

the canister surface and around 60°C at the rock after about 2 years. The average temperature gradient was then about 0.34°C per cm radial distance.

The evolution of the wetting process at mid-height of the canister in the wettest hole is shown in Fig. 3.39, in which the upper curve set represents the recorded relative humidity (RH), which is a measure of the degree of water saturation, while the lower set represents the temperature. The drop in RH signalled by a sensor adjacent to the canister logically indicates that drying took place.

The RH plottings give the impression that nearly complete water saturation (92–95%) had been reached after about one year. This is in contrast with the recorded evolution of the swelling pressure as indicated by Fig. 3.40. The homogeneous distribution of water according to the RH measurements should correspond to a uniform distribution of the pressure but the obvious variations in pressure shown in the pressure diagram indicate that complete saturation and maturation – microstructural homogenization – of the buffer had not taken place even after 2.5 years. The

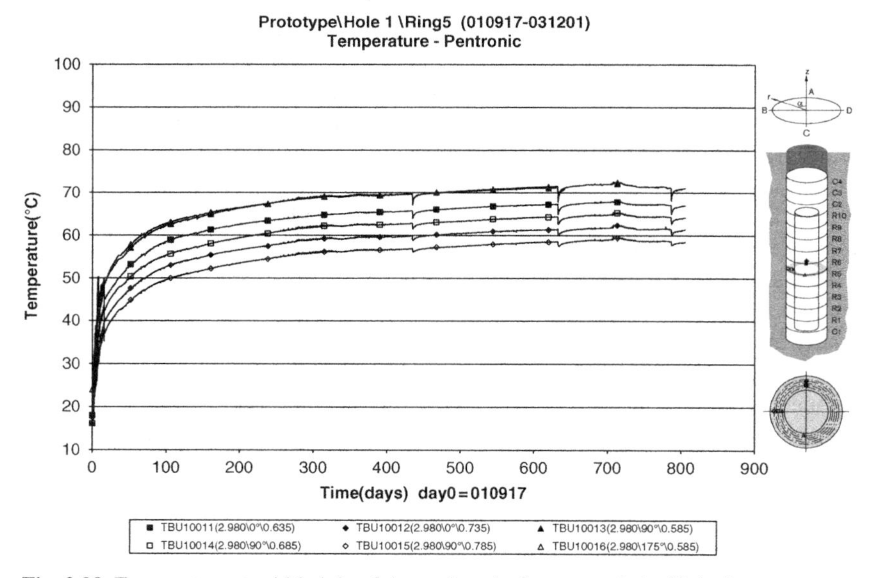

Fig. 3.38 Temperature at mid-height of the canister in the wettest hole (Hole 1)

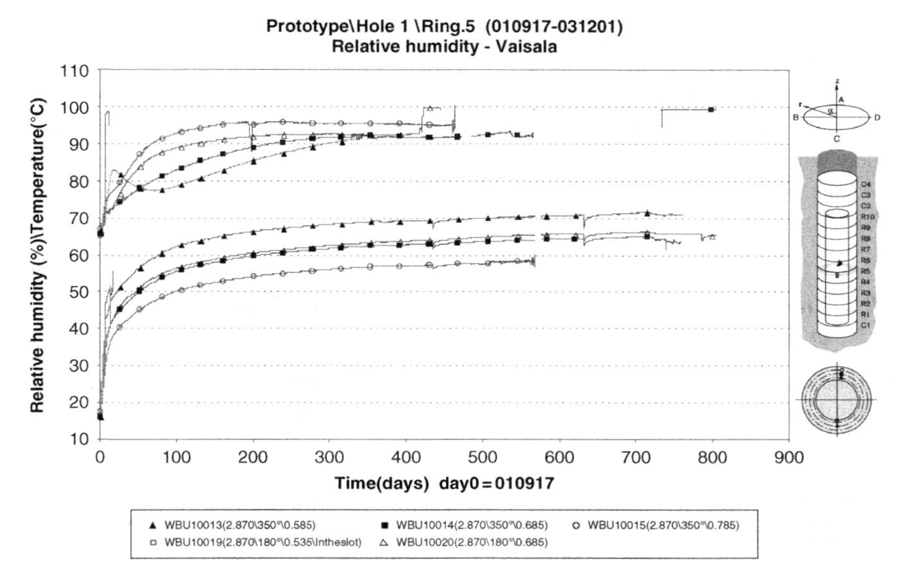

Fig. 3.39 RH and temperature distributions in the buffer in the wettest hole (Hole 1). The data can be approximately taken as the degree of water saturation. The *upper* curve set shows the RH readings and the *lower* set gives the temperature

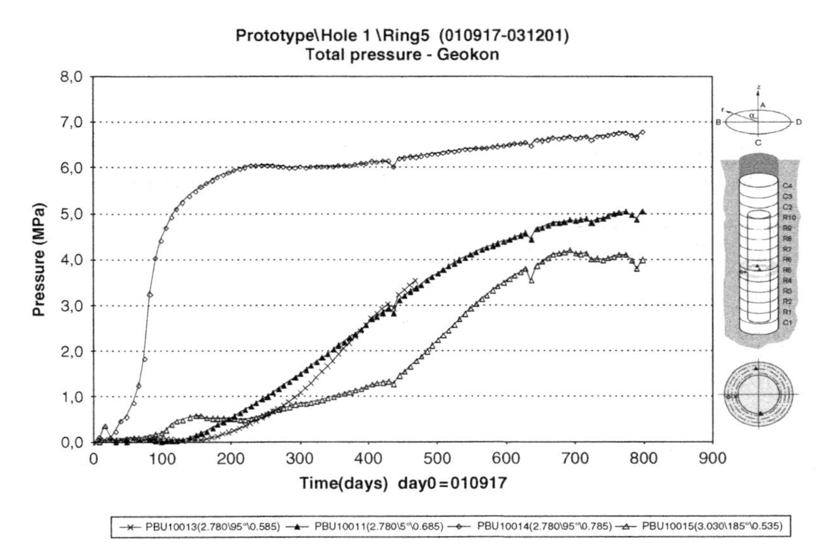

Fig. 3.40 Evolution of total pressure at mid-height of the buffer in the wettest hole. The highest pressure, about 6.7 MPa, was reached after about 2 years while the lowest (4 MPa) was signalled by a cell close to the canister

discrepancy was partly caused by delay in homogenization by creep, expansion and consolidation but even more by a process found in many other field- and bench-scale buffer experiments, namely that the cables connecting the RH gauges to the recording units served as water conductors and caused local wetting and early saturation of the clay where the gauges were located. The pressure recordings in fact also suffered from such irrelevant moistening and it is expected that the true growth of swelling pressures was in fact even slower than shown in Fig. 3.40.

3.5.9.2 Theoretical Simulation of the Evolution of the Buffer by Using the Codes

The codes were used for simulating the evolution of the buffer in the CROP project and the three most accurate ones gave the predicted data in Tables 3.6, 3.7 and 3.8, which also give the actually recorded data [29].

Table 3.6 Recorded and predicted temperature in degrees centigrade at the canister and rock surfaces at mid-height in the wettest hole after 1 and 2 years from start of heating

Location	Recorded	COMPASS(UWC)	CODE_BRIGHT(CIMNE, Enresa)	THAMES(JNC)
Canister	1 y = 69 2 y = 72	1 y = 70 2 y = 72	1 y = 70 2 y = 72	1 y = 87 2 y = 92
Rock	1 y = 56 2 y = 60	1 y = 56 2 y = 59	1 y = 57 2 y = 60	1 y = 71 2 y = 76

Table 3.7 Recorded and predicted degree of saturation in percent at the canister surface at mid-height in the wettest hole after 1 and 2 years from start of heating

Location	Recorded	COMPASS (UWC)	CODE_BRIGHT (CIMNE, Enresa)	THAMES(JNC)
Canister	1 y = 90–100 2 y = 90–100	1 y = 96 2 y = 100	1 y = 95 2 y = 97	1 y = 79 2 y = 99
Rock	1 y = 90–100 2 y = 90–100	1 y = 98 2 y = 100	1 y = 99 2 y = 99	1 y = 94 2 y = 100

Table 3.8 Recorded and predicted pressure in MPa at the canister surface at mid-height in the wettest hole after 1 and 2 years from start of heating

Location	Recorded	COMPASS (UWC)	CODE_BRIGHT (CIMNE, Enresa)	THAMES (JNC)
Canister	1 y = 1.0 2 y = 4.0	1 y = 0.8 2 y = 3.2	1 y = 3.0 2 y = 5.1	1 y = 4.7 2 y = 6.2
Rock	1 y = 6.0 2 y = 6.7	1 y = 2.8 2 y = 3.9	1 y = 5.0 2 y = 7.2	1 y = 6.4 2 y = 7.2

Comparison of predictions and recordings led to the following major conclusions [2]:

- The best agreement between predictions and measurements was obtained for the temperature evolution. One of the codes overestimated the temperature for the first two years, while the others gave accurate data. There are indications that the thermal conductivity of the buffer was higher than assumed and that the heat transfer was assisted by some undefined mechanism like convection through vapour flow. In conclusion: The temperature evolution can be predicted with acceptable accuracy.
- The hydration rate is a difficult matter. If the recorded data had been correct one could have concluded that all but one of the codes gave adequate information. However, the recordings overestimated the hydration rate because of water migration along the cables to the moisture sensors, meaning that the code that predicted too low temperature gave the most correct simulation of the hydration rate. All the simulations were in fact unreliable.
- Further discrepancies were found respecting the evolution of swelling pressure, which should follow the same patterns as the hydration and hence give the best fitting of recorded and predicted pressure for the code that gave the lowest hydration rate. However, the table shows that it was the least reliable one. In fact, the discrepancy must have been even larger since water migration along pressure cell cables gave too early response.

The evolution of pressure in the buffer in "dry" deposition holes is even more difficult to predict and it is totally unknown whether water saturation will take a few tens or hundreds of years for buffer in deposition holes that are not intersected by any 4th order disconinuity. This may imply, as in argillaceous clay, that the clay undergoes permanent changes in the form of contraction of the stacks of lamellae.

3.5.10 How do Predicted and Recorded Canister Movements in the Clay Agree?

3.5.10.1 General

Modelling of the evolution of the buffer using the codes referred to has not included movement of the canisters, which is expected to have the practically important effect indicated in Fig. 3.7. Nor has any large-scale field or mock-up test been made for measuring it. However, small scale tests and theoretical modelling using boundary element technique have been made and we will examine them here [17]. A small-scale test involved recording of the movement of a model canister embedded in dense smectite-rich clay at constant temperature of 21.5–22.0°C (Fig. 3.41). The clay was placed in compacted form in the load cell and saturated with distilled water for 3 months yielding nearly complete saturation at a density of 2000 kg/m^3. The

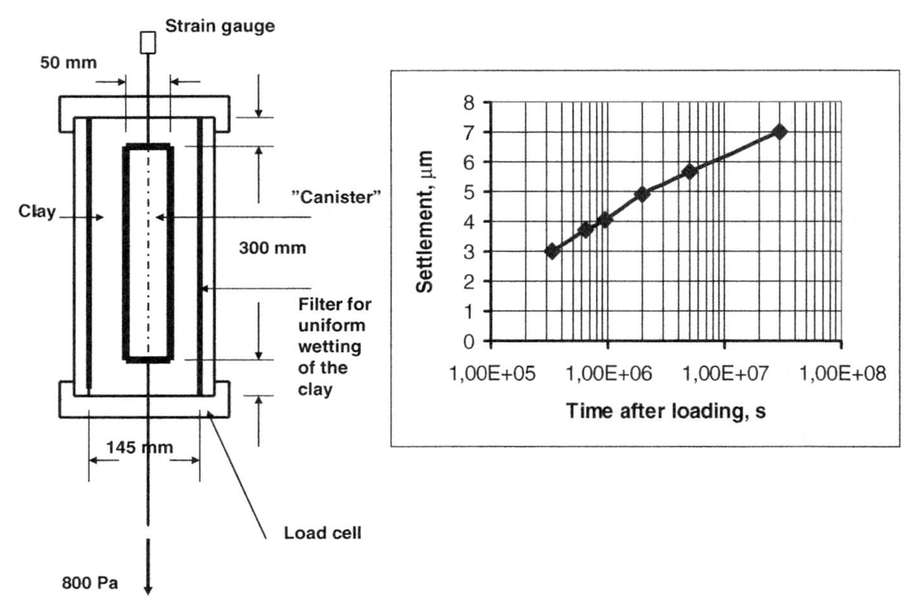

Fig. 3.41 Model test with loaded canister confined in a tight cell and exposed to a load of 800 N simulating the weight of a canister with a weight of 80 kg.

diagram shows that the settlement was less than 8 μm after 3 months under a constant load of 80 kg with access to water, and that it was approximately proportional to log time. One realizes that if the load on the buffer below and around the lower part of the canister makes it sink, and the expansion upwards of the upper part of the buffer lifts the canister, it will be exposed to tension.

For defining the problem we need to consider the physical nature of the stress/strain relation, which requires that the microstructural constitution is considered and that relevant physical relationships between stress and strain utilized.

3.5.10.2 Stress and Strain

Soil mechanics employs a special basic stress principle, the so-called effective pressure concept proposed by Terzaghi some 90 years ago. It states that the effective, or grain, pressure is the difference between the total pressure and the pressure in the porewater. For most soils the shear strength and volume changes depend only on the effective stress. For high-plastic clays this is not correct, however, primarily because it is difficult to imagine how grain pressure is really transferred at particle contacts, which are believed to be of the type shown in Fig. 3.42, i.e. via electrical double-layers and merged hydrate layers on the basal surfaces of the stacks of lamellae. True mineral/mineral contacts do not exist unless the pressure is several hundred MPa.

The Terzaghi concept of consolidation and expansion has a quite obvious physical meaning for soils containing non-expanding clay mineral particles, like illite

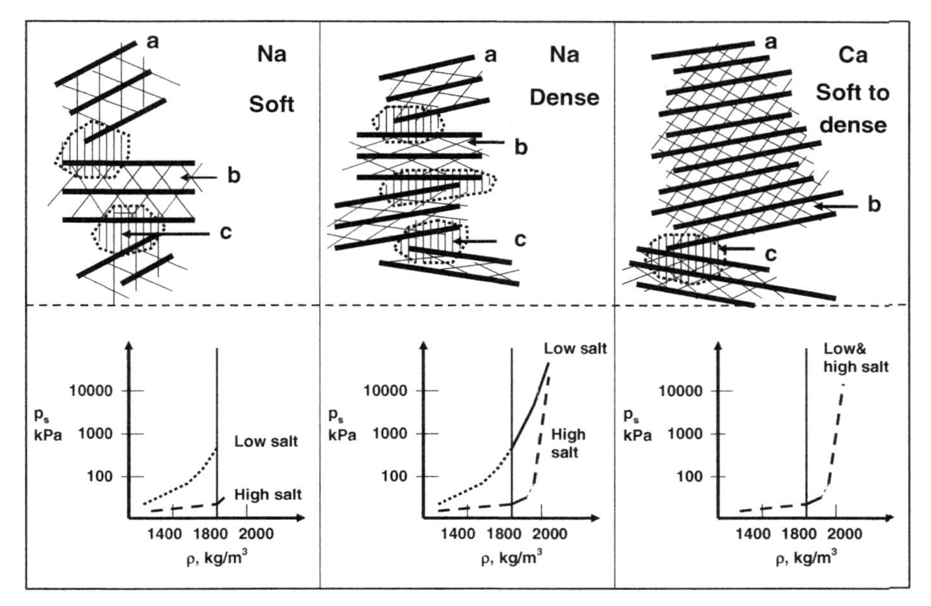

Fig. 3.42 Schematic pictures of interacting stacks of lamellae and influence of density at water saturation, expressed in g/cm^3 (1 g/cm^3 equals 1000 kg/m^3) and salinity for Na and Ca montmorillonite clay on the swelling pressure. a) lamella, b) interlamellar space, c) contact region with interacting electrical double-layers [4]

and kaolinite. Thus, an increased total pressure Δp of water saturated clay of these latter types generates a pore pressure of exactly the same magnitude but if water is allowed to be drained the pressure Δp is successively transferred to the particle network, which is thereby compressed and densified. The same process takes place in smectite-rich clay leading to a closer distance between some of the stacks of lamellae and closer spacing of the lamellae in some stacks, by which the repulsion between the lamellae and stacks of lamellae increases. Unloading has the opposite effect, i.e. water is taken up in the interlamellar space and on the basal surfaces of the stacks of lamellae by the very strong hydration potential, yielding expansion. The difference between the expandable and non-expandable clays is that the grain contact surface is much smaller for the latter clays than for the expandable ones, which in fact invalidates the effective stress concept.

All changes of the microstructural constitution are associated with shear-induced slip within and at the contacts of adjacent stacks of lamellae. Such slip takes place under constant volume conditions or in conjunction with consolidation in all soils and the accumulated time-dependent strain is termed creep. For smectite clay, lacking true mechanical friction between particles, creep is stronger than for non-expandable clays but has the same character of stochastically distributed slip occurring where energy barriers are overcome. This brings to stochastical mechanics as simplified below.

3.5.10.3 Creep

The empirically derived relationship in Eq. (3.3) is being used in current buffer design work for SKB for describing time-dependent strain as a function of the principal stress state and stress conditions at failure:

$$\dot{\gamma} = \dot{\gamma}_o e^{\alpha \frac{(\sigma_1 - \sigma_3)}{(\sigma_1 - \sigma_3) f_e} (-\alpha) \frac{(\sigma_1 - \sigma_3)_0}{(\sigma_1 - \sigma_3) f_e}} \left(\frac{t}{t_r} \right)^n \quad (3.3)$$

where:

t = time after stress change
t_r = reference time (10^5 s)
$(\sigma_1 - \sigma_3)_0$ = reference deviator stress $[0.5\,(\sigma_1 - \sigma_3)]$
$(\sigma_1 - \sigma_3)_{fe}$ = deviator stress at failure
γ = creep rate
γ_o = creep rate at time t_r
n and α = parameters derived from laboratory tests

Equation (3.4) expresses the shear strength q as a function of the mean effective stress p:

$$q = ap^b \quad (3.4)$$

where:

$a = q$ for $p = 1$ kPa
b = inclination of curve in log$p$/log$q$ diagrams.

The experimental background is very limited but it is claimed that triaxial tests have yielded a-values between 2.8 and 5.5 depending on the porewater chemistry and type of adsorbed cation while b appears to have a constant value, 0.77.

Equations (3.3) and (3.4) are completely empirical and not based on any conceptual microstructural model and are therefore not of general validity. For deriving reliable theoretical expressions of creep and creep rate as functions of stress and clay density the true involved mechanisms must be identified, which we will briefly comment here.

Shearing of any material takes place through the activation of barriers on the microstructural scale to slip and the barriers are represented by bonds of various kinds, like those emanating from weak Van der Waals forces, via hydrogen bonds to strong primary valence bonds. They form a spectrum of the type shown in Fig. 3.43.

Like for most materials the spatial differences in barrier heights forms a spectrum of interparticle bond strength representing a first-order variation, a second being represented by the interaction of differently sized particle aggregates, "peds" (Fig. 3.44). They behave as strong units that remain intact for small strain but yield

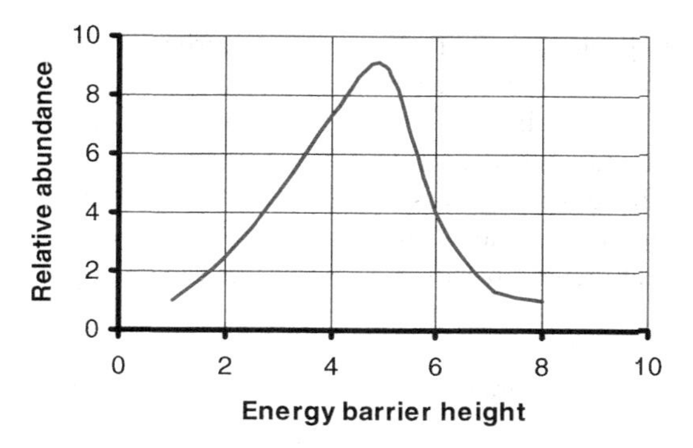

Fig. 3.43 Schematic energy barrier spectrum [18]. The lowest energy barriers represent weak hydrogen bonds and the highest primary valence bonds

at large strain and contribute to the bulk strength by generating dilatancy. The energy spectrum is hence not a material constant but changes with strain and thereby with time. An appreciable fraction of the strain-induced microstructural changes are preserved but exfoliated smectite stacks reorganize and cause self-healing by forming gels of different density depending on the available space and rate of strain. This means that local bond breakage is balanced by formation of numerous new bonds that makes the altering microstructural network stay coherent for low and moderate bulk shear stresses, while higher shear stresses cause irrepairable changes of the network leading to bulk failure at a certain critical strain.

The response of the structure to a macroscopic shear stress is that the overall deformation of the entire network of particles changes by disintegration, translation and rotation of weaker aggregates while larger are less affected and stay strong. The break-down of weak aggregates transforms them to a laminated structure of flaky particles. This microstructural organization is believed to be the reason for the Newtonian rheology of smectite clay that has undergone large strain in one or two directions.

The theoretical basis for modelling creep is provided by thermodynamics, which has led to derivation of analytic expressions for macroscopic creep under constant volume conditions in contrast to the empirical expressions that are commonly used

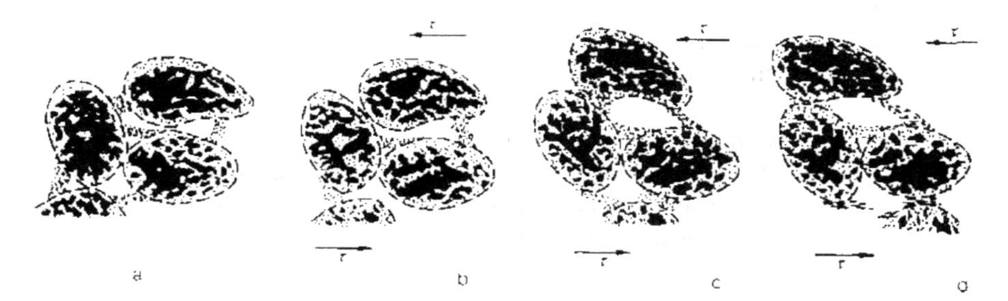

Fig. 3.44 Consecutive stages in the evolution of shear strain of microstructural network of clay particles. a) before loading, b) instantaneous shear and formation of slip units due to application of a shear stress τ, c) formation of slip domains accompanied by healing and breakdown, d) failure [18]

in geotechnical practice, and the derivation of analytical expressions of creep strain can be summarized as follows.

For a clay element subjected to a constant deviator stress one can assume that the number of energy barriers of height u is $n(u,t)\delta u$ where δu is the energy interval between successive jumps of a unit, and t the time, the entire process being stochastic. The change in activation energy in the course of evolution of strain means that the number of slip units is determined by the outflux from any u-level into the adjacent, higher energy interval and by a simultaneous inflow into the interval from u-δu [19].

Each element of clay contains a certain number of slip units in a given interval of the activation energy range and displacement of such a unit is taken to occur as the shifting of a patch of atoms or molecules along a geometrical slip plane. In the course of the creep the low energy barriers are triggered early and new slip units come into action at the lower energy end of the spectrum in Fig. 3.43. This end represents a "generating barrier" while the high u-end is an "absorbing barrier". A changed deviator stress affects the rate of shift of the energy spectrum only to higher u values provided that the shearing process does not significantly reduce the number of slip units. This is the case if the bulk shear stress does not exceed a certain critical value, which is on the order of 1/3 of the conventionally determined bulk strength. It implies that the microstructural constitution remains unchanged and that bulk strain corresponds to the integrated very small slips along interparticle contacts. In principle this can be termed "primary creep".

For low shear stresses, allowing for "uphill" rather than "downhill" jumps one gets for the rate of change of $n(u,t)$ with time:

$$\delta n(u,t)/\delta t = \nu[-n(u+\delta u,t)exp[-(u+\delta u)/kT] + n(u,t)\exp(-u/kT)] \quad (3.5)$$

where:

δu = width of an energy spectrum interval
ν = vibrational frequency (about E11 per second)
t = time
k = Boltzmann's constant
T = absolute temperature.

Using Eq. (3.5) and introducing Feltham's transition probability parameter to describe the time-dependent energy shifts and that each transition of a slip unit between consecutive barriers gives the same contribution to the bulk strain one gets the bulk shear strain rate as in Eq. (3.6) with $t<t_o$ as boundary condition:

$$d\gamma/dt = B(1 - t/t_o) \quad (3.6)$$

The appropriate constant B and the value of t_o depend on the deviator stress, temperature and structural details of the slip process. The creep can hence be expressed as in Eq. (3.7):

$$\gamma = \alpha t - \beta t^2, (t<a/2\beta) \tag{3.7}$$

meaning that the creep starts off linearly with time and then dies out.

For higher bulk loads, the strain on the microstructural level yields some irreversible changes associated with local breakdown and reorganization of structural units. Still, there is repair by inflow of new low-energy barriers parallel to the strain retardation caused by the successively increased number of slip units being halted by meeting higher energy barriers. This type of creep can go on for ever without approaching failure. Following Feltham the process of simultaneous generation of new barriers and migration within the transient energy spectrum lead to the expression for the creep shear rate in Eq. (3.8):

$$\frac{\mathrm{d}\gamma}{\mathrm{d}t} = \mathrm{A}\int_{u_1}^{u_2} n(u)\ \nu(u,t)\mathrm{d}u = \mathrm{A}\int_{u_1}^{u_2} n(u,t)\exp(-u/kT)\mathrm{d}u \tag{3.8}$$

The implication of this expression in which A is a constant, is that the lower end of the energy spectrum mainly relates to breakage of weak bonds and establishment of new bonds where stress relaxation has taken place due to stress transfer from overloaded parts of the microstructural networks to stronger parts, while the higher barriers are located in more rigid components of the structure.

Feltham [19] demonstrated that for thermodynamically appropriately defined limits of the u-spectrum the strain rate appertaining to logarithmic creep takes the form in Eq. (3.9):

$$\mathrm{d}\gamma/\mathrm{d}t = \mathrm{B}T\tau/(t + t_o) \tag{3.9}$$

where B is a function of the shear stress τ. t_o is a constant of integration which leads to a creep relation closely representing the commonly observed logarithmic type implying that the creep strain is proportional to $\log(t+t_o)$. The significance of t_o is understood by considering that in the course of applying deviatoric stress, at the onset of the creep, the deviator rises from zero to its nominal, final value. A u-distribution exists at $t = 0$, i.e., immediately after full load is reached, which may be regarded as equivalent to one which would have evolved in the material initially free from slip units, had creep taken place for a time t_o before loading. Thus, t_o is characteristic of the structure of the prestrained material [20].

This model implies, for moderate deviator stresses that allow for microstructural recovery, that the creep strain is as illustrated in Fig. 3.45. If there is successive retardation of the creep rate according to the logarithmic time law it is expected that failure will not be caused even after very long period of time.

Further increase in deviator stress leads to what is conventionally termed "secondary creep" in which the strain rate tends to be constant and giving strain that is proportional to time. Following the same reasoning as for the lower stress cases one can imagine that creep of critically high rate makes it impossible for microstructural self-repair: comprehensive slip changes the structure without allowing reorganization, which yields a critical strain rate which unevitably leads to failure (Fig. 3.46).

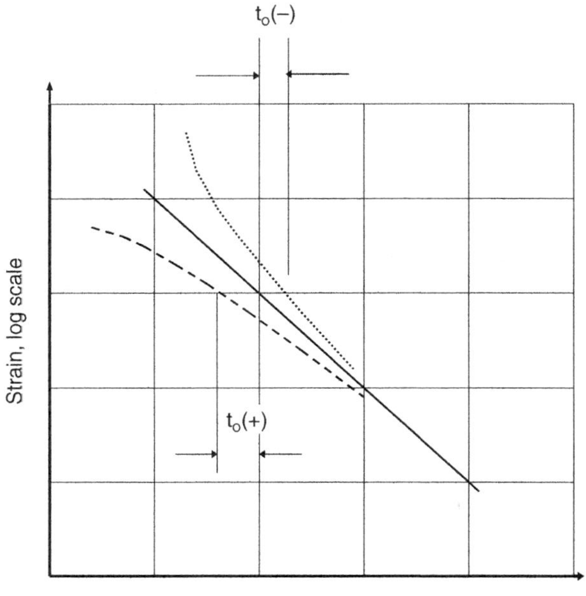

Fig. 3.45 Generalization of creep curves of log time type

For smectite clay shear-induced formation of slip units, consisting of dispersed stacks of lamellae, is much more comprehensive than in illitic and kaolinite-rich in clays. The successive increase in the number of slip units in fact implies an important self-sealing ability that causes attenuation of the creep rate even for high shear stresses as indicated by the diagrams in Figs. 3.47 and 3.48, representing creep testing of smectitic clay. Such experiments in fact indicate that smectite-rich clays behave as viscous fluids at large strain and not at all as predicted by common soil mechanics. Such a case is shown in Fig. 3.49.

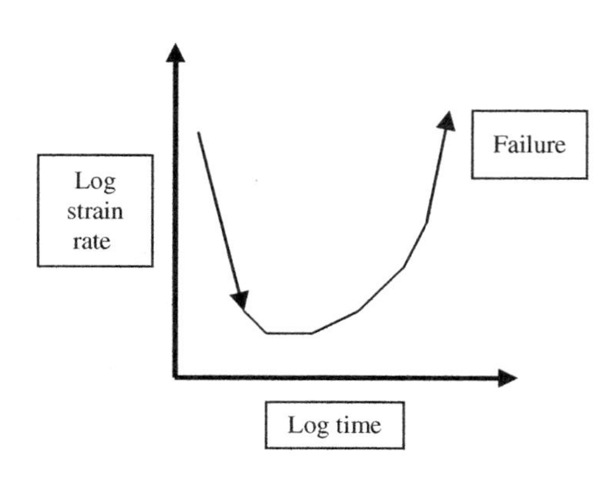

Fig. 3.46 Evolution of creep for critically high stresses

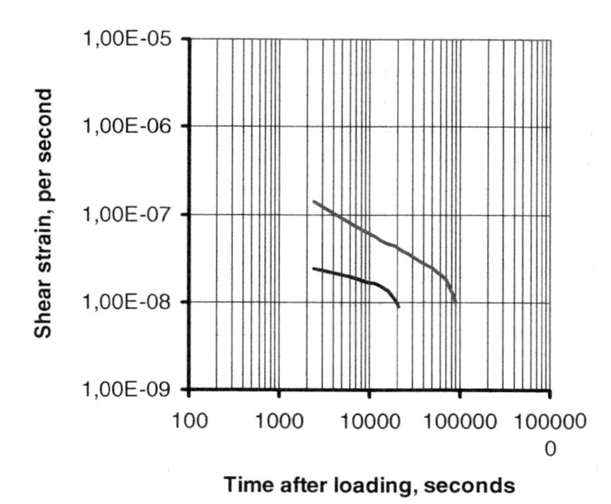

Fig. 3.47 Creep strain of smectitic clay for the average shear stresses 11, 23 and 39 kPa (upper curve). Density at saturation with distilled water 1940 kg/m^3

The strain rate evolution shown in Fig. 3.49, i.e. a nearly constant strain rate, was recorded after increasing the average shear stress to a level that was expected to yield instant failure, is believed to be a strain-hardening effect caused by successive dispersion of the initial dense particle aggregates to form an anisotropic matrix of aligned and partly disintegrated stacks of lamellae oriented in the direction of the maximum shear stress. This caused an increasing specific area with most of the shear resistance provided by a steadily increasing but ultimately stable number of quickly formed and broken hydrogen bonds like in viscous gels of colloidal matter.

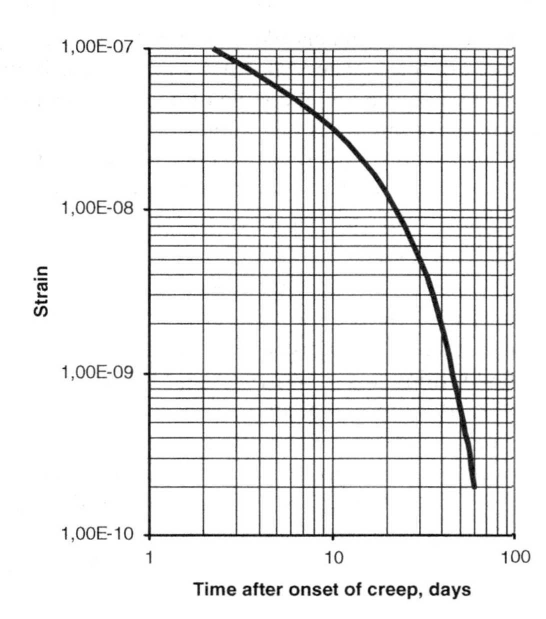

Fig. 3.48 Creep strain of smectite-rich clay (MX-80) with a density at saturation with distilled water of 1500 kg/m^3. Double shear apparatus, average shear stress 6 kPa

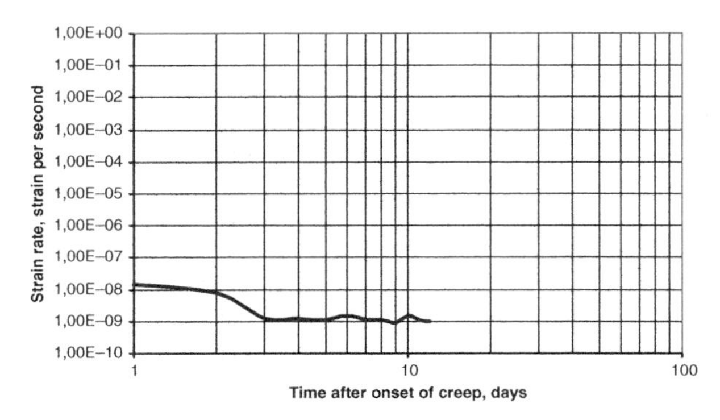

Fig. 3.49 Creep of the same sample as in Fig. 3.48 by increasing the shear stress to 23 kPa. The creep rate attenuated but turned into a constant, relatively low strain rate without leading to failure in the two week long test performed at room temperature

This suggests that smectite clays may not undergo failure like illitic clays, i.e. by slip along discrete failure planes, but deform to yield zone failure using the soil mechanical term.

3.5.10.4 Conclusive Remarks Concerning the Stress/Strain Behaviour of Buffer and Backfills

If the buffer embedding canisters were confined by rigid rock all movements in the mechanically closed system would be small and negligible. However, the pressure that will be exerted by very dense, smectite-rich buffer on the overlying compressible backfill in tunnels like in the KBS-3 V concept, will displace it and cause the tension of the canisters shown in Fig. 3.7. The boundary between buffer and backfill may be displaced by a couple of decimetres and can lift the rock close to the floor of blasted tunnels so that subhorizontal fractures open and dramatically increase the hydraulic conductivity of the rock below the tunnel floor. For TBM-excavated tunnels with less damaged floor the risk is smaller but still existing.

Considering the impact of the heavy load, about 25 tons, on the buffer clay in SKB's and POSIV's concepts one would think that the increase in effective pressure at the base of the canister would be negligible compared to the effective pressure represented by the swelling pressure of the buffer. The increase would be only a few percent and classical consolidation theory would give a settlement of only a few millimetres. However, shear strain will dominate over compression and, as indicated by BEM-calculated creep, it may well lead to a settlement of a few tens of millimetres in a few thousand years, which, in combination with the upward movement of the upper part of the canister, can cause the aforementioned tension and damage of canisters of SKB type. Tension stresses in a HIPOW canister would cause no problems.

A major conclusion from these considerations is that problems with canister tightness can not be ruled out even in a short term perspective. If the buffer below midheight canisters undergoes weakening by increase in void size, or shrinkage by chemical impact or gas attack, the settlement may increase and thereby tension caused in the upper, colder part if the buffer embedment maintains its physical properties. This emphasis the need for investigating the long-term constitution of the buffer as we will do next.

3.5.11 Long-Term Function of Buffer Clay

3.5.11.1 General

We see from the assessment of the performance of KBS-3 V canisters that the very important waste-isolating function of the buffer must last for at least 100 000 years and this brings the long-term function of the buffer in focus. It primarily concerns the stability of smectite-rich clays, which is determined by thermodynamically controlled reactions, implying that mineral transformations take place that depend on the stability conditions of interacting minerals and porewater under the prevailing temperature and temperature gradient conditions. However, also a number of other factors affect the constitution of the buffer and one can identify the following processes of expected importance:

- Long-term exposure to heat and groundwater of partly and fully water saturated buffer clay contacting metal canisters.
- Exposure to gas emanating from corroded canisters.
- Exposure to high pH by contacting cement.
- Exposure to radiation.

3.5.11.2 Conversion of Smectite to Non-expanding Minerals

Pytte et al. [21], considering initially only time and temperature, put together literature data as a support of the hypothesis that only these two parameters are of importance for modeling smectite-to-illite (S/I) conversion (Table 3.9). The required potassium was assumed to emanate from degraded feldspars or released from various high-temperature reactions.

Later, other mineralogical changes than K-feldspar dissolution have been identified and a more complete list of mineral alteration associated with S/I conversion has been proposed:

- Decomposition of mica.
- Decomposition of feldspars.

Table 3.9 Summary of time/temperature constraints on complete S-to-I reaction

Time range, years	Temperature range, °C
E0-E3	250–400
E6-E7	200–500
E7-E7.5	90–230
E8.5	<80

- Loss of kaolinite.
- Formation of chlorite and chlorite interlayers in S/I stacks.
- Loss and formation of quartz, cristobalite and amorphous silica.
- Loss of calcite and other carbonates.

The following total reaction affecting smectite has been proposed:

$$S + (F_k + Mi) = I + Q + Chl \tag{3.10}$$

where:

S = Smectite, F_k = K-feldspar, Mi = K-mica, I = Illite, Q = Quartz, Chl = Chlorite.

Quartz has been assumed to originate from decomposed feldspars and from silicons lost from the SiO_4 tetrahedrons in montmorillonite in conjunction with conversion of this mineral to beidellite as a first step in conversion to illite.

Subsequent investigations conducted in the US and in other parts of the world, particularly in conjunction with studies of the formation and accumulation of hydrocarbons, especially those related to contact metamorphism, gave evidence of S/I conversion at relatively low temperatures. A classical investigation of this type was published by Janet Hoffman and John Hower [22]. They introduced the scheme of metamorphic grade index for pre-greenschist facies pelitic rocks in Fig. 3.50 and suggested the relation of grades with temperatures according to Table 3.10.

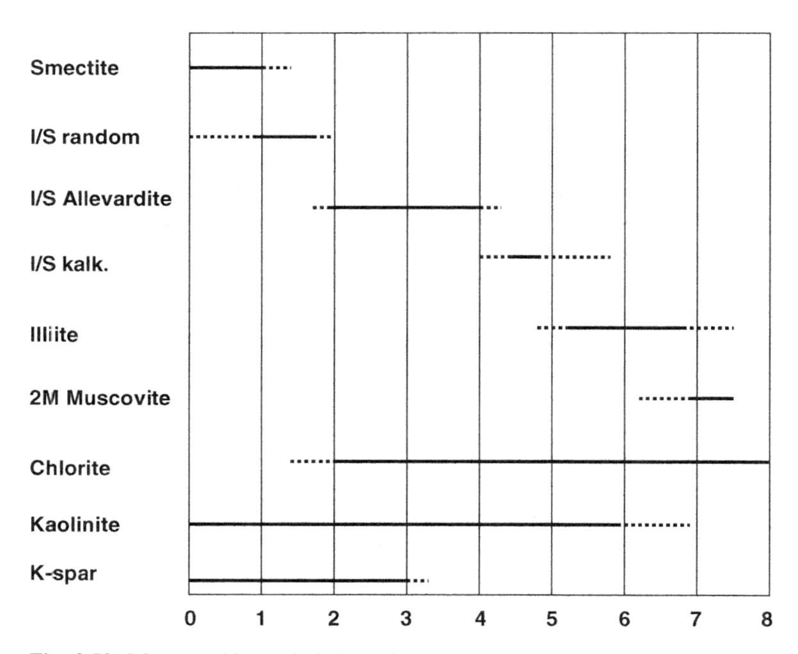

Fig. 3.50 Metamorphic grade index minerals according to Hoffman and Hower. (Allevardite is ISISIS..., and Kalkberg ISIIISIII)

Table 3.10 Temperatures for metamorphic grade changes with indication of reaction times as inferred from the age of the deposits [22]

Grade	Upper mesozoic – Lower tertiary temperature, °C	Pliocene-pleistocene temperature, °C
1	60	60
2	100	140
3	120	–
4	175	–
5	200	–
6	–	–
7	300	–

One concludes from Table 3.10 that temperatures as low as 60°C can cause largely complete conversion of smetite to illite in a sufficiently long period of time, and this was also the basis of Pytte's conversion model that is presently favoured by some THMC modellers and formulated in Eq. (3.10).

$$-dS/dt = \left[Ae^{-U/RT(t)}\right]\left[(K^{+}/Na^{+})^{m}S^{n}\right] \tag{3.11}$$

where:

S = Mole fraction of smectite in *I/S* assemblages	T = Absolute temperature
U = Activation energy	t = Time
R = Universal gas constant	m,n = Coefficients

Use of Eq. (3.10) with appropriate parameter values has given diagrams like the one in Fig. 3.51 showing the rate of illitization for the activation energy of 27 kcal/mole to yield conversion from smectite to illite. The graph shows that heating to 50°C would cause only insignificant loss of smectite in E6 years, while about 100% of the original smectite will turn into illite in this period of time at 100°C. A typical buffer temperature history for SKB canisters with spent fuel is: (a) 100–150°C in the first hundred years, (b) followed by a 500 year period with an average temperature of 50–100°C, and (c) less than 50°C in the subsequent 1000 years, leading to conversion of 15% of the initial smectite content to illite in 1500 years.

The problem is that the activation energy is not known with any certainty and that the access to potassium is difficult to define and derive. For an activation energy of around 20 kcal/mole the model would imply very significant degradation of the buffer by conversion to illite in a hundred years. This would be disastrous and totally ruin the plans of using smectite of montmorillonite type as buffer.

What has then been made to find out what the activation energy really is, and above all, whether the theoretical conversion model is at all correct? Not much, although there have been some attempts, primarily by finding examples from nature of smectite clays having been exposed to temperatures and temperature gradients that resemble those in repositories.

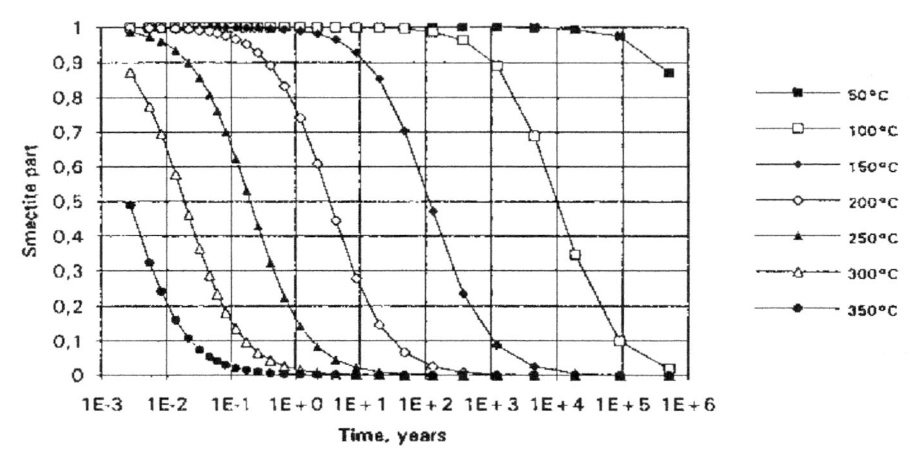

Fig. 3.51 Diagram showing predicted conversion of smectite to illite for the activation energy 27 kcal/mole

3.5.11.3 Natural Analogues

An almost classical case comprises the so-called meta-bentonites at Kinnekulle in southwestern Sweden. They are of Ordovician age and make up a series of up to 2 m thick beds (Fig. 3.52) that have been exposed to at least 30 MPa pressure in the latest glaciation period. The present content of montmorillonite in expandable form in the bentonite bed is about 25%, the rest being illite, quartz, feldspars, chlorite, and carbonates. Analyses of samples taken from the bed gave the water content values (w) in this figure, indicating that the clay has expanded from a denser state reached during the maximum ice load some tens of thousands of years ago [23].

The temperature history, derived from conodont analysis and calculation of the temperature evolution after intrusion of Permian basalt is shown in Fig. 3.53. The average temperature gradient was 0.02°C/cm in the first hundreds of years.

3.5.11.4 Laboratory Experiments

Various hydrothermal tests of smectites with different porewater compositions indicate that the conversion of montmorillonite to non-expanding minerals and formation of precipitates may still not be as simple as presumed and the matter appears to require more research. Here, a few carefully performed tests will be referred to for showing the state-of-art of this branch of clay mineralogy.

A joint SKB/ANDRA experiment in the early nineties was made to investigate the effect of gamma radiation on montmorillonite-rich (MX-80) clay with a dry density of 1650 kg/m^3 [24]. The sample was exposed to a radiation dose of about 3E 7 Gy at one end that was sealed by an iron plate, while at the opposite end the sample was confined by a filter through which it was in contact with weakly brackish water with Na as dominant cation contained in a large vessel. The solution had a very low potassium content (<10 ppm). The radiated end was heated to 130°C during the

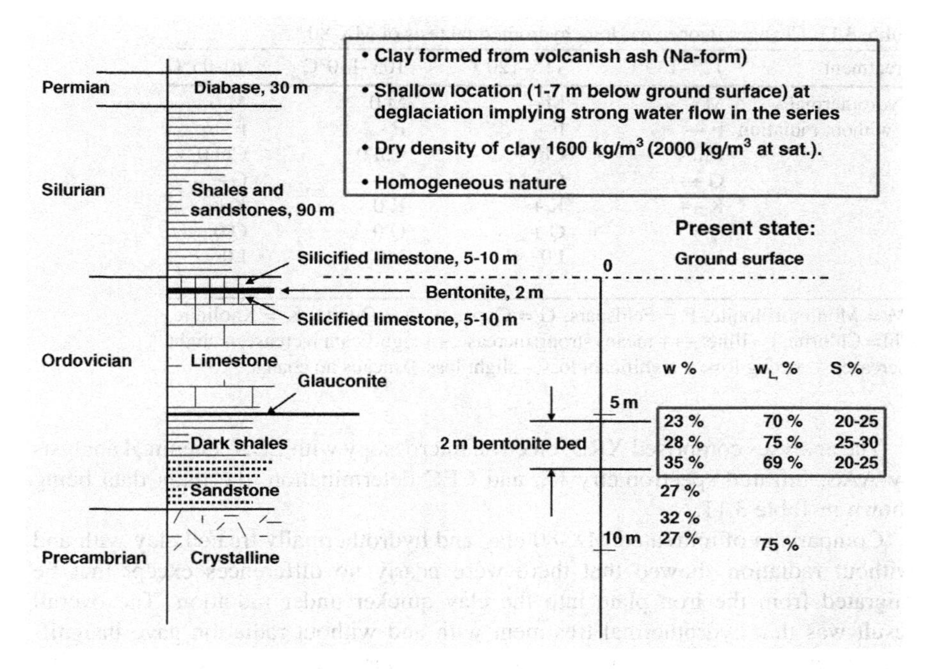

Fig. 3.52 The Kinnekulle stratigraphy. The *right* part describes the geotechnical/mineralogical data evaluated from analysis of 10 m cores taken from the present ground surface. w_L denotes the Atterberg liquid limit, which is a measure of the smectite content ("S") that is 20–30% of the mineral mass in the thick bentonite bed [23]

1 year experiment and the opposite to 90°C (thermal gradient about 6°C/cm). The solution was pressurized to 1.5 MPa. The adsorbed radiation dose was 3972 Gy/h at the hottest contact, around 700 Gy/h at half length of the sample, and 456 Gy/h at the coldest end. A parallel test of the same duration was conducted without radiation.

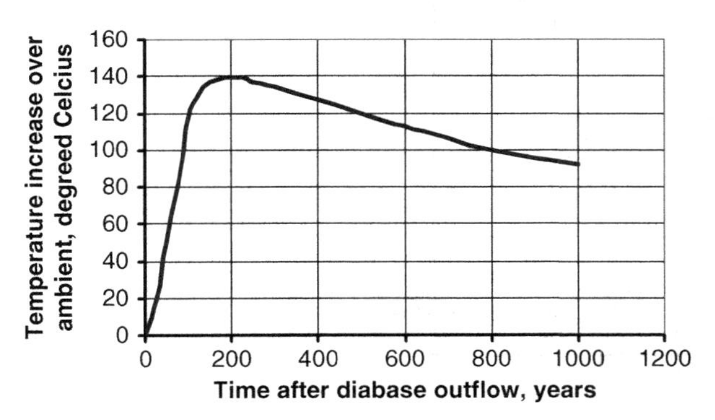

Fig. 3.53 Temperature history of the Kinnekulle bentonite as evaluated from finite element calculations and conodont analyses [23]

Table 3.11 Changes in one year long hydrothermal tests of MX-80

Treatment	125–130°C	115–120°C	105–110°C	90–95°C
Hydrothermal without radiation	M -	M –	M 0	M 0
	F —	F –	F -	F -
	Chl +	Chl +	Chl 0	Chl 0
	G ++	G +++	G +	G+
	K ++	K +	K 0	K -
	Q +	Q +	Q 0	Q 0
	I +	I 0	I 0	I 0

M = Montmorillonite, F = Feldspars, G = Gypsum, Q = Quartz, K = Kaolinite, Chl = Chlorite, I = Illite. +++ means strong increase, ++ significant increase, + slight increase, — strong loss, – significant loss, - slight loss. 0 means no change.

The analyses comprised XRD, electron microscopy with EDX, chemical analysis by AAS, infrared spectrometry IR, and CEC determination, the main data being shown in Table 3.11.

Comparison of untreated MX-80 clay and hydrothermally treated clay with and without radiation showed that there were nearly no differences except that Fe migrated from the iron plate into the clay quicker under radiation. The overall result was that hydrothermal treatment with and without radiation gave insignificant chemically induced changes. This is supported by the CEC data showing that untreated MX-80 had CEC = 99 meq/100 g and the most harsly treated clay 93 meq/100 g. However, creep tests gave witness of significant stiffening by cementation (Fig. 3.54). Thus, the shear strain of samples exposed to 130°C was nearly one thrid of the 90°C sample.

In the late eighties a series of one year long laboratory tests were performed as part of the Stripa Project [25]. The clay was confined in teflon-coated cells placed

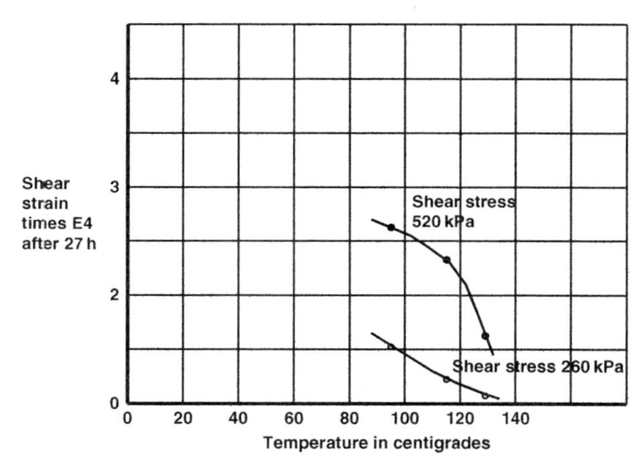

Fig. 3.54 Shear box testing of MX-80 clay from hydrothermal experiment without radiation. Normal stress 6 MPa. The accumulated strain was 2E-4 to 27E-4 in 27 hours [24]

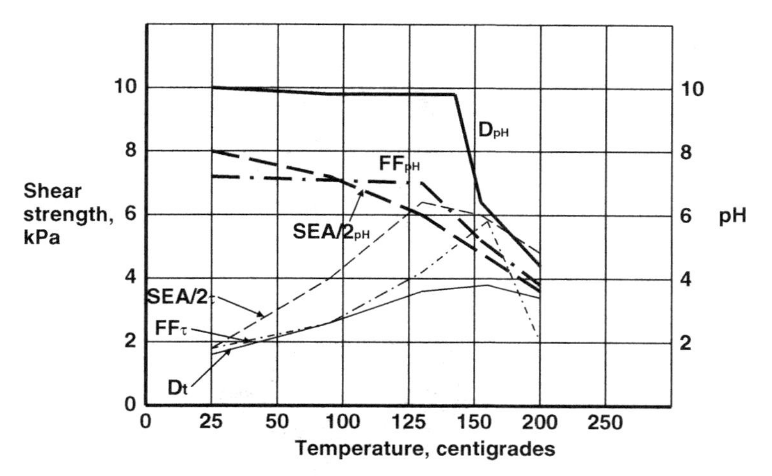

Fig. 3.55 Drop in pH and increase in shear strength at termination of hydrothermal tests [25]

in larger vessels with initially distilled water (D), solution (10 000 ppm) of $CaCl_2$ (FF), and 50% artificial ocean water (SEA/2). Figure 3.55 shows that considerable changes took place. Thus, pH dropped by about one unit from 10 at room temperature for distilled water and 7–8 for salt solutions to 4–8 by heating to 130–150°C, which means that dissolution of any carbonate in the buffer must have taken place. Very acid conditions were created by heating to 200°C, which naturally had a strong dissolving impact on all minerals.

The diagram shows that the shear strength, determined by vane testing, doubled or trebled by heating to 130–160°C but dropped at higher temperatures, presumably by dissolution and loss of solid matter. It is believed that the strengthening was due to precipitation of cementing silicious compounds. The concentration of silicons increased from around 10 mg/l to more than 90 mg/l in the pore fluid at the termination of the tests. Figure 3.56 illustrates the assumed processes induced by heating to 90°C and higher temperature: 1) contraction of stacks of lamellae under a sufficiently high effective (swelling) pressure yielding larger and more interacting voids, and 2) precipitation of silicious matter causing cementation. The two processes combine to increase the hydraulic conductivity and the shear strength.

Mineralogical analyses by XRD confirmed that calcite was largely dissolved at 90°C and completely dissolved at 130°C. K- and Na-feldspars and also quartz were slightly attacked at 90°C and 130°C but largely dissolved at 160°C. They were nearly completely dissolved at 200°C. The heights and areas of the characteristic XRD peaks of montmorillonite were not significantly altered indicating only slight attack.

A three year Mock-up tests, simulating SKB's concept KBS-3 V on half scale, was performed at the Technical University in Prague early in this century giving detailed information on the performance of the Czech buffer candidate "RMN" under repository-like conditions [26]. The test is referred to here because it indicates how smectite clay rich in montmorillonite and iron performs under such conditions.

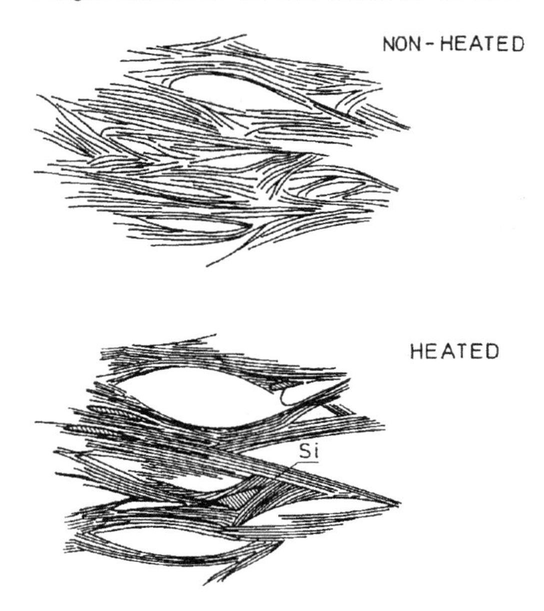

Fig. 3.56 Microstructural changes in hydrothermally treated MX-80 clay [25]

The rock was represented by a steel tube with 80 cm diameter to which a filter was attached for uniform wetting of the buffer, which consisted of highly compacted blocks of clay with 75% Ca montmorillonite, 10% finely ground quartz and 5% graphite powder (Fig. 3.57). The clay blocks having a dry density of 1800 kg/m^3, had a radial thickness of 18 cm and a water content of 7%. The 50 mm gap between the steel tube and the heater was filled with loose "RMN" granules. Samples could be taken through 3.5 cm pipes that intersected the tube and were subsequently sealed with equally dense clay plugs. The constant 95°C temperature at the heater surface gave a temperature of 45–48°C at the outer boundary of the clay buffer at mid-height (temperature gradient 2°C/cm). The filter was saturated with moderately brackish "granitic" water under 60 kPa pressure. Nearly complete saturation was reached after 2 years.

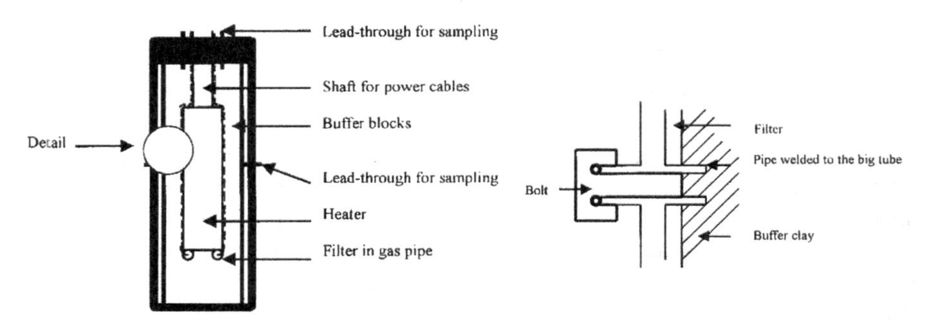

Fig. 3.57 The Czech Mock-up. General arrangement with central heater surrounded by buffer blocks and granules

Table 3.12 Geotechnical data of samples M1 to M4 in oedometers. The data ρ_{sat} = density at water saturation, and hydraulic conductivity = K, represent equilibrium reached after about 40 days [27]

Sample	Distance from heater, cm	T, °C	ρ_{sat}, kg/m^3	p_s, kPa	K, m/s
M1	16–18	45–47	1800	430	1.6E-11
M2	12–14	54–56	1945	650	1.9E-11
M3	6–8	67–69	1910	355	2.3E-10
M4	0–2	85–90	1925	310	2.1E-10

The heating, wetting and development of swelling pressure were recorded by rich instrumentation, and samples were extracted two to three times per year for checking the wetting rate. Samples taken at the termination were used for mineralogical analyses and ordinary geotechnical investigations summarized in Table 3.12.

It is obvious that a large part of the clay buffer had undergone significant changes. The ratio of the swelling pressures for the M2 and M4 samples, which had practically the same density, was 650/310=2.09. The coldest 50 mm zone had the same conductivity as virgin clay with the same density, while for the clay 10–18 cm from the heater it was about 10 times higher than of equally dense untreated clay. The clay closer than 10 cm from the heater was up to 100 times more permeable than unheated clay with the same density. The swelling pressure showed the corresponding pattern, i.e. a marked drop for the samples taken close to the heater.

Mineralogical characterization of the original clay was made by use of TEM with EDX and CSD, i.e. Coherent Scattering Domain data for determining the thickness of particles or stacks of lamellae. The Koester diagram in Fig. 3.58 shows the charge distribution in the various layers in untreated RMN clay to be compared to the distribution of charge after termination of the experiment.

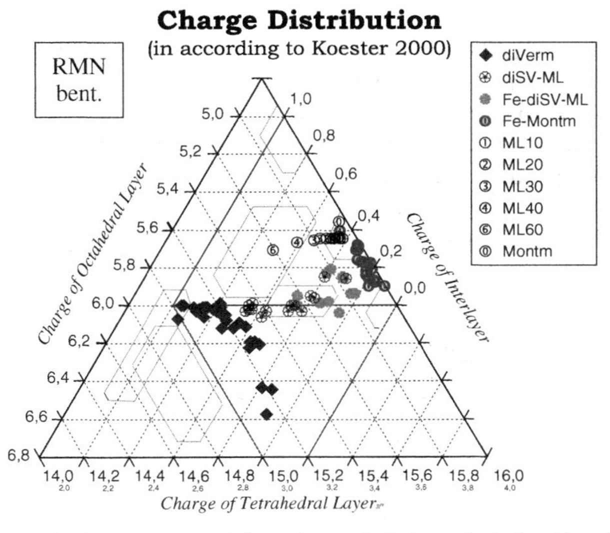

Fig. 3.58 Distribution of charge in tetrahedral, octahedral and interlayers in M4. ML means mixed-layer minerals, the figures represent percentages [26]

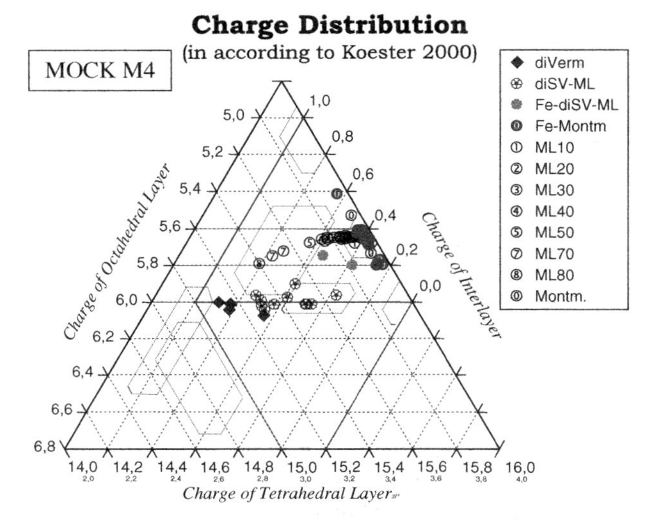

Fig. 3.59 Distribution of charge in tetrahedral, octahedral and interlayers [27]

Figure 3.59 shows that very obvious changes in charge distribution took place in M4. It was concluded that significant dissolution of the Fe-montmorillonite and complete disappearance of the intergrowth of illite and kaolinite had taken place, suggesting migration and precipitation of released elements at different distances from the heater. TEM-EDX photos of M4 showed that desiccation fissures assumed to have been formed before resaturation took place remained, demonstrating that the clay was not sufficiently expandable to self-heal. The XRD analyses showed obvious reduction in montmorillonite content in comparing M3 and M4, and obvious increase in illite particle thickness in samples M3 and M4 as compared with M2 (cf. Table 3.13).

A reliable verification of the significant changes in physical properties of the larger part of the RMN buffer because of mineral conversion and cementation is given by Fig. 3.60. It shows the change in Atterberg liquid limit, w_L, which is an excellent measure of the hydration potential of clays. For unheated RMN clay w_L is about 170% and this figure dropped by about 30% in the hottest part of the buffer and by about 10% at the periphery of the buffer.

The main results from the described study of samples from the Mock-up test were:

Table 3.13 Mineralogical changes in the Mock-up test [27]

Mineral	Sample M2	Sample M3	Sample M4
Montmorillonite	Slight dissolution	Moderate dissolution	Signif. dissol., low crystallinity
Illite	High disorder, minor illite	Significant illite formed or reduced disorder	
Talc and kaolinite	Low disorder, particle growth	Increased disorder or neoformation	

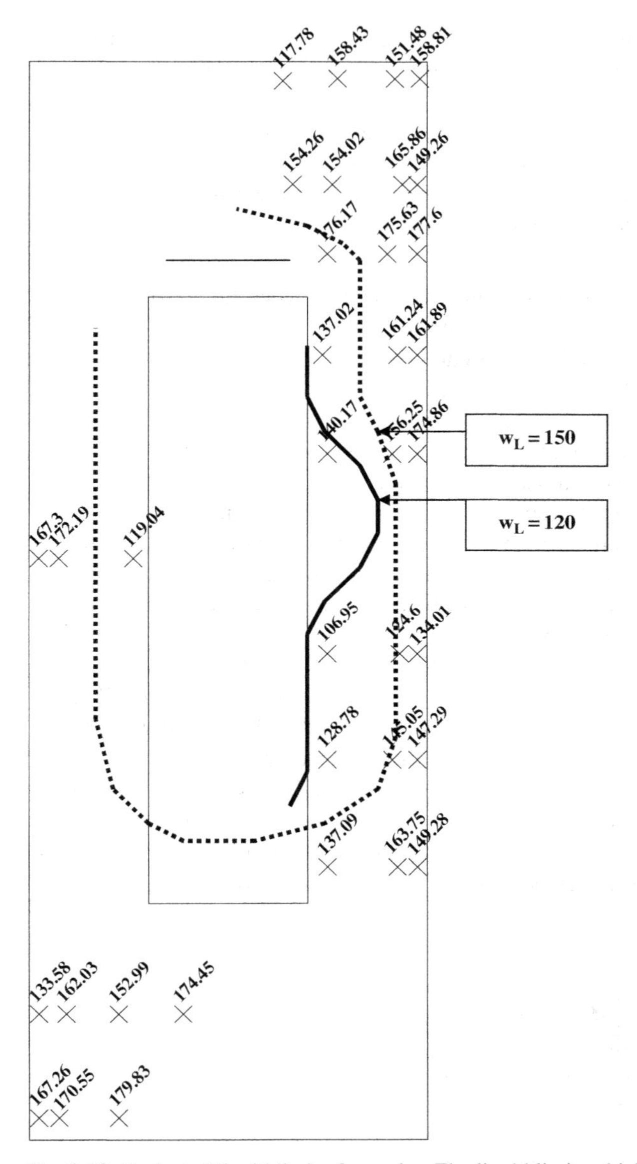

Fig. 3.60 Evaluated liquid limit of samples. The liquid limit, which is at least 170% of untreated RMN clay, had dropped to about 150% over the larger part of the buffer clay and to less than 120% near the heater. At mid-height heater the larger part of the clay w_L ranged from 110 to 120%

- The swelling pressure had dropped by more than 50%, in the larger part of the buffer. The hydraulic conductivity had increased by 10–100 times.
- Intergrowth of illite and kaolinite is obvious in the "dioctahedral vermiculite" in untreated RMN clay. In these aggregates illite dominates over kaolinite. In sample M4 most of the illite/kaolinite intergrowths had dissolved.

- Fe set free by the dissolution of the Fe-montmorillonite can have formed iron complexes causing cementation in the entire buffer mass, especially in the most heated part (M4).
- Replacement of octahedral Al by Fe can have caused a drop in coherence of the montmorillonite crystals promoting easier dissolution.
- Formation of illite in the hottest part of the buffer (sample M4) may have been associated with uptake of K from dissolved vermiculite.

Further data have been obtained from a full-scale 5 year field experiment at SKB's underground laboratory at Äspö with very dense MX-80 clay buffer surrounding a full-size copper-lined KBS-3 canister in an 8 m deep, 1.75 m diameter deposition hole in granite. It was evaluated in 2007 with respect to the performance of the clay [28]. A filter had been placed at the walls of the hole for providing the clay with groundwater, which had a total salt content of about 6000 ppm with a Na/Ca ratio of about unity. The canister contained electrical elements for simulating the heat produced by the highly radioactive content of such canisters and its surface temperature was maintained at 85°C for a couple of years and at successively lower temperatures later in the experiment. The average radial thermal gradient was 1°C/cm. The evolution of the swelling pressure at the buffer/canister contact at mid-height canister is shown in Fig. 3.61.

A representative sample numbered "R8:225: Canister" was taken from the nearness of the heater and investigated by oedometer testing for finding out if these properties deviated from those of equally dense virgin MX-80 clay and what possible chemical and mineralogical changes that can have taken place.

Virgin MX-80 clay, saturated and percolated with distilled water and having the same density as a sample R8:225 from the vicinity of the canister, has an average hydraulic conductivity that is less than one hundredth of that of this sample. In contrast, there was no discrepancy between the swelling pressure of this sample and

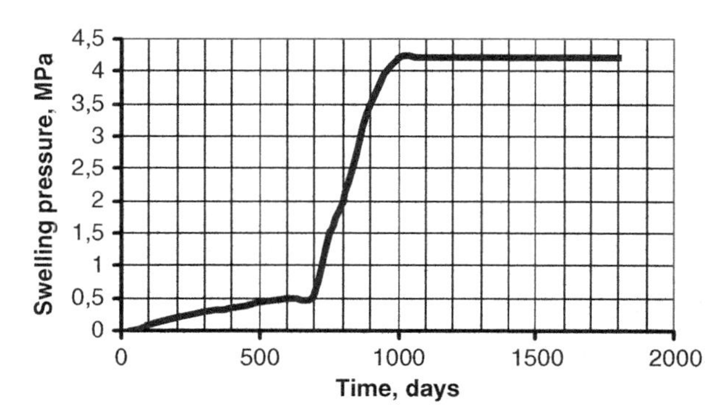

Fig. 3.61 Development of swelling pressure at mid-height canister. The flat part of the curve represents the initial phase of hydration and compression of the pellet filling as well as closure of joints between the clay blocks

Table 3.14 Summary of physical data of the sample adjacent to the canister at mid-height, and of virgin MX-80 clay with the same density, i.e. 1970 kg/m^3 (1540 kg/m^3 dry density) saturated with distilled water or artificially prepared ocean water (3.5% TDS)

Sample	R8:225: Canister, parallel to canister	R8:225: Canister, perpendicular to canister	Virgin MX-80 dist. water	Virgin MX-80 ocean water
Hydraulic conductivity, m/s	2E-11	2E-11	8E-14	2E-13
Swelling pressure, MPa	4.2	4.3	4.0	3.5–4.0

that of virgin MX-80. One hence concludes that while the swelling pressure of the clay adjacent to the canister does not indicate any physico/chemical changes, the hundred-fold increase in hydraulic conductivity and the reduced dispersibility of this clay caused by the exposure to hydrothermal conditions certainly demonstrate such changes. Table 3.14 summarizes the data from testing of samples from the field experiment.

The main results from the mineralogical analyses are summarized in Table 3.15. The data were derived from comprehensive (>100 specimens) particle-wise TEM-EDX-measurements.

Major findings were:

- MX-80 clay contains two general types of montmorillonite: i) montmorillonite with a normal charge as end member of IS-ml series, and ii) low-charge montmorillonite as end member of diVS-ml series. The ratio IS:diVS varies from 1/20 to 2/3.
- The hydrothermal treated clay has undergone substitution of Al^{3+} by Fe^{3+} in the octahedral layer causing higher lattice stresses because of the larger ion radius of Fe^{3+} than of Al^{3+} and reducing the resistance of the montmorillonite to dissolution. The change in chemical composition of the montmorillonite is illustrated by the derived formulae:

Table 3.15 Parameters from mineral formula per [$(OH)_2$ O_{10}] from TEM-EDX-analyses

Parameter	Virgin MX-80	R8:225: Canister
Phases Traces:	Montmorillonite (low-charge montmorillonite), diVS-mlIS-mlPSV-ml	Montmorillonite (low-charge mont-morillonite), diVS-mlIS-mlKSV-ml"albite"
Smectite Layers (S%)		
S% in IS-ml phases	85%	95%
S% in diVS-ml phases	90%	80%
Interlayer charge (average of all)	0.222	0.305

IS-ml – illite-smectite mixed layer phases; diVS-ml – dioctahedral vermiculite-smectite mixed layer phases; PSV-ml – pyrophyllite-smectite-dioctahedral vermiculite mixed layer phases; KSV-ml – kaolinite-smectite-dioctahedral vermiculite mixed layer phases.

a) montmorillonite as end member of IS-ml series in virgin MX-80:
$Ca_{0.04}\ Mg_{0.09}\ K_{<0.01}\ Al_{1.64}\ Fe^{3+}{}_{0.13}\ Mg_{0.23}\ Ti_{<0.01}\ (OH)_2\ Si_{3.96}\ Al_{0.04}O_{10}$
b) montmorillonite as end member of IS-ml series in hydrothermally treated clay:
$Ca_{0.07}\ Mg_{0.02}\ Na_{0.04}\ K_{0.07}\ Al_{1.46}\ Fe^{3+}{}_{0.23}\ Mg_{0.25}\ Ti_{0.03}\ (OH)_2\ Si_{3.98}\ Al_{0.02}O_{10}$

- The hydrothermally treated clay has undergone step-wise alteration from normally charged montmorillonite to a low-charge montmorillonite by replacement of original Mg by Al. The composition of the montmorillonite as end member of diVS-ml series in the hydrothermally treated clay has the following form:

  c) $Ca_{0.02}\ Mg_{0.04}\ Na_{0.01}K_{0.03}\ Al_{1.62}\ Fe^{3+}{}_{0.16}\ Mg_{0.17}\ Ti_{0.04}\ (OH)_2\ Si_{3.99}\ Al_{0.01}O_{10}$
- Montmorillonite is the dominating mineral phase in both virgin and hydrothermally treated clay. However, mineralogical changes induced by the hydrothermal conditions are obvious as indicated by an increase in smectite layer frequency in the IS-ml phases of the heated clay (95%) compared to the typical number for virgin MX-80 (85%), and an associated drop in smectite layer frequency in the diVS-ml phases from 90% of the virgin MX-80 to 80% in the heated clay.
- The lower interlayer charge of R8:225 than of virgin MX-80 means, according to classical double layer theory, that it should have higher swelling pressure, which is in agreement with the diagrams in Fig. 3.62 and with the recorded swelling pressure of R8:225 (Table 3.14).

An effect of the exposure to hydrothermal conditions that is of importance in modelling microstructural and rheological evolution is the altered particle morphology shown in Fig. 3.63. It demonstrates that the typical appearance of interwoven thin stacks of montmorillonite lamellae has been locally changed to a mass of kaolinite-like, discrete particles.

The particles in Fig. 3.63 are believed to have been formed in the softest parts of the heterogeneous microstructure, i.e. the channels described earlier and illustrated schematically in Fig. 3.36. In freshly prepared buffer clay of MX-80 type channels contain homogeneous clay gels with a density that is low in large channels and high in narrow ones. Salt porewater causes coagulation of the gels, which increases their conductivity and thereby the bulk couductivity [10], and if the smectite particles are converted to illite the conductivity would be increased furthermore and cause the more than hundredfold rise in bulk conductivity that the hottest sample underwent (cf. Table 3.14).

Putting together the results of the described laboratory and field tests and keeping in mind the experience from the Kinnekulle case one comes to the following major conclusions:

- In all the tests montmorillonite remained the major clay mineral in the buffer even at temperatures of up to 150°C. However, the fact that natural analogues with temperature histories similar to that of repositories have undergone significant loss of montmorillonite demonstrates that buffer clay will also have its content of this mineral reduced. The rate of conversion of the montmorillonite is in fact

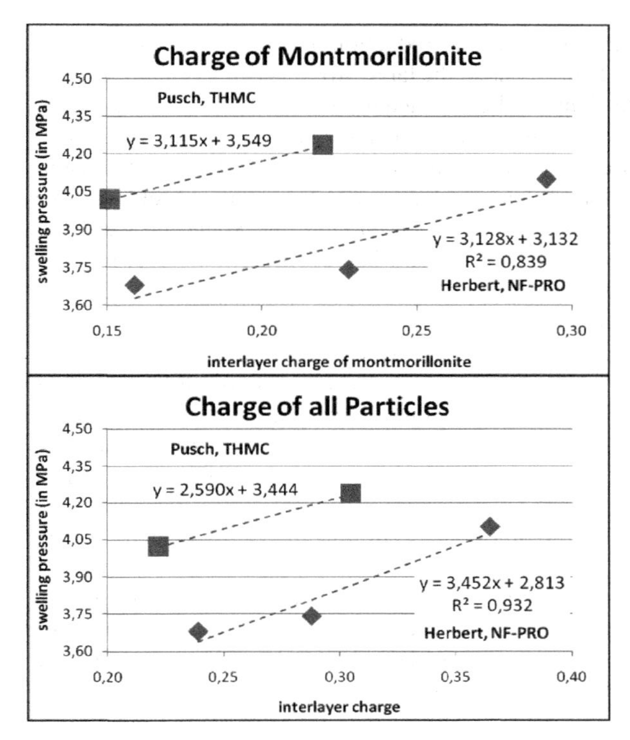

Fig. 3.62 Comparison of swelling pressure vs. charge for MX-80 clay showing the same trend of increased pressure for increased charge in different experiments using different techniques

Fig. 3.63 Partially hypidiomorphic, pseudohexagonal particles in sample R8:225

not yet known, for the Kinnekulle case the reduction by 50–75% may have taken place in about 1000 years. In this context the large difference in thermal gradient– less than 0.02°C/cm for the Kinnekulle case and more than 1°C/cm for the experiments – should be considered. It may imply much quicker changes in a repository.

- Mineralogical changes were similar in all the experiments. The most obvious phenomenon was dissolution of the accessory minerals calcite and feldspars but also of quartz and montmorillonite. The released elements Si and Fe were precipitated in the course of the experiments or when cooling took place, forming cementing agents that significantly reduced the expandability. This speaks in favour of using very pure montmorillonite clays for preparing buffers.
- In the early stage of evolution desiccation will take place in a large part of the buffer and frequent fissures and growth of initial voids will appear in the matrix of contracted stacks of lamellae. Precipitation of Si and Fe can weld the lamellae and stacks together and prevent expansion when water later enters this part, giving rise to permanent increase in hydraulic conductivity and drop in swelling pressure.
- The observed silicification in the limestones confining the bentonite beds at Kinnekulle agrees with the basic smectite-to-illite conversion model that specifies that silica is released from silicious minerals, primarily feldspars, and precipitated where lower temperature prevails, yielding cementation. At Kinnekulle, cementation by silica precipitation (Fig. 3.64) took place to at least 5 m distance from the bed and this is also the calculated effect using Grindrod/Takase's geochemical model [29].

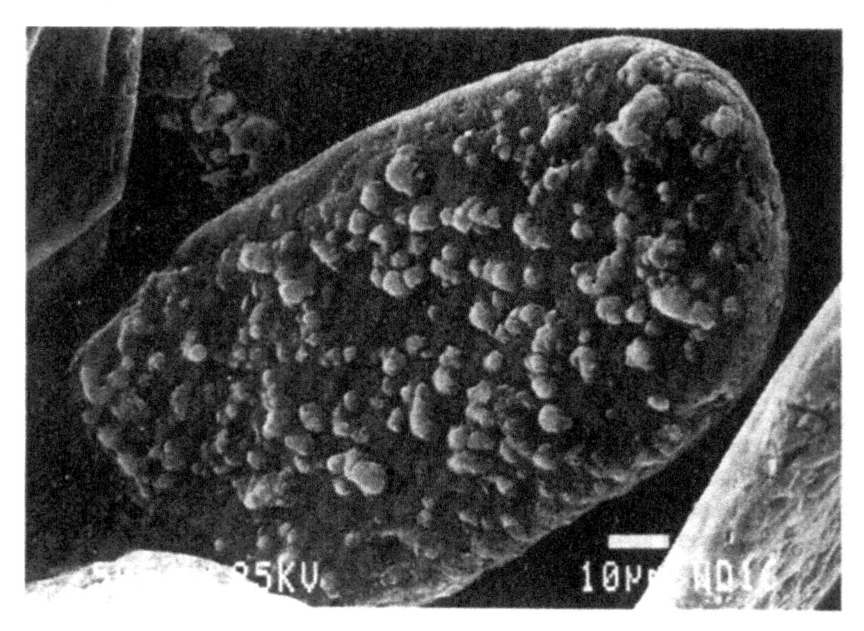

Fig. 3.64 Silica precipitated on a silt particle in Kinnekulle bentonite

It is important to realize that the conditions under which the tests ran can have saved the buffer from stronger degradation by offering unlimited access to water and by limiting the testing time. In practice, the majority of the buffer in deposition holes will be supplied with very little water and this water may have much higher TDS than used in the experiments. Continued R&D is hence a necessary prerequisite for certifying that montmorillonite buffer will perform acceptably in a long-term perspective. The most obvious threat to acceptable long-term performance of montmorillonite buffers is cementation, which hence requires particular attention.

3.5.11.5 Geochemical Modelling

The need for safer predicition of the fate of smectitic clays exposed to hydrothermal conditions has caused intense work, primarily by the oil- and gas-industry for developing lasting drill muds and this work together with the R&D of the Organizations has given further insight in this problem area. Comprehensive work has been made at the Universities of Strasbourg and Greifswald and we will summarize the experience gained, which has led to development of conceptual models and geochemical codes. We will start by referring to a few major evolution processes identified in natural sedimentary smectitic clays.

Neoformation of illite is expected to take place at certain concentrations of silica (H_4SiO_4), aluminum and potassium, yielding crystal nuclei in the form of laths. When precipitation takes place the potassium concentration drops locally and the concentration gradient thus formed brings in more potassium by which the process continues. Geochemical codes tend to indicate that illite should be formed from smectites in a certain "window" of phase diagrams of silica, aluminum and potassium, but they do not seem to be able to indicate whether the conversion takes place via mixed-layer mineral stages or by dissolution/neoformation. Illite formation in this way can take place parallel to conversion of smectite via mixedlayer S/I minerals.

As to the role of iron we saw from the experiments with the Fe-rich Czech smectite clay with Fe located in octahedral positions in the crystal lattice that Fe may promote solubility. It can not be excluded that reactions between clay and contacting iron and steel canisters can yield both cation exchange to Fe in the clay and precipitation of iron compounds not only when oxygen is still available in the system but also under aerobic conditions. Reduction of the structural iron in buffer material is most likely and Fig. 3.65 describes possible mechanisms.

A general conclusion from all tests and attempts to model the chemical evolution of montmorillonite buffer is that the buffer will undergo two major degrading processes under the hydrothermal conditions that will prevail in most HLW repositories:

- Practically important dissolution of montmorillonite starts at temperatures exceeding about 60°C but is assumed to be very slow at temperatures lower than 80–90°C. It depends on the concentration of silicons in the porewater, which

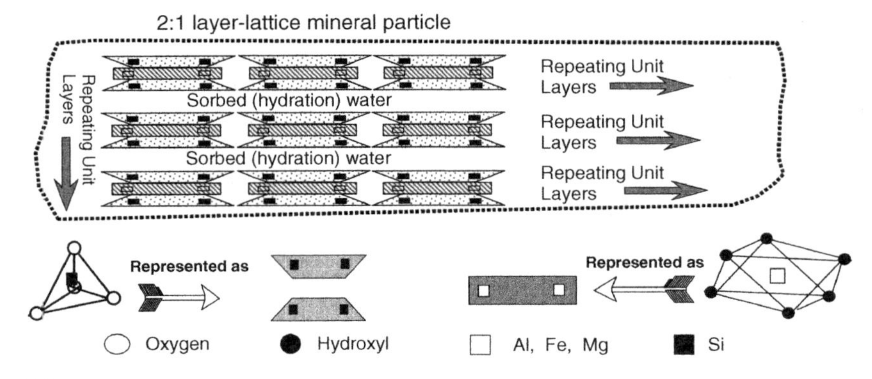

☐ Fe -- common transition metal; Can exist as divalent or trivalent positive in octahedral sheet, or trivalent form in tetrahedral sheet.

<u>Oxidation-Reduction of Octahedral Fe</u> :-- due to biological activity, alternate wetting and drying cycles, redox-type chemicals in soil solution. This will significantly alter short-range forces next to clay layers.

<u>Reduction of structural Fe^{3+} to Fe^{2+}</u> :-- will result in (a) a lower SSA, high degree of layer collapse, (b) decreased water-holding capacity, (c) reduced swelling, and (d) better stacking of particles.

Fig. 3.65 Schematic showing the repeating layers of a 2:1 layer-lattice mineral particle (montmorillonite) and what might happen in a reducing environment [10]

in turns depends on the dissolution of accessory minerals and precipitation of silicious compounds, like illite. For formation of illite potassium must be available and therefore controls the formation of this mineral. Buffers with very little feldspars, especially those containing potassium, and calcite are preferable for long-term stability. The lack of longlasting tests under controlled conditions makes predictions uncertain.

- Precipitation of silicious compounds yields illite to an exent that is controlled by the access to potassium, and other crystalline or amorphous matter. They are formed in the most porous parts of the clay matrix, the channels, which thereby contain minerals with less hydration potential than montmorillonite and represent permanent, pervious paths.

3.5.12 Impact on Buffer and Backfills of Chemical Reactions with Other Barriers

3.5.12.1 Interaction of Canister Material and Buffer Clay

A number of experiments and theoretical calculations have been performed for investigating mutual chemical interaction of buffer and metal canisters, but we will confine ourselves here to look at a recent study of the interaction of copper metal and three potential buffer candidate clays for use in crystalline and argillaceous rock, i.e. montmorillonite-rich bentonite (MX-80), saponite-rich clay (GeoHellas, Greece), and mixed-layer montmorillonite/muscovite (Friedland Ton).

The clays were prepared by compacting air-dry powder of the clays in hydrothermal cells of stainless steel with one end made of copper. The steel end was kept at 35°C by circulating 1% $CaCl_2$ solution through filters consisting of a coarse part and a fine-porous part contacting the clay, while the copper end, separated from the steel by an O-ring and located 50 mm from the opposite end, was kept at about 95°C (Fig. 3.66).

The temperature gradient of slightly less than 15°C per cm clay was maintained throughout the saturation period, which lasted for about 3 weeks, and in the subsequent hydrothermal period. Samples of the solution were extracted twice, the first time after 3 weeks and the other at the end of the 8 week long hydrothermal treatment.

At the end of the tests the samples were extruded into oedometer cells in conjunction with sectioning them into three parts for determining the hydraulic conductivity and swelling pressure for different distances from the hot end.

The test results of the physical measurements showed that the dry density was different in the earliest and latest water-saturated parts of the samples, demonstrating that compression of the last-mentioned, most heated part had taken place. MX-80 had undergone considerable changes, particularly in the form of a very significant loss of swelling pressure and a rise in hydraulic conductivity by 100 times in the hottest part. The central and cold parts were also significantly changed with

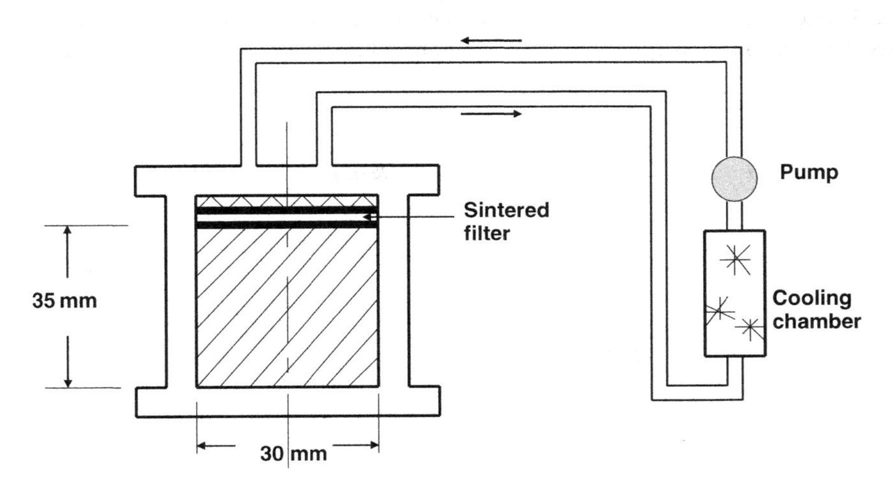

Fig. 3.66 Test arrangement with the cold end kept at 35°C by circulating $CaCl_2$ solution through the upper filter and the hot end of copper at 95°C

respect to conductivity but retained their expandability. The physical properties of the saponite changed much less than those of MX-80. Thus, the swelling pressure remained high in the hottest part and was roughly the same as in untreated saponite clay in the central and cold parts. The hydraulic conductivity of the hot part was about 70 times higher than that of untreated saponite while it was raised insignificantly in the central and cold parts. The swelling pressure of the mixed-layer Friedland Ton had dropped significally in the hottest part but only slightly in the central and cold parts. The hydraulic conductivity of the hot part was about 50 times higher than of untreated FIM clay but remained unchanged in the central and cold parts.

The mineralogical changes can be summarized as follows:

- MX-80

Most of the montmorillonitic particles underwent mineralogical alteration yielding two series of mixed layer phases: an illite-smectite mixed layer series (IS-ml) and a dioctahedral vermiculite-smectite mixed layer series (diVS-ml). These results are in agreement with those obtained in the large-scale tests. The difference between the two diVS-ml series indicates obvious trends for Al, Fe and Mg in the composition of the octahedral layer especially for the cold parts: Al decreased with increasing content of vermiculitic, while Fe and Mg showed reverse behaviour (Fig. 3.67).

The XRD diagrams show that part of the montmorillonitic phases collapsed in the hot region to 10 Å spacing, which contributed to the higher density of the material in the hot region. Montmorillonite in the colder parts was mainly unchanged except for a weak tendency of the Si concentration to have increased in the central and colder parts. The clay had undergone slight dissolution and minor precipitation of silicious

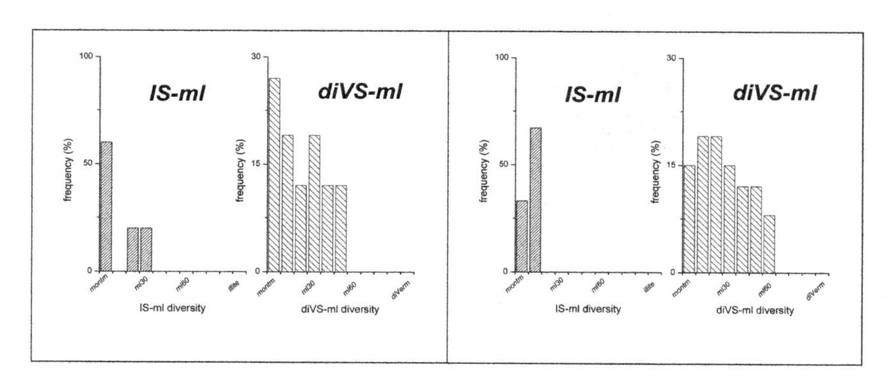

MX80 hot	MX80 cold

IS-ml = illite-smectite mixed layer phases; diVS-ml = dioctahedral vermiculite-smectite mixed layer phases; diVerm = dioctahedral vermiculite (interlayer charge deficient illite)

Fig. 3.67 TEM-EDX: Diversity of mixed layer phases for MX80 hot (*left group*) and MX80 cold (*right group*)

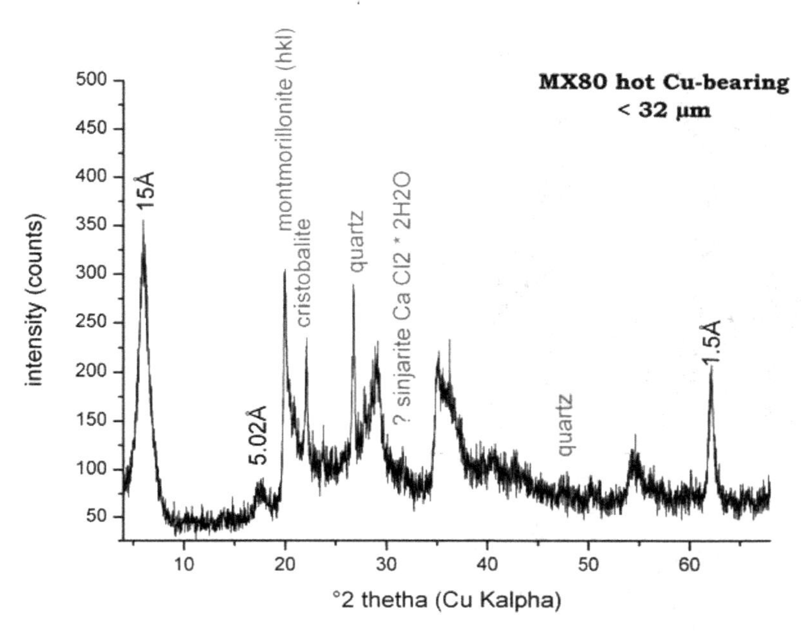

Fig. 3.68 XRD-powder diffractogram of the hottest part of the MX80 sample. Major minerals are Ca-montmorillonite (15 Å), quartz, and cristobalite. Sinjarite (2.83 Å) was neoformed

material can have taken place causing cementation and partial loss of expandability. The chloride mineral sinjarite (2.83 Å) was neoformed in the hottest part, i.e. up to 10 mm distance from the copper plate (Fig. 3.68).

Copper had dissolved and entered the clay to a significant extent in the hot part. The concentration gradient indicates that the migration is diffusive and curve fitting gives an approximate diffusion coefficient of E-12 m^2/s (Fig. 3.69).

- Saponite

The mineralogical changes were very small and had the form of Ca replacing initially sorbed Na and Mg. The sum of the Na and Ca concentrations was almost 50% higher in the coldest part than in the hot and central ones. The concentrations of Fe, Al, Mg and Si dropped slightly in the direction from the hot to the cold end. The dissolution of minerals was much less significant than in the MX-80 clay. Copper did not enter the clay at all (Fig. 3.69).

- Friedland Ton

Ca replaced the initially sorbed Na nearly completely in the hottest part and also to a high degree in the central and coldest parts. The content of Mg sorbed by the clay was partly replaced by Ca or Fe set free in conjunction with slight dissolution of iron complexes. The largely unchanged Fe, Al and Si contents show that the clay mineral structure remained nearly unchanged throughout the clay. Copper had

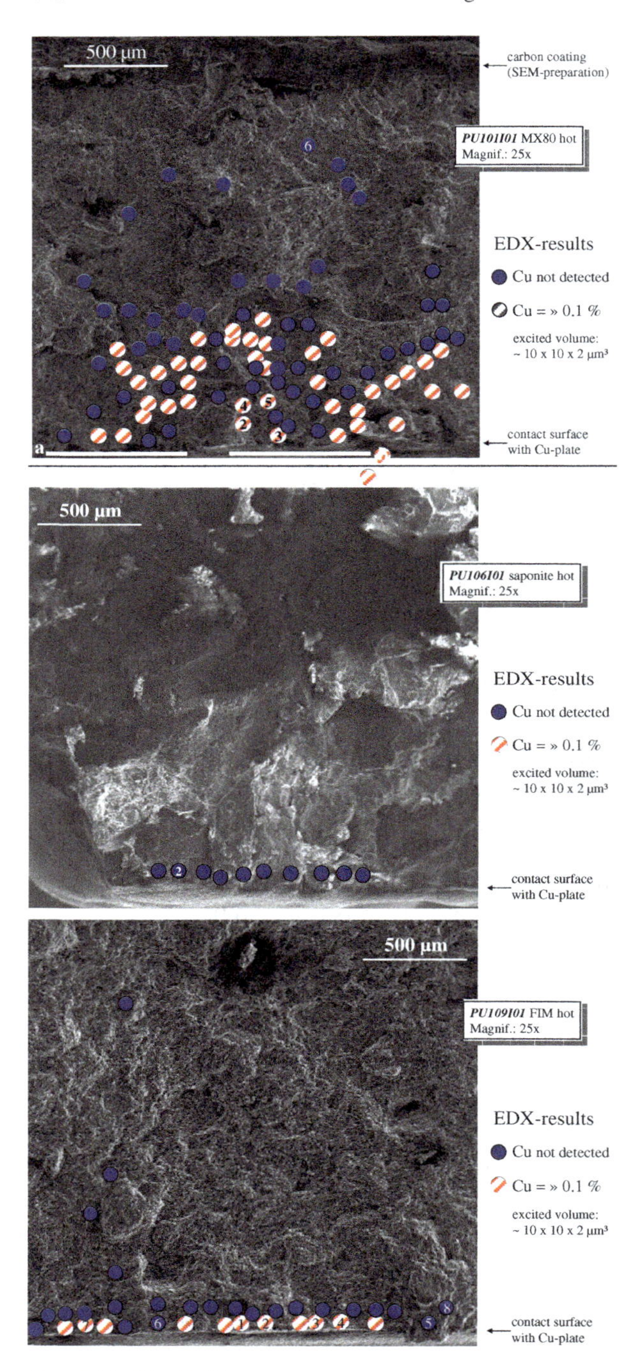

Fig. 3.69 SEM EDX analysis showing migration of Cu from the hot end of the cells into the respective clay. *Upper*: MX-80 showing significant effect. *Center*: Saponite showing no migration of Cu at all. *Lowest*: Slight, local migration of Cu in Friedland Ton

dissolved and entered the clay to an insignificant extent in the hot part but migration of this element had occurred even into the colder parts as concluded from chemical analysis.

One concludes from this study that the saponite clay suffered much less from the hydrothermal treatment than the montmorillonite clay and that the mixed layer Friedland Ton was intermediate in this respect. Figure 3.69 shows that copper migration was obvious in the MX-80 clay and, to a much smaller extent, in the Friedland Ton sample. The migration in the clay is believed to have had the form of surface as well as pore diffusion from the metal contact of 1–2 mm in two months. Since the saponite clay sustained the hydrothermal treatment and did not interact to a measurable degree with the copper it qualifies as buffer candidate for use in HLW repositories. This conclusion is in agreement with earlier findings from deep oil drilling and some national buffer studies [10,30].

The copper metal in contact with MX-80 clay showed significant pitting corrosion scars as demonstrated by the photos in Fig. 3.70.

Fig. 3.70 Micrograph of the corroded copper plate contacted with MX-80 clay showing a 200 μm corrosion scar with a depth of up to 11 μm

3.5.12.2 Interaction of Concrete and Backfills of Expandable Clay

This matter is of particular importance for the long-term performance of concrete plugs and smectitic backfills. The construction principles, preferably shotcreting in several layers for temporary plugs, and on-site casting of permanent concrete plugs, are known and tested through full-scale tests in the ESDRED project, but the long-term stability and chemical interaction with adjacent smectitic backfills remain to be determined. The matter has been in focus for a long time since the high pH of cement materials is known to be a threat to smectite minerals.

The ECOCLAY project [31] comprised batch leaching tests with bentonite mixed with synthetic cement porewater solutions and determination of chemical and microstructural changes at the cement/bentonite interface. This and other investigations have shown that if montmorillonite is taken to have the composition ($Al_4Si_8O_{20}(OH)_4$), the expected reaction with Portland cement should be $Ca(OH)_2$ + $Al_4Si_8O_{20}(OH)_4 > 2CaAl_2Si_4O_{12}, 6\,H_2O$, the reaction product corresponding to the zeolite mineral chabazite. Depending on the composition of the cement porewater other zeolites can be formed, all of them being excellent cation exchangers with high hydraulic conductivity. Theoretically, attack by OH^- on smectite minerals can yield $Al(OH)_4$, $Mg(OH)_2$ and $SiO(OH)_3$, which can all serve as cementing agents and reduce the expandability of the clay. Hydrothermal tests have shown that illite has been formed as well because of the potassium content of the degrading cement, a process that seems to require a temperature of 40°C for yielding a practically important reaction rate.

A simple and conservative chemical reaction model for estimating the rate of chemical degradation of concrete with Portland cement has been proposed and applied in practice for the licensing of the Swedish LLW repository at Forsmark. It is based on the presumption that $Ca(OH)_2$ in the cement paste is dissolved section-wise from the clay/concrete contact and that the dissolved components migrate by diffusion into the clay [32]. This gives a calcium concentration gradient that drives the process. It is assumed that the solubility of calcium in equilibrium with calcium silicate is 2E-3 kmole Ca/m^3. The concrete in the 50 m high silo wall is expected to be converted to amorphous products in the voids of the ballast (aggregates) to a distance from the clay contact of 10 cm after 50 years, to 13 cm distance after 100 years and to 30 cm distance after 500 years. The smectite clay is correspondingly transformed to an extent that is controlled by mass ratios and stoechiometry. Tests confirm that the controlling process at the chemical interaction between Portland cement and smectite is release of potassium from the cement and diffusion of calcium ions into the clay. Considerable microstructural changes were found in the reaction zone but their impact on the bulk hydraulic conductivity has not been certified. A certain insignificant conversion of smectite minerals to illite caused by potassium given off from the cement could be identified. The reaction zone was found to be 1 cm after one year at room temperature.

Other studies have verified that zeolites can be formed in a few months in batch tests with KOH/NaOH/$Ca(OH)_2$ at 90°C and after several months at lower temperatures. Interstratified smectite/illite (S/I) with up to 15–20% illite was found when the solution contained much K^+. Uptake of Mg^{2+} in the montmorillonite crystal lattice yielded the smectite species saponite.

The most important conclusions from all the studies are:

- High-alkali cement degrades quicker than low-alkali cement. The latter degrades by destruction of the CAH gel
- Released elements and water start to migrate from the fresh cement paste to the smectite in the first few hours after establishment of contact.
- The cement paste is dehydrated and its voids become wider.
- Water migrates from the smectite to the dense cement matrix.
- The solidified cement paste fissures.
- pH=12.6 is a critical value for significant changes of contacting smectite.
- Ca migrates from the cement to the clay causing ion exchange and change in the microstructure of the clay by coagulating softer parts.

The obvious risk of significant destruction of montmorillonite-rich backfill in contact with ordinary concrete as suggested by the outcome of the earlier tests, has initiated studies of the chemical interaction of low-pH cement and mixed-layer clay (Friedland Ton), which is being proposed as backfill in SKB repositories of KBS-3 V type. These studies, being part of the EU project LRDT, comprised compaction of Friedland Ton in cells to 1400 and 1600 kg/m^3 dry density with holes prepared in the clay for filling of low-pH cement water [33]. The experiments were made by use of cells shown in Fig. 3.71. Clay powder was compacted in the cells to a density corresponding to a dry density of 1450 kg/m^3. The central perforated tube containing the cement water was surrounded by a thin filter for preventing clay particles to migrate into the solution that was contained in the tube. The solution was replaced at the end of each test period, which lasted for 2 months, and analyzed with respect to pH and the concentration of major elements. After 5 months specimens

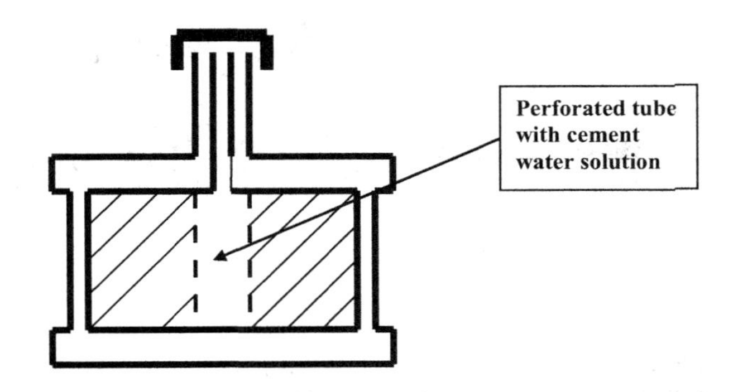

Fig. 3.71 Cell for clay/cement-water experiments

were extracted from the clay samples for examination using X-ray diffraction and electron microscopy, the rest of the clay samples being used for determination of the hydraulic conductivity.

XRD powder diffractograms were obtained from the inner (internal, closest to solution) and outer part (external) of the clay annulus (Fig. 3.72). No changes could be seen when comparing the two parts and no change compared to untreated Friedland Ton could be identified either. It hence seems that low-pH cements has a very limited impact on this clay, indicating that they would serve acceptably in repositories.

It was concluded from all the XRD analyses that no changes in the mineralogical composition could be identified and that the expected neoformation of zeolites like e.g. phillipsite or analcime could not be detected. The amount of possibly neoformed minerals was below the detection limit of XRD, i.e. $<1\%$. For the oriented, air dried specimens of the clay fraction <2 μm one could observe a change of the position of the broad (001) reflection of the muscovite-montmorillonite mixed-layer mineral from 1.2 to 1.4 nm (12 to 14 Å), meaning that exchange of Na by Ca had taken place. SEM electron microscopy of the clay reacted with Merit 5000 cement water (Fig. 3.73) and with Spanish Electroland cement water show a network of fine aggregates coated with amorphous precipitates.

Determination of the hydraulic conductivity at percolation with distilled water gave the data in Table 3.16.

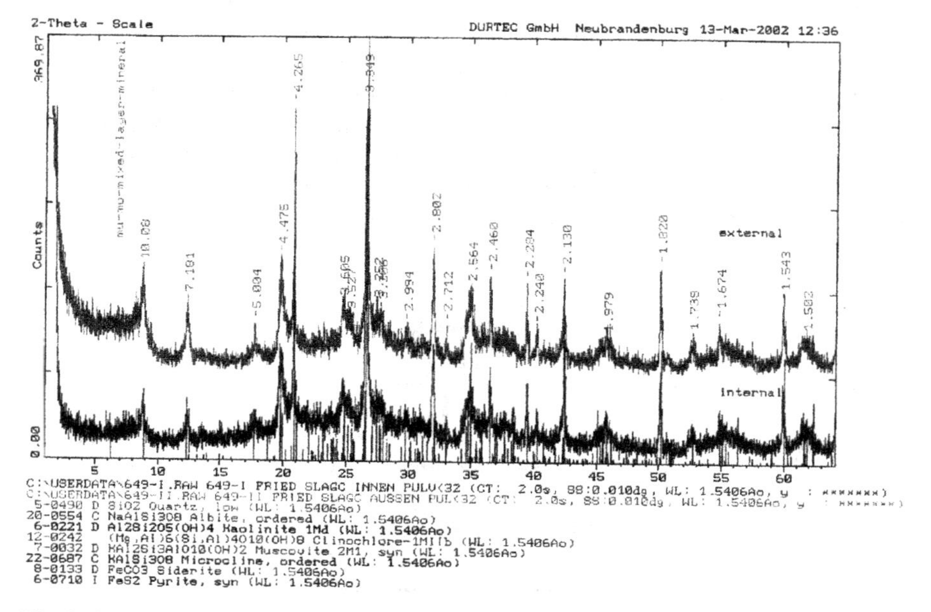

Fig. 3.72 XRD diagram of Friedland clay powders from outer ("external") and inner ("internal") part of clay reacted with porewater of slag cement (Merit 5000)

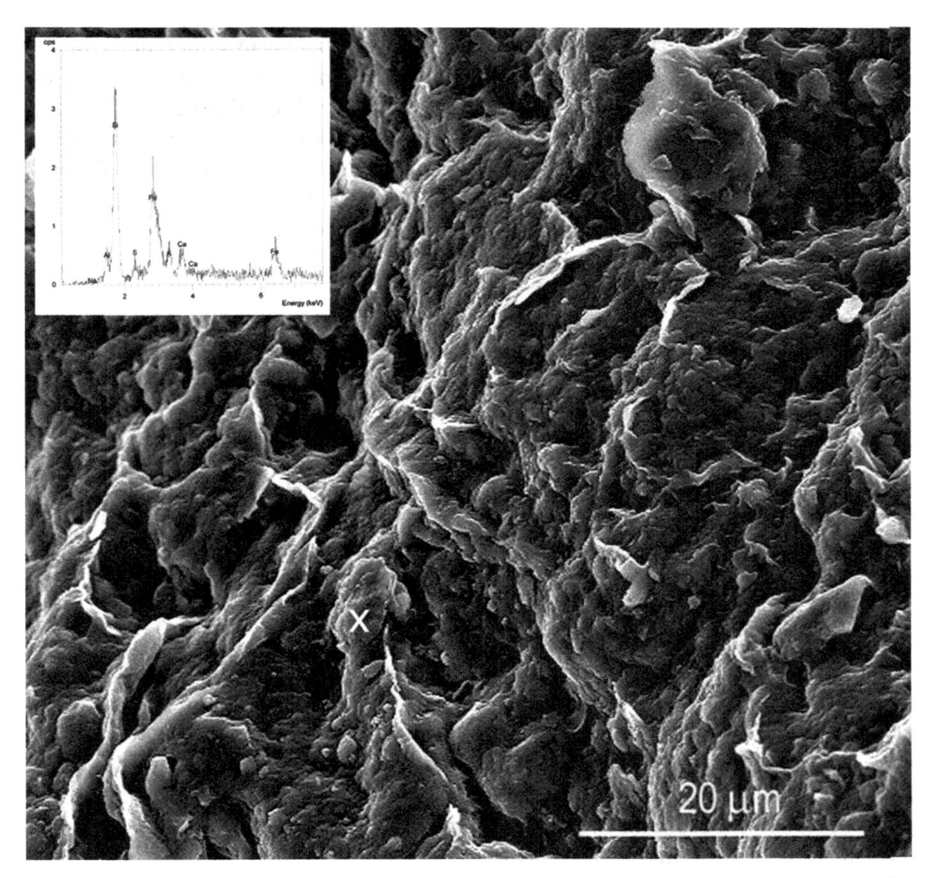

Fig. 3.73 Typical microstructural constitution of Friedland Ton reacted with cement water. At the upper right a rounded amorphous silicious compound is present

One finds that the saturation with cement water did not make the clay more permeable. On the contrary, its hydraulic conductivity was reduced to about ¼ of that of virgin clay, the most probable reason being clogging of voids by amorphous hydrosilicate compounds formed by slight cement-water attack on the clay minerals. Slight dissolution was in fact manifested by release of small amounts of Mg

Table 3.16 Evaluated hydraulic conductivity of the clay samples treated with cement-water and of virgin Friedland Ton [33]

Cement water	Density at saturation, kg/m^3	Hydraulic conductivity of clay treated with cement water, m/s	Hydraulic conductivity of virgin clay, m/s
Merit 5000, pH = 9.4	1850	1.3E-11	7.0E-11
Electroland, pH = 8.1	1900	9.0E-12	4.0E-11

and Al from the clay samples, but the insignificant change in potassium content in the solutions shows that illitization was insignificant. The average amount of dissolved smectite was found to be 12 g of smectite per liter of pore solution, i.e. 1.2% under the closed conditions that prevailed in the tests. Considering the mass ratio of cement and clay in a repository it was concluded that a concrete wall with a thickness of several decimeters will not have any practically important impact on contacting backfill of dense Friedland Ton in any time perspective.

3.5.13 Other Processes of Importance to the Function of Buffer Clay

3.5.13.1 Impact of Gas Under Pressure

Gas can be of importance in two respects: (1) Hydrogen gas from corrosion of steel canisters or from the iron component of canisters of the kind presently favoured by SKB can create channels that serve as paths for quick release of radionuclides from the canisters to the biosphere via fracture systems, and (2) Water vapour formed in the hotter parts of the buffer in conjunction with water saturation can pressurize and displace water from colder parts of the buffer, forming channels.

Considering the matter of hydrogen gas first, a number of tests have been made for finding out how and at what pressure gas can penetrate smectite clay confined between filters. These tests have shown that pressurizing it through a filter causes displacement of a small fraction, i.e. 0.02 to 0.2%, of the porewater before gas passes through the clay (Fig. 3.74). Several of the studies seem to indicate that the gas pressure has to exceed a threshold value for proceeding through the clay that is on the same order of magnitude as the swelling pressure.

However, no tests have been conducted for a sufficiently long time to make sure if a low pressure can yield breakthrough after sufficiently long time and whether step-wise pressure increase with very small pressure increments gives the same results

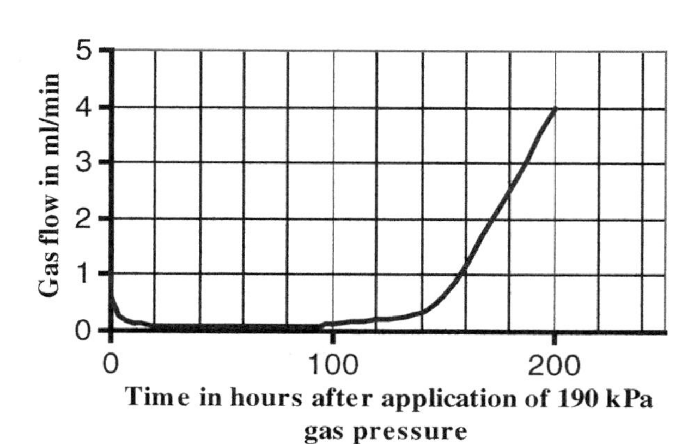

Fig. 3.74 Nitrogen gas movement in bentonite clay with 1680 kg/m^3 density at saturation with seawater [10]

as large increments or more rapid pressurizing. The results from some tests are contradictory in many respects and there is no convincing conceptual model of the involved processes [34]. This is because no credible microstructural model has been proposed as a basis of outlining the evolution of gas penetration. A hint of how this could be made is by considering the microstructural model in Fig. 3.36, which consists of a network of channels of varying size and content of clay gels with a density that is related to the size. The variation in density of the clay gels means that their shear strength varies correspondingly and this determines the route taken by advancing gas. Thus, gas makes its way through the channels offering least resistance, displacing and compressing clay gels and porewater in a finger-like pattern. Using this conceptual model one can explain why stepwise pressurizing with small increments can lead to a higher critical gas pressure than for larger increments: the clay gels find time to consolidate under the applied pressure by which they form dense plugs that require higher pressure to break. For larger pressure steps or quick pressurizing there is not enough time for the gels to stabilize and they hence offer less resistance to the advancing gas. Use of the theoretical 3Dchan model would need to be significantly extended to include also the gel consolidation process so for the time being only conceptual understanding of the mechanisms involved in gas penetration can be obtained.

3.5.13.2 Vapour Effects

In 1985 a laboratory study on the influence of hot water vapor on the expandability of smectite clay was reported by Rex Couture [35]. He showed that while reaction of bentonite with liquid water at temperatures ranging between 100 and 200°C produces minimal loss in swelling capacity of Na bentonite, reaction with water vapor at 150 to 250°C results in rapid irreversible loss of most of the expandability. Couture concluded that the loss of expandability, which was also found to be associated with a dramatic increase in hydraulic conductivity, is related to the influence of pressurized water vapor but gave no explanation of the mechanism.

The conditions in the deposition holes of a KBS3 repository imply that steam will be produced in the buffer clay near the hot canisters. When the joints between the large buffer blocks become tight by absorption of water supplied by the surrounding rock, a closed system of interconnected voids is formed around the canisters (Fig. 3.75). This space is filled with air and water vapor and the relative humidity will be 100% at a late stage of wetting when water has moved in from the rock and the wetting front is close to the hot canister surface. Hence, depending on whether vapor can escape from the buffer clay and what the temperature will be in the vapor-filled space, the buffer will be exposed to steam pressurized to 75 to 150 kPa. These conditions are hence similar to the hydrothermal conditions that prevail in the final phase of orogenetic processes, in which precipitation of clay minerals and zeolites is known to take place. It is therefore possible that the effect that Couture noticed may be due to precipitation of cementing minerals.

The impact of hot water vapour on the montmorillonite-rich MX-80 clay was investigated in the nineties in a study by Geodevelopment AB and NAGRA focusing

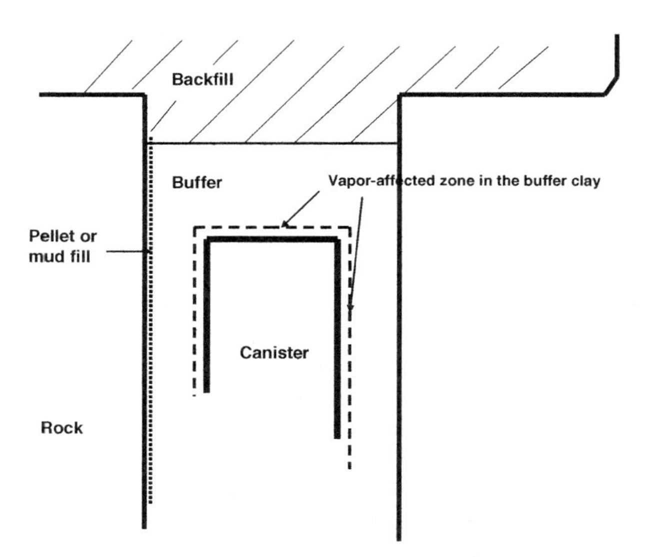

Fig. 3.75 Vapour-filled zone in the buffer clay close to canisters

on whether changes in expandability and hydraulic conductivity will take place at all under experimental conditions that are representative of those in a KBS-3 V deposition hole [36]. A second aim was to identify the physico/chemical processes that lead to such changes if they take place.

Isostatically compacted MX-80 samples with a water content of 7.5% and a dry density of 1900 kg/m had been stored for about 2 years under ordinary room atmosphere conditions and samples were trimmed to cylindrical form for fitting into oedometers serving as vapour pressure cells (Fig. 3.76). The temperature was maintained at 90, 100 and 110°C in separate tests and the steam pressure measured. The vapor treatment lasted for 30 days and small specimens then taken for microstructural analysis while the remainder was saturated with distilled water for determining swelling pressure and hydraulic conductitivity.

At the opening of the test cells it was noticed that the upper surface of all the samples had risen significantly and was crumbly, brittle and fractured. The net dry density after the subsequent trimming to 20 mm thickness and hydration for determining the swelling pressure and hydraulic conductivity was 1190–1350 kg/m^3. The swelling pressure was recorded continuously in the hydration and percolation phases and was found to grow much slower than for virgin MX-80 clay. In fact the swelling pressure had not reached its maximum value even after 3 months. Figure 3.77 shows the evolution of swelling pressure for the sample that matured most quickly.

The figure demonstrates that the maturation rate was low, an interesting fact being that untreated MX-80 clay with this density and saturated with distilled water has a maximum swelling pressure of 700 kPa, which indicates that the clay sample exposed to the most severe vapor treatment (110°C) showed no sign of sign

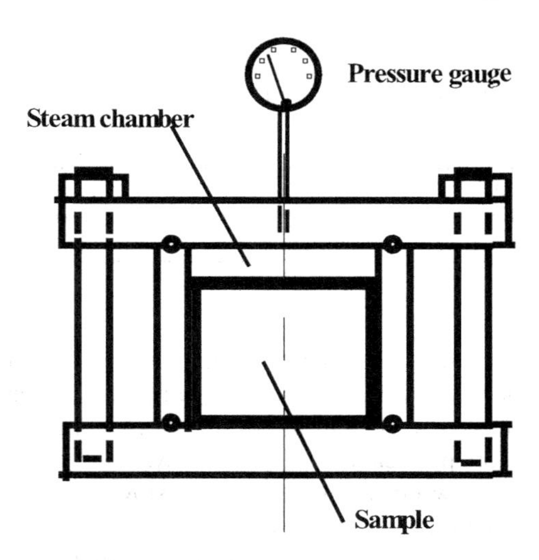

Fig. 3.76 Schematic picture of steam chamber with 30 mm diameter and 40 mm height. The chamber contained a small metal cup with water that evaporated and created vapor at a pressure that was recorded

of degradation. However, the determination of hydraulic conductivity gave another picture. Thus, we see from Table 3.17 that all the clay samples had a hydraulic conductivity that was more than 100 times higher than that of virgin MX-80 clay.

The explanation of the remarkable increase in hydraulic conductivity of the cooled clay is believed to be the same as for the hydrothermal experiments, i.e. cementation of stacks of smectite lamellae by precipitated silica stemming from dissolved accessory and smectite minerals. In the 110°C moist environment the initially relatively dry clay absorbed water and a network of partly expanded stacks of smectite lamellae had been formed as if the clay had taken up liquid water in the oedometer. However, pressurized vapour remained in larger voids, the remnants from the original compaction-generated granular structure that could not be closed as in liquid water. Elements, primarily silicons, were released by dissolution of

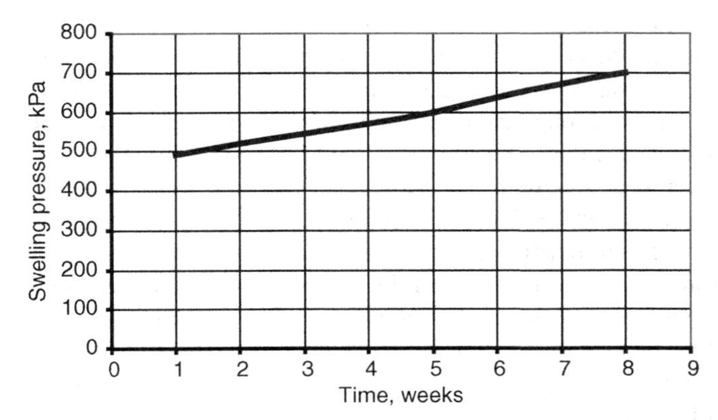

Fig. 3.77 Evolution of the swelling pressure of a clay MX-80 sample with dry density 1190 kg/m^3 exposed to 110°C vapour for 30 days

Table 3.17 Evaluated hydraulic conductivity from tests using distilled water

Vapor Temperature, °C	Density, kg/m^3	Hydraulic conductivity of vapour-treated clay, m/s	Hydraulic conductivity of virgin clay, m/s
90	1800	6.1E-11	4E-13
100	1850	2.2E-11	2E-13
110	1750	1.7E-11	7E-13

accessory and smectite minerals and they precipitated and cemented the particle network with wide voids together. In the subsequent oedometer tests the cementation prevented closure of the wider voids, which remained and made the clay much more permeable than virgin clay.

How does this fit with the slow rise in swelling pressure recorded in the percolation tests? The explanation is believed to be that cementation bonds were broken successively by expanding components of the particle network that were only insignifically cemented. The delay in the evolution of the swelling pressure was hence probably caused by the resistance to expansion, which was still not large enough to close larger, channel-forming voids.

The overall conclusion from the vapour experiments and the hydrothermal tests described earlier in this chapter is that a considerable part of buffer clay will retain most of its expandability but loose a significant part of its tightness. This is expected in a short term perspective if the conditions are favourable for quick and uniform wetting, while stronger degradation of the buffer may take place in several hundred years or more in rock that provides only little water for hydration.

3.5.13.3 Microbial Effects

Microbial species detected in the groundwater in deep vault repositories include aerobic and microaerophilic heterotrophic microorganisms, and anaerobic iron-reducing and sulphate-reducing bacteria. The capability of microorganisms such as facultative anaerobic bacteria, fungi and even anaerobes in reducing iron is well known but their performance in dense buffer is not well known. The presence of water in the buffer material allows both acid-base reactions and also oxidation-reduction reactions. The latter can be abiotic and/or biotic. In respect to biological issues, microorganisms are significant participants in catalyzing redox reactions. The activity of electrons is a significant factor in oxidation-reduction reactions since transfer of electrons between the clay minerals and biotic issues can cause reduction of structural Fe^{3+} in the octahedral and tetrahedral sheets to Fe^{2+}, which can significantly alter the short-range forces between the lamellae and result in: (a) a lower specific surface area, (b) a high degree of layer collapse, (c) decreased water-holding capacity, (d) reduced swelling capability, and (e) stacking of the layer sheets.

The matter of microbial processes has received much attention because of the possibility that microbes active on the repository level can bring radionuclides with them to the ground surfaces or give rise to organic debris that can be brought up by flowing water. A number of comprehensive investigations have been conducted for finding out whether repository rock and clay-based engineering barriers can host microbes and offer conditions for them to multiply [37]. Some ten years ago it was estimated that the energy state of the porewater controls the conditions for biological cells including bacteria and spores to survive, implying that the competition between smectite and microbes respecting access to water is the key issue. Later, it has been argued, referring to microstructural models, that the size of the interparticle voids and the strength of the surrounding particle network determine the fate of microbes [38]. Both microstructual modelling and laboratory tests have shown that microbes and spores can hardly move in highly compacted smectite clay with a density at saturation of more than about 1800 kg/m^3.

3.5.13.4 Influence of Radiation

A number of laboratory experiments have indicated that γ radiation has two major effects, (1) radiolysis of the porewater, by which the radical OH and various other products like negatively charged O_2 are produced, and (2) breakdown of the crystal lattice of clay minerals [4]. The first process yields decomposition of the porewater that can result in gas accumulation at the surface of HLW canisters, while the second mechanism yields mechanical disintegration of the crystallites by which the average particle size is reduced and the crystal lattice fragmented. At present the general understanding is that strong gamma radiation will not cause substantial degradation of smectitic buffer as exemplified by the Saclay tests referred to earlier in this chapter. This study showed that thermally induced mineral transformations like dissolution of feldspars and neoformation of chlorite and quartz occurred but that the large majority of the smectite content remained unchanged.

Porewater from the canister-embedding smectite clay will enter leaking canisters and cause dissolution of radioactive waste and released radionuclides will diffuse through the water into the clay. Positively charged elements will migrate into the clay yielding cation exchange, which means that they get adsorbed and expose the crystal lattice to α radiation [4]. Plutonium and americium are radionuclides that can be of concern and laboratory tests have shown that montmorillonite that has been saturated with either of these elements, yielding around 5×10^{18} alpha doses per gram of clay, is completely destroyed and converted to an amorphous, silicious mass. Using a relevant value of the number of alphas per mass unit of waste and considering a single hole representing a manufacturing defect or resulting from "pitting" corrosion of a KBS3-type canister, one finds that such degradation may take place in a cone-shaped zone all the way through the clay to the rock in about 10 000 years. The amount of buffer that can be altered is very small, however, and the net effect on its physical performance is concluded to be insignificant.

3.5.13.5 Influence of Electrical Potentials

Electrical potentials have been considered because they may generate currents that can have an impact on the chemical stability of HLW canisters and affect water flow in repository rock as well as in engineered barriers. At shallow depth currents flow parallel to the ground surface because the electrical conductivity of air is very low compared to that of soil and rock. If the conductivity is constant along any plane in the earth the natural current flow is none, while variations in conductivity cause natural currents that can be oriented in different directions. Man-made current flow can have both horizontal and vertical components depending on the distance to the sources and the depth to which measurements are made.

In crystalline rock numerous vertical and sub-vertical conductive fracture zones and mineralized zones can perturb current flow dramatically. If the current flows perpendicular to the zones electrical charges are set up at their boundaries, by which the electric field can become strongly enhanced or reduced and strong vertical currents generated. If the current flows parallel to the zones the electric field is more or less unperturbed, but the resulting current in such conductive zones may become magnified by the ratio of the electrical conductivity of the zone and that of the adjacent crystalline bedrock. Pipelines and cables with high electrical conductivity will act as conducting fracture zones, i.e. they can collect currents from the surroundings, resulting in enhanced corrosion. It is obvious from these facts that the structural constitution of the repository rock is a major factor in the generation of natural electrical currents and potentials and that conductive components like cables can serve as short-circuiting conductors that can undergo corrosion and have a significant impact on recorded data. Assuming that electric currents follow the major water conductors significant electrical potentials may be set up at these conductors while the potentials formed across the large space with deposition tunnels may not be high. Water-bearing fractures that intersect deposition holes are expected to serve as electrical conductors both because of the water content and because they are usually coated with chlorite and clay minerals.

A phenomenon that has not yet attracted much interest is the electrical potential that is formed in the buffer in the course of the wetting and that may have an impact on the performance of instruments placed in the buffer. Such a potential exists between the copper canisters and the earth connection in SKB's underground laboratory at about 400 m depth and may have three effects: (1) The flow of water in the buffer in the maturation period is counteracted meaning that the recorded wetting rate is lower than predicted by use of Darcy models, (2) Metal gauges placed in the buffer act as electrodes and may undergo corrosion and the recordings are thereby affected, and (3) Canister corrosion may accelerate.

Electrical currents in nature can be measured by means of magneto telluric (MT) measurements, i.e. simultaneous recording of the electric and the magnetic fields. The electromagnetic field derives from basically two sources: (1) the natural field due to large scale current systems in the ionosphere generated by interaction of the Earth's static magnetic field and charged particles from the sun, and (2) man-made current systems in power cables and grounding points and large consumers of electric power.

A number of conclusive observations have been made from field studies:

- The amplitudes of the horizontal magnetic field components are very similar, whereas the amplitude of the vertical magnetic field is considerably lower except during magnetic storm events.
- The electric fields are very similar in signature, but the amplitude ratios of 6 have been found, which indicates that the total electric field is strongly polarized due to lateral changes in the electrical conductivity close to the surface.
- Daily variations may be a few nTesla.
- DC voltages between borehole measuring positions can represent electric fields of about –250 mV/km. For an average approximate distance between the boreholes of about 250 m these fields correspond to voltage differences of about –60 and 20 mV, respectively.
- The inhomogeneous nature of the electric field contrasts to the homogeneity of the magnetic field.

Figure 3.78 shows the results from measurements in SKB's URL where four full-scale canisters surrounded by dense smectite clay are located in 8 m deep deposition holes on the 400 m level. The measurements show that the cables have a potential that is higher or lower than net-ground. An obvious fact is that the potential is different in holes with no instrumentation in the buffer and in holes where the buffer is instrumented. It is hence obvious that the instrumentation played a role for the accuracy of the data and that electronic gauges measuring temperature and pressure may give incorrect values and undergo corrosion.

A phenomenon related to electro-osmosis occurs in the buffer surrounding the copper-lined canisters in the deposition holes. It has an initial degree of water saturation of 50–70% and complete water saturation requires several years in fracture-rich

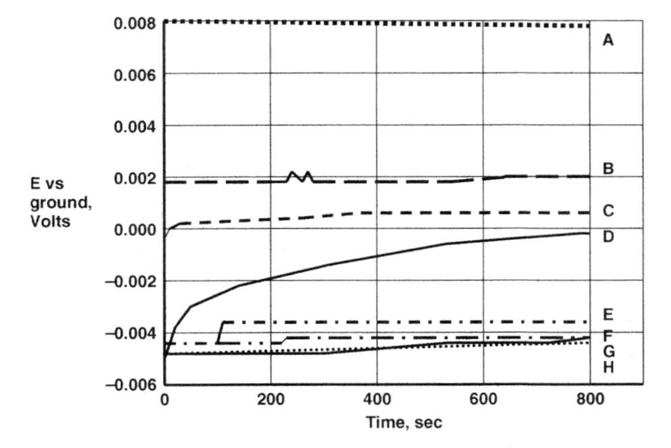

Fig. 3.78 Results from measurements of electrical potentials vs. ground in power cables to electrical heaters in KBS-3 V canisters about 400 m depth in the Äspö URL. For one of the canisters different cables give very different values after different periods of time (D)

rock to more than a hundred years in fracture-poor rock. The process is commonly considered as a purely physical process but is in fact associated with creation of local electrical potentials that have an impact on the rate of water saturation and possibly also on the corrosion rate of the copper lining. Starting with the simple case of flow in porous media one has, by combination of the Darcy and Kozeny/Carman flow laws, the expression for the hydraulic conductivity *K*in Eq. (3.12), [39]:

$$K = n^3/(1-n)^2 bT^2R^2 \tag{3.12}$$

where:

b = pore shape factor (2.5)
T = tortuosity ($\sqrt{2}$)
R = particle surface area per unit volume of solids
n = porosity

By considering the variation in void geometry discussed earlier one finds the general relationship between porosity and average flow rate to be of the type in Fig. 3.79.

Following Olsen one has, from Onsager's general phenomenological relationships for irreversible thermodynamic processes, the following equations:

$$q_w = L_{11}(P/L) + L_{12}(E/L) \tag{3.13}$$

$$q_e = L_{21}(P/L) + L_{22}(E/L) \tag{3.14}$$

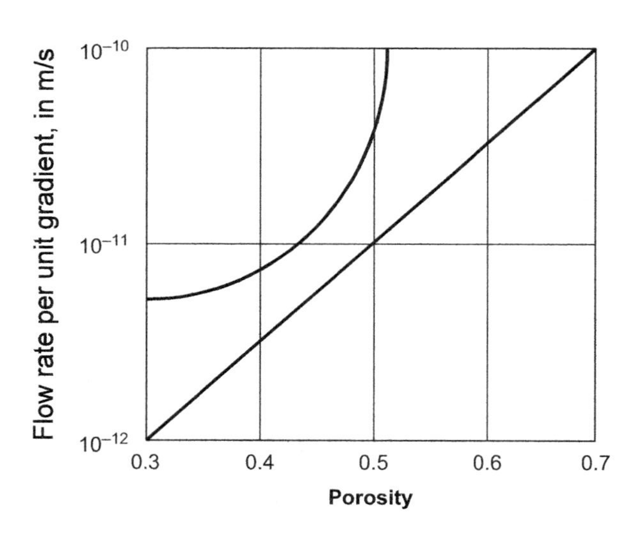

Fig. 3.79 Generalized difference between predicted (*straight line*) and recorded (*curve*) relationship of flow rate and porosity for illite percolated by saline Na-dominated water [39]

where:

L = element length
q_w = water flow rate
q_e = electrical current flow rate
P/L = hydraulic gradient
E/L = electrical potential gradient
$L_{12} = L_{21}$
$L_{11}, L_{12}, L_{21}, L_{22}$ = phenomenological coefficients

Equation (3.13) implicitely expresses the hydraulic conductivity and it is obvious that if the influence of electrokinetic coupling is negligible or absent, the electrical flow potential term vanishes and the expression gets the form of Darcy's law, i.e. Eq. (3.15):

$$q_w = L_{11}(P/L) \tag{3.15}$$

The influence of electrical coupling to hydraulic flow can be derived from these equations, which yields Eq. (3.15):

$$(q_2 - q)/q = L_{12}/L_{11}(E/P) \tag{3.16}$$

The maximum retardation in flow rate due to electrokinetic coupling occurs when the percolate is electrolyte-free. According to Olsen it is less than 3.3% for dense illite clay in Na form and assuming that the effect is proportional to the cation exchange capacity one would assume that the corresponding reduction is about 10% for smectite-rich buffer and would hence be of little concern in practice. However, the relationship may be much more important for the smectites because of the very narrow voids, the very large particle surface area and the electrical double-layers. This brings us to consider the role of the microstructure in electro-osmotic water transport and focus on how migration of cations, dragging water molecules with them, takes place. It is expected that electro-osmotic water flow can be quantified by using microstructural models that are detailed enough to define the size of microvoids ranging from fractions of one micrometer to a few tens of micrometers like the one referred to earlier in this chapter [16] and the earlier GMM model [40,41].

In principle, there are three different mechanisms for cation migration, cf. Fig. 3.80:

1) Surface diffusion – migration of cations in the electrical double-layer.
2) Matrix diffusion – diffusive migration of cations in the interlamellar space. (Only cations because anions are hindered by Donnan exclusion).
3) Pore diffusion – diffusive migration of cations and anions.

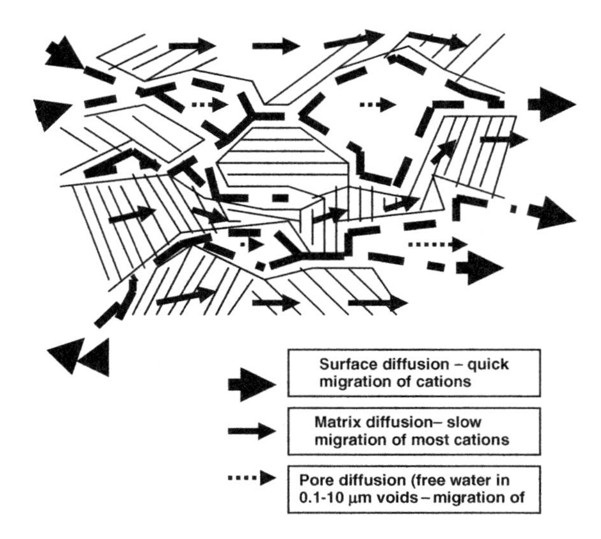

Fig. 3.80 Smectite clay microstructure with dense particle aggregates shown as hached areas. Surface diffusion takes place along the exposed free surface of aggregates of stacks of smectite lamellae. Matrix diffusion occurs within the aggregates, primarily by cation diffusion in the interlamellar space, while pore diffusion takes place in voids of various types

References

1. SKB, 2003. Planning report for the safety assessment SR-Can. SKB, Stockholm.
2. Svemar C, 2005. Cluster Repository Project (CROP). Final Report of European Commission Contract FIR1-CT-2000-20023, Brussels, Belgium.
3. Grauer R, 1986. Ueber moegliche Wechselwirkungen zwischen Verfuellmaterial und Stahlbehaelter in Endlager C. Interner Bericht 86-01, NAGRA, Baden, Schweitz.
4. Pusch R, 1994. Waste disposal in rock. Developments in Geotechnical Engineering, 76. Elsevier Publ. Co. ISBN: 0:444-89449-7.
5. Munier R, Follin S, Rhén I, Gustafson G, Pusch R, 2001. Projekt JADE, Geovetenskapliga studier. SKB R-01-32. SKB, Stockholm.
6. Sandström R, Wu R, 2007. Origin of the extra low creep ductility of copper without phosphorous. SKB TR-07-02.
7. Lönnerberg B, Larker H, Ageskog L, 1983. Encapsulation and handling of spent nuclear fuel for final disposal. SKB/KBS Technical Report 83-20. SKB, Stockholm.
8. Martin Burström, 2007. Personal communication
9. SKB, 1999. SR 97 – Post-closure safety. SKB Technical Report TR-99-06. SKB, Stockholm.
10. Pusch R, Yong RN, 2006. Microstructure of smectite clays and engineering performance. Taylor & Francis, London and New York. ISBN10: 0-415-36863-4.1
11. Kehres A, 1983. Isotherms de deshydratation des argiles. Energies d'hydratation – Diagrammes de pores surfaces internes et externes. Dr. Thesis, Université Paul Sabatier, Tolouse, France.
12. Tardy Y, Lesniak P, Duplay J, Proust R, 1980. Energies d'Hydratation des Argiles. Application a l'Hectorite. Bull. Mineral., Vol. 103 (pp. 217–223).
13. Gueven N, Huang W-L, 1990. Effects of Mg^{2+} and Fe^{3+} substitutions on the crystallization of discrete illite and illite/smectite mixed layers. Int. rep. Dept. Geosciences Texas Tech University, Exxon Production research Co, Houston, Texas.
14. Svemar C, 2005. Prototype repository project. Final Report of European Commission Contract FIKW-2000-00055, Brussels, Belgium.
15. Pusch R, 2001. Selection of THMCB models. Äspö Hard Rock Laboratory, Prototype Repository. Int. Progr. Report IPR-01-66. SKB, Stockholm.

16. Pusch R, Moreno L, Neretnieks I, 2001. Microstructural modelling of transport in smectite clay buffer. In Proc. Int. Symp. on Suction, Swelling, Permeability and Structure of Clays. K Adachi and M Fukue (Eds). Rotterdam/Brookfield: A A Balkema.
17. Pusch R, Adey R, 1986. Settlement of clay-enveloped radioactive canisters. Appl. Clay Science, Vol. 1 (pp. 253–365).
18. Pusch R, Feltham P, 1980. A stochastic model of creep of soils. Géotechnique, Vol. 30, No. 4 (pp. 497–506).
19. Feltham P, 1968. A stochastic model of creep. Phys. Stat. Solidi, Vol. 30 (pp. 135–146).
20. Pusch R, 1984. Creep in rock as a stochastic process. Engineering Geology, Vol. 20 (pp. 301–310).
21. Pytte AM, Reynolds RC, 1989. The thermal transformation of smectite to illite. In Thermal History of Sedimentary Basins. N D Naeser and T H McCulloh (Eds). New York: Springer-Verlag (pp. 133–140).
22. Hoffman J, Hower J, 1979. Clay mineral assemblages as low grade metamorphic geothermometers. Application to the thrust-faultet disturbed belt of Montana, USA. SEPM Special Publ. No. 26.
23. Pusch R, Madsen F, 1995. Aspects on the illitization of the Kinnekulle bentonites. Clays and Clay Minerals, Vol. 43, No. 3 (pp. 261–270).
24. Pusch R, Karnland O, Lajudie A, Decarreau A, 1993. MX-80 exposed to high temperatures and gamma radiation. SKB Technical Report TR-93-03. SKB, Stockholm.
25. Pusch R, Karnland O, Hökmark H, Sandén T, Börgesson L, 1991. Final report of the Rock Sealing Project – Sealing properties and longevity of smectitic clay grouts. Stripa Project Technical Report 91-30. SKB, Stockholm.
26. Pacovsky J, Svoboda J, Zapletal L, 2005. Saturation development in the bentonite barrier of the Mock-up CZ geotechnical experiment. Clay in Natural and Engineered Barriers for Radioactive Waste Confinement – Part 2. Physics and Chemistry of the Earth, Vol. 32/8-14. Elsevier Publ. Co. (pp. 767–779).
27. Pusch R, Kasbohm J, Pacovsky J, Cechova Z, 2005. Are all smectite clays suitable as "buffers"? Clay in Natural and Engineered Barriers for Radioactive Waste Confinement – Part 1. Physics and Chemistry of the Earth, Vol. 32/1-7. Elsevier Publ. Co. (pp. 116–122).
28. Pusch R, Kasbohm J, Thao HM, 2007. Evolution of clay buffer under repository-like conditions. Proc. Int. Workshop on THMCB processes, Lund 2007. Applied Clay Science (In print).
29. Grindrod P, Takase H, 1993. Reactive chemical transport within engineered barriers. In: Proc. 4th Int. Conf. on the Chemistry and Migration Behaviour of Actinides and Fission Products in the Geosphere, Charleston, SC USA, 12-17 Dec. Oldenburg Verlag 1994 (pp. 773–779).
30. Linares J et al., 1989. Investigacion de bentonitas como materiales de sellado. U.E.I. Fisicoquimica y Geoquimica Mineral Estacion Experimental del Zaridin (CSIC), Granada, Spain.
31. Huertas F et al., 2000. Effects of cement on clay barrier performance, ECOCLAY project. Final report Contract No F14 W-CT96-0032, European Commission, Brussels.
32. Pusch R, 1982. Chemical interaction of clay buffer materials and concrete. Technical Report SFR 82-01. SKB, Stockholm.
33. Pusch R, Zwahr H, Gerber R, Schomburg J, 2003. Interaction of cement and smectite clay – theory and practce. Appl. Clay Science, Vol. 23 (pp. 203–210).
34. Horseman, ST, Harrington JF, 1997. Study of gas migration in MX-80 buffer bentonite. Nat. Envir. Research Council, British Geol. Survey. Report WE/97/7.
35. Couture RA, 1985. Steam rapidly reduces the swelling capacity of bentonite. Nature, Vol. 318 (p. 50).
36. Pusch R, Bluemling P, Johnson L, 2003. Performance of strongly compressed MX-80 pellets under repository-like conditions. Appl. Clay Science, Vol. 23 (pp. 239–244).
37. Pedersen K, 1995. Survival of bacteria in nuclear waste buffer materials. The influence of nutrients, temperature and water activities. SKB Technical Report TR 95-27. SKB, Stockholm.

38. Pusch R, 1999. Mobility and survival of sulphate-reducing bacteria in compacted and fully water saturated bentonite – microstructural aspects. SKB Technical Report TR-99-30. SKB, Stockholm.
39. Olsen HW, 1961. Hydraulic flow through saturated clays. Dr Thesis, Civ. Eng., MIT, USA.
40. Pusch R, Karnland O, Hökmark H, 1990. GMM – A general microstructural model for qualitative and quantitative studies of smectite clays. SKB Technical Report TR 90-43. SKB, Stockholm.
41. Pusch R, Muurinen A, Lehikoinen J, Bors J, Eriksen T, 1999. Microstructural and chemical parameters of bentonite as determinants of waste isolation efficiency. Final Report EC Contract No F14 W-CT95-0012, European Commission, Brussels.

Chapter 4
Repository Concepts for HLW Including Spent Fuel and Waste with Long-Lived Radionuclides

4.1 Major Principles of Storing HLW

The use of engineered barriers, i.e. canisters, buffers, and backfills, is common to all repository concepts irrespective of the geological conditions. The concepts and design principles vary very much, however, because of the different hydraulic and mechanical performances of the geological media and the strongly varying chemical conditions in them. While the function of crystalline rock is primarily to provide mechanically stable confinement of the radioactive waste, argillaceous rock offers tightness. The former is conductive and the latter mechanically weak. For salt, the conditions are different in several respects: deposition tunnels and rooms are not stable at all because of the extreme creep potiential, while the tightness is even better than of argillaceous rock, and one may say that it can be too good. Thus, chemical reactions between salt and waste containers yield hydrogen gas that may not be readily released but stay at the interface and form gas bubbles with such high pressure that large-scale rupture and transport of contaminated brine to the biosphere can take place.

In this chapter we will briefly describe proposed concepts, focusing on advantages and disadvantages. We will start by considering crystalline rock for which there are detailed design and construction plans as well as complete performance analysis of one concept, the KBS-3 V repository model, which we will use a basic reference case in this chapter.

4.2 Repository Concepts

4.2.1 The Complete Repository

A repository can have the following underground space:

- Adit in the form of a helical ramp from the ground surface to the repository level, some 400–500 m below.

R. Pusch, *Geological Storage of Highly Radioactive Waste*,
DOI: 10.1007/978-3-540-77333-7_5, © Springer-Verlag Berlin Heidelberg 2008

- Transport shafts from the ground surface to the repository level (skip, personnel, ventilation).
- Ventilation shafts from the ground surface to the repository level.
- Central area at the repository level from where the ramp and main shafts for transporting material and personnel extend.
- Transport tunnels at the repository level connected to deposition tunnels and to the central area.
- Deposition tunnels or holes.

These features are illustrated in Fig. 4.1, which represents the Swedish repository concept KBS-3 V.

We have made acquaintance with the KBS-3 V concept in Chap. 3 and seen that it represents a rather simple and robust method of confining HLW but also identified issues that need to be taken into consideration in the detailed design and planning of the construction work. They are common to all sorts of repositories in crystalline rock and have to do with the presence of major discontinuities in the form of fracture zones in the rock. Thus, a repository should be suitably adaptated to the rock structure, implying that canister deposition tunnels and holes must not be located in or very near major discontinuities that are water-bearing and can undergo tectonically induced shear strain, a matter that will be discussed further in Chap. 6. The need for such adaptation is obvious when trying to incorporate repositories with deposition tunnels of several hundred meter length in the typical fractal-like structure model

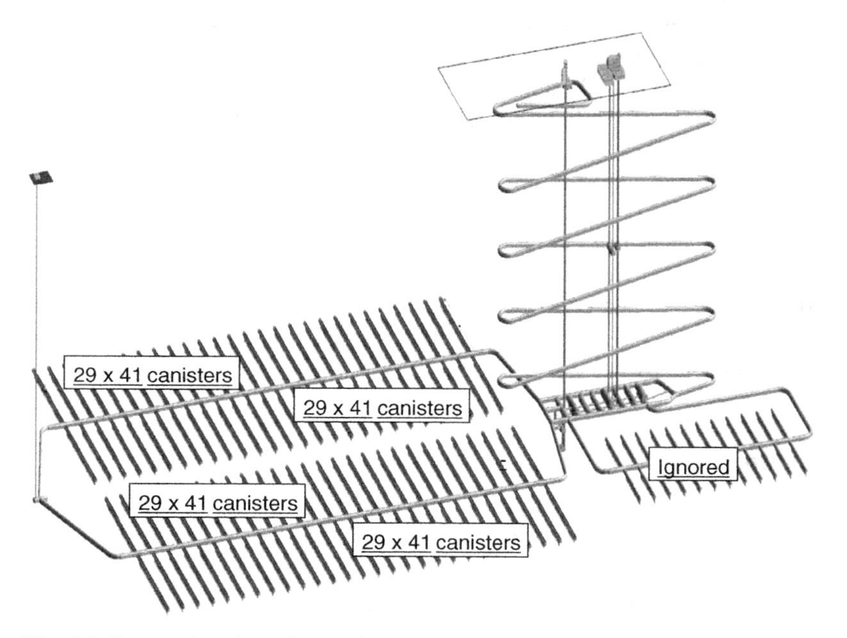

Fig. 4.1 Perspective view of a one-level repository representing the KBS-3 V concept (the part termed "ignored" is a possible pilot or experimental repository), [1]

of crystalline rock. Thus, assuming patterns with common spacing and persistence of 3rd order discontinuities – i.e. minor water-bearing fracture zones – one finds that a number of planned positions of deposition holes have to be abandoned since they would be located in such zones or too close to them. Since a high degree of utilization of the rock mass is desired for economical and some other reasons, great care must be taken in designing a HLW repository with respect to orientation and spacing of deposition tunnels and holes.

Applying the principle that the deposition tunnels must not be intersected by 2 nd order discontinuities, i.e. major fracture zones, they will still be intersected by minor fractures zones, i.e. those of 3rd order. Optimal orientation and spacing of KBS-3 V deposition tunnels can be of the type shown in Fig. 4.2, which again demonstrates that a significant number of planned locations of deposition holes must be abandoned because of the discontinuities. The presence of those of 3rd order causes much trouble not only by reducing the utilization for locating deposition holes but also because of the water inflow in the tunnels in the backfilling phase and because they interact hydrulically with the EDZ and thereby serve as effective flow

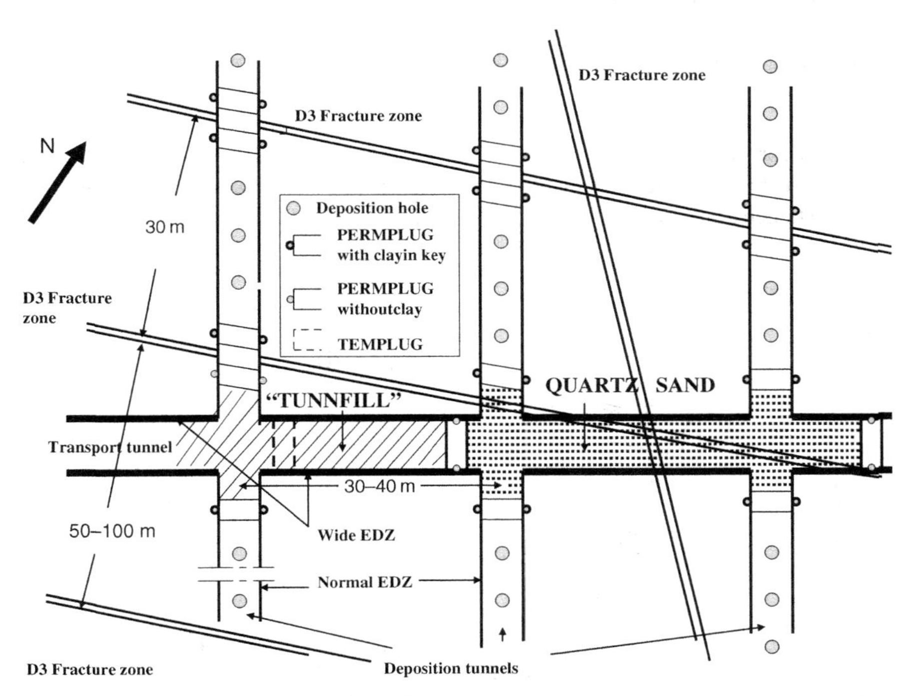

Fig. 4.2 Schematic plan view of a KBS-3 V repository with a transport tunnel connected to deposition tunnels in rock with typical frequency of 3rd order fracture zones that have to be accepted in the repository but not allowed to intersect deposition holes. Tight plugs are keyed into the rock for isolating the waste-containing parts of the deposition tunnels from fracture zones and from transport tunnels. Where transport tunnels are intersected by water-bearing fracture zones they are suitably backfilled with cement-stabilized quartz sand

paths of nuclide-contaminated water. These two issues have to do with backfilling and plugging that we will look into in some detail.

4.2.2 Alternative Concepts

Why put just one canister in a deposition hole? Why vertical holes? Why not very deep holes? These questions have been raised and encouraged designers to consider other concepts than KBS-3 V and similar types. Figure 4.3 shows four principles that have been proposed by the Organizations, except for the new Case C, which may have the highest potential for future use. The four principles can be described as follows:

- Steep holes with two or more canisters. An extreme version is the so called "Very Deep Hole" (VDH).
- Wide rooms with deposition holes with single canisters. Inherited from disposal of chemical waste.
- Inclined deposition holes with single canisters. New version of the KBS-3 V concept (KBS-3i).
- Very long holes. Swiss and Spanish concepts further developed by SKB (KBS-3H).

Except for the VDH concept they all imply construction of tunnels at depth and the same principles in adaptating the respective repositories to the macroscopic rock structure hence apply. We will consider them in detail when dealing with buffers and backfills.

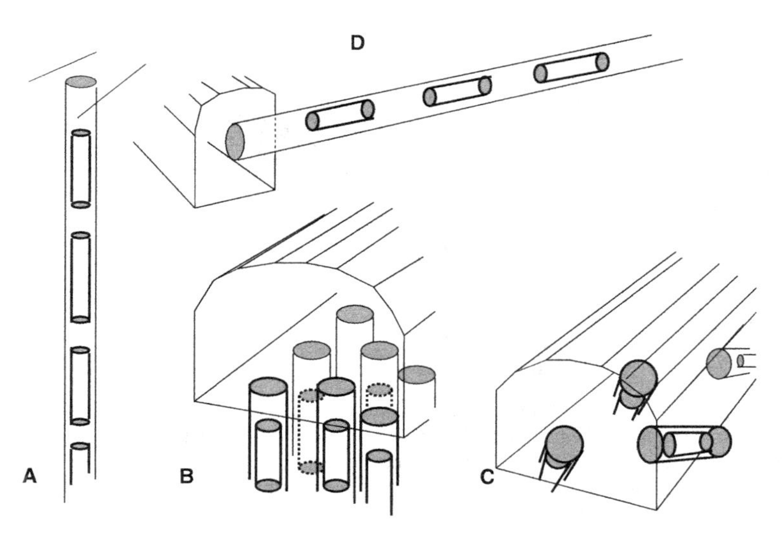

Fig. 4.3 Four alternatives to the KBS-3 V concept. Small cylinders in bigger ones represent HLW canisters surrounded by buffer clay and placed in the deposition holes

4.3 Canisters, Buffers and Backfills – The Engineered Barriers

4.3.1 General Principles

Our reference case is the KBS-3 concept with its major engineered barriers, the canister, surrounded by very dense smectite clay in deposition holes and the tunnels being backfilled with smectitic clay of moderate densite. For the VDH and VLH concepts canisters and buffer make up the EBS in the deposition holes. In all concepts transport tunnels, central areas, shafts and ramps will be backfilled, at least partly, with smectitic backfill. In this chapter we will consider proposed materials and techniques for placement of these components.

4.3.2 Canisters

SKB's development of canisters has been in focus in the evolution of the KBS-3 V and KBS-3H concepts with the aim of finding long-lasting versions that can sustain significant shear strain. Most of the other Organizations propose use of canisters of iron or steel that are not expected to be intact for more than 10 000 years. We have seen in Chap. 3 that there will be chemical interaction of canisters and buffer clay and noticed what impact it will have on the buffer and will not deal specifically with this matter in the present subchapter. However, placement of canisters will be considered here since it is associated with practical difficulties that have led to the development of a special component, the supercontainer, that we will have a good look at.

4.3.3 Buffers

The buffer should have the following functions:

- To be less permeable than the surrounding rock and to be so tight that ion transport takes place by diffusion and not by water flow.
- To keep the canisters in place and protect them by homogenising the stress field.
- To mechanically stabilise the holes or rooms to limit convergence and long-term settlement of overlying rock.
- To act as a transport barrier for radionuclides and a barrier for colloids.
- To provide a suitable geochemical environment.
- To ensure low corrosion rates of both canisters and waste forms.

We saw in Chap. 3 that several types of smectite-rich clay can be used for preparing buffer clay and that the density determines its physical properties. Some of the Organizations have taken for granted that montmorillonite-rich clay ("bentonite") should be used as buffer and have defined a density span for reaching a minimum swelling pressure on the rock in the deposition holes for guaranteeing tight contact between buffer and rock, and a maximum swelling pressure for avoiding overstressing of the rock and canisters. It is currently proposed that the density of the fully water saturated and matured smectite-rich buffer should be higher than about 1850 kg/m^3 but lower than about 2150 kg/m^3.

In order to provide the necessary functions over time it is required that the buffer does not undergo such mineralogical changes, desiccation, or cementation, that the requirements are no longer fulfilled. Some national programmes envisage that the temperature of the bentonite should always stay below 100°C (e.g., Sweden, Belgium, Spain and France) while other national programmes allow higher temperatures. Thus, the Swiss concept specifies a maximum temperature below 100 to 110°C for a significant part of the bentonite but up to 150°C near the canister. Smectite-rich, dense clay with 40–60% water saturation has a low thermal conductivity and it drops further under the thermal gradient that prevails in the deposition hole. Some concepts imply plans to add some material with higher thermal conductivity (e.g. graphite and silica sand) or to wet the buffer to increase the thermal conductivity of the buffer material.

4.3.4 Backfills

Backfill is the term for material that must be placed in tunnels, rooms and shafts that do not contain nuclear waste. It is needed to provide support to the rock since collapse of such rooms can dramatically change the groundwater flow through the repository. For some of these rooms, and for the tunnels in a KBS-3 V repository the principle should be that the backfill does not have to be tighter than the surrounding rock. Sealing off the EDZ at strategic sites, i.e. where water-bearing fracture zones are intersected, will create local stagnant hydraulic regimes between the plugs (Fig. 4.2).

4.3.5 Plugs

In the 1980s and 1990s considerable effort was put in construction and testing of plugs since it was realized that they will be required for temporary or permanent isolation of deposition tunnels from transport tunnels, central areas and shafts etc. One carefully performed test comprised construction and testing of a plug at about 360 m depth in granitic rock in SKB's underground laboratory at Stripa. It represented a case with two concrete bulkheads connected by tension bolts and a steel tube for allowing passage past the plug (Fig. 4.4). The purpose was to evaluate the performance of the approximately 2.2 m long concrete bulkheads equipped with inset gaskets (O-rings) of highly compacted bentonite (HCB) of approximately 0.5 m length and depth and with a dry density of 1600 kg/m^3. A post-tensioning system was used for taking the force from the pressurized sandfill between the bulkheads, the sand being used for simulating a fracture zone with up to 3 MPa water pressure. The

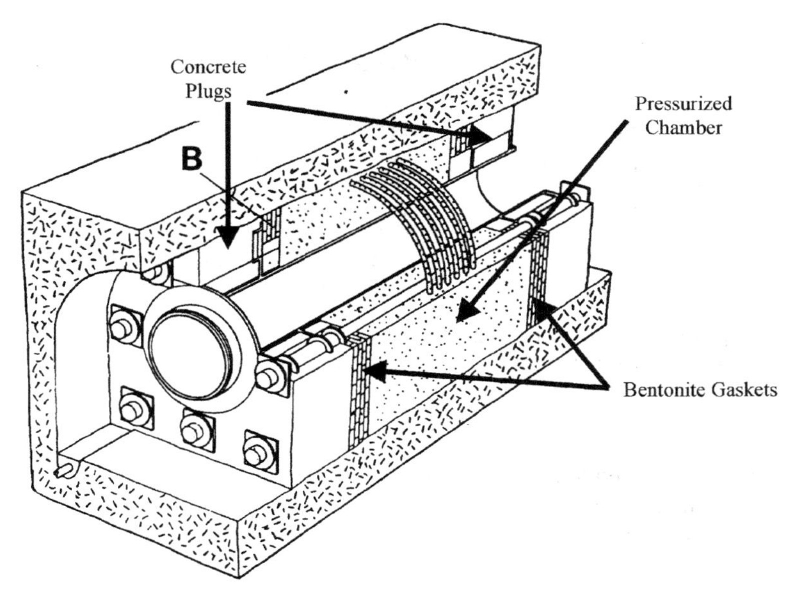

Fig. 4.4 Layout of the tunnel plugging test [2]

project had a bonus in giving evidence of the existence of EDZ in blasted tunnels and providing information of how it can be sealed.

The monitoring of flow past the tunnel plug was observed to decrease with time and this was attributed to the gradual closure of some of the interface flow paths as the result of the swelling of the bentonite and invasion of shallow fractures by dispersed clay. It was estimated that with no bentonite gasket present, the seepage rate past the concrete bulkhead would have been in the order of 1000 l/hr at a hydraulic head of 3 MPa. With the bentonite gasket in place the seepage past the bulkhead at 3 MPa hydraulic head was reduced from 1000 to approximately 75 l/hr. Strong leakage at the rock-concrete interface on the downstream end was observed in the first few days at relatively low hydraulic pressure (100 kPa) but this changed and after a month no seepage could be visually detected. A major conclusion from the experiment was that while plugs with clay gaskets can be made practically tight, water flows along them within the rock, thereby verifying that the EDZ is a major hydraulic conductor. Post-grouting by injection of clay grout through seven strategically placed boreholes from the outside of one of the bulkheads gave effective sealing of a few discrete fractures of 4th order. The bulkhead served to support the rock for preventing "grout (hydraulic) fracturing".

In the described experiment five 50 mm tensional bars, connecting the two plugs, were used to resist the water pressure that corresponded to an axial force of 7000 t (70 MN). Alternatively, pressures of this magnitude can be resisted by keying concrete plugs into the surrounding rock, which would also cut off the EDZ as indicated in Fig. 4.5. Such a plug was also constructed in SKB's URL at Äspö and effective

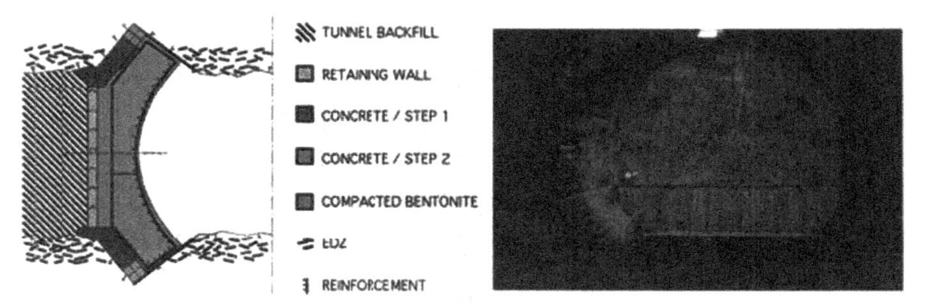

Fig. 4.5 Permanent plug ("Permplug" in Fig. 4.2) of concrete cast on site in two steps for minimizing thermally induced strain. The plug shown was constructed at about 400 m depth in crystalline rock and leaked despite the careful construction work [3]

tightness was obtained by equipping the concrete in the recess with "O-rings" of highly compacted smectite clay seals as in the figure. For cutting off the blast-generated EDZ the seal must reach about 1 m into the rock. This particular plug leaked through fractures in the concrete formed by shrinkage and thermal impact in the hardening phase.

Simpler design and construction can be used for temporary plugs as indicated in Fig. 4.6. They can be made by shotcreting. The slot, which is preferably made by

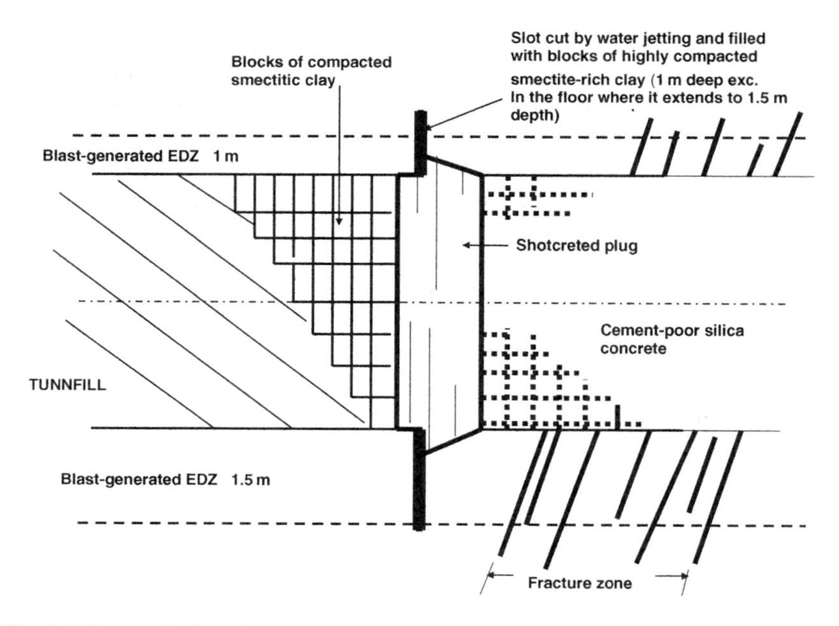

Fig. 4.6 Proposed plug of permanent or temporary type with water-jetted slots in the rock filled with effectively sealing clay blocks for cutting off the EDZ

water jetting, can be filled with blocks of highly compacted blocks of smectite clay for ensuring tightness.

The slot and recess can be prepared well in advance of the arrival of the tunnel backfill front for minimizing the construction time of the cement-rich concrete plug. On the side where the water-bearing fracture zone intersects the tunnel, cement-poor silica concrete is placed by shotcreting or casting using a movable form.

4.3.6 Grouts

Construction of repositories in crystalline and argillaceous rock can be very difficult where tunnels, rooms, and shafts intersect water-bearing fracture zones and grouting will have to be made for reducing inflow of water [4]. Development and use of grouts has been in focus of the Organizations for decades and various types have been tested. The goal is to develop and use techniques for short-term use in the construction phase and for long-term tightening of rock around deposition holes. For the latter there is not yet any reliable solution, while techniques for temporary tightening and stabilizing rock in the construction phase are at hand. However, several failed sealing projects in ordinary tunnel construction projects demonstrate that careful planning and practical experience are required.

The importance of sealing and stabilizing rock by grouting justifies the following, more detailed examination of techniques and results.

4.3.6.1 Grouting for Tightening Rock in Deposition Holes and Around Plugs

Comprehensive international research and development have been performed for finding suitable grout materials and injection techniques [5,6,7]. Systematic search was made that comprised cement and clay materials and also chemical grouts and bitumen. The chemical grouts all have a very low viscosity and were therefore of primary interest. They were silicate compounds (dilute sodium silicate, with and without ethylacetate or formaldehyde), acrylamide, lignosulfonates ($Na_2Cr_2O_7$, $2H_2O$), phenoplasts (phenol + aldenhyde and NaOH as catalyst), amonoplast (urea + formaldehyde + acid catalyst), polyurethanes (polyisocyanate + polyethers + glycols), acrylate, epoxy (resin base + catalyst). The conclusion concerning them was that silicates are subject to syneresis and have poor mechanical stability. All the other chemical grouts were found to be neurotoxic or hazardous for other reasons as described in the comprehensive international literature and were therefore abandoned and blacklisted in the Stripa Project [5]. Surprisingly, some of them appeared again some years later in attempts to stabilize tunnels like the big railway tunnel through the horst named Hallandsås in Sweden but were prohibited for medical reasons.

The finally selected grout types in SKB's R&D work and used in the big field experiments at Stripa were cement and MX-80 smectite clay. It was well recognized that the water content of Portland cement grout must be fairly high for making it sufficiently fluid, which gives poor sealing and risk of bleeding [4,6,7]. The penetration into fractures is poor even if very fine-grained cement material is used and the project therefore included a study of whether the use of an earlier developed "dynamic" injection technique can improve penetration into narrow fractures. This technique, which involves application of a constant grout pressure with superimposed oscillation, turned out to be effective for cement grouts since the effective (grain) pressure is lowered, while for thixotropic clay grouts it worked less well [4]. The effectiveness of the technique for cementitious grouts deserves further development of the equipment while the theoretical hydrodynamic grout flow model is valid in its present form [4,8,9]. The shear resistance as a function of shear rate and viscosity-related parameters in "dynamic" injection has the following form:

$$\tau = m[(d\gamma/dt)/d\gamma_o/dt)]^n \quad (4.1)$$

where

τ = shear stress
$d\gamma/dt$ = shear rate
$d\gamma_o/dt$ = normalized shear rate
m = "viscosity"
n = exponent.

This grout flow model gave predicted penetration depths that were in good agreement with measured penetration in tests with artificial fractures in the form of thin gaps between very stiff steal beams (Fig. 4.7). The frequency varied in the interval 20–80 Hz and the static "backpressure" between 100 and 1500 kPa. This diagram shows the interdependence of the viscosity parameter m, the fracture aperture and the penetration depth for a backpressure of 1.5 MPa and the frequency 40 Hz, which represent optimum conditions. Under these conditions the penetration depth in fractures with smaller geometrical aptertures than 100 μm is about 2 m for the most fluid grouts (m about 0.1) while it will not exceed 4 dm in a fracture with 20 μm aperture. Without the superimposed oscillation the grout would reach significantly less than 50% of these penetration depths.

Despite the improvement provided by the dynamic injection the efficiency of grouting is still problematic because of insufficient tightening and difficulty in predicting how grout moves into real fractures:

1. Fractures are undulating and have varying aperture that puts a limit to the penetrability.
2. Fractures have coating minerals that are easily disintegrated, like chlorite, and move ahead of the grout and clog the fractures where the aperture is small.
3. The grout pressure must exceed the piezometric pressure by about 1 MPa, which can widen the fracture and open other fractures that are not sealed by the grout.
4. Cement grouts have a particle size that can hardly be reduced to less than 10 μm even by very effective grinding.

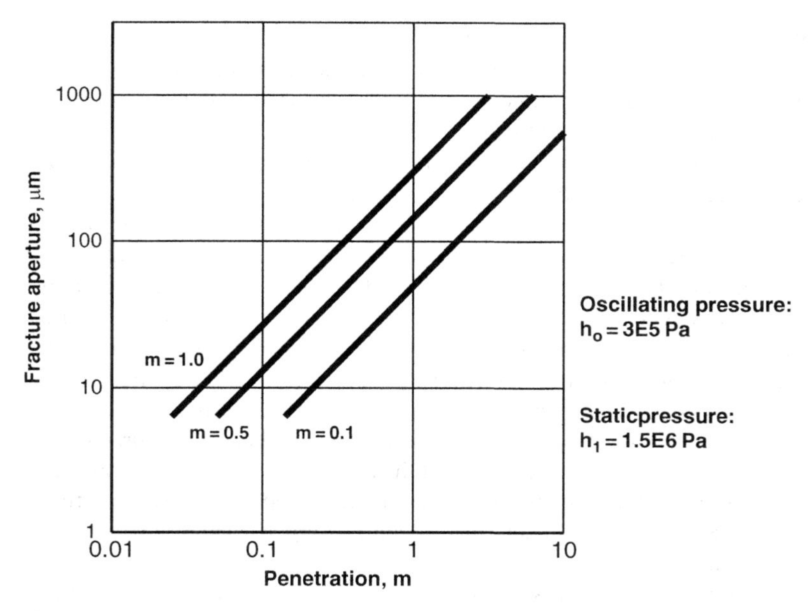

Fig. 4.7 Theoretical relationship between penetration depth and viscosity of grouts and the aperture of plane-parallel fractures for high backpressure (1.5 MPa) and average oscillatory frequency (40 Hz), [9]

5. Clay grouts must be low-viscous to be injectable, meaning that the density can not be higher than 1100 to 1200 kg/m^3 for sealing fractures with a geometrical aperture of 100–200 μm.

The most important conclusions from all these experiments were:

- Post-grouting, i.e. grouting from the interior of excavated rooms, is much less effective than pre-grouting of rock to be excavated.
- Very permeable rock can be tightened to a lower average hydraulic conductivity than fairly tight rock. Thus, while rock with an initial average conductivity of K = E-7 m/s can be tightened to K = E-10 m/s, rock with K initially equal to E-9 m/s can be only slightly tighter by the same grouting effort.
- Clay grouting is effective but the sensitivity to piping and erosion of clay injected in fractures requires that the water pressure is lowered by drainage of the rock mass and that that outer end of the grout holes are mechanically sealed with low-pH cement. Clay grouts consisting of mixtures of smectite clay and very fine silica particles are more piping-resistant and seem to be preferable.

There has been particular interest in clay grouting because of the excellent chemical compatibility of rock and smectite grouts and continuation of the research work in the Stripa Project has been made in AECL's URL in Canada. The work was focused on sealing EDZ in a tunnel in highly stressed granite and the contact between plugs (bulkheads) of concrete and highly compacted blocks of smectitic clay in the tunnel by injecting clay grouts [10]. The tunnel was excavated by controlled drilling and blasting techniques and the excavation technique, causing redistribution of *in-situ* stress around the tunnel, led to the development of an EDZ of variable extension like in the Stripa Project. The plugs were keyed into the rock wall of the tunnel for taking axial forces by 4 MPa water pressure and for cutting off the EDZ, the recesses being excavated with a mechanical technique using line sawing and perimeter reaming. The clay grouting was made as an additional measure to interrupt the connectivity of EDZ at the bulkhead. The conclusion from these activities was that the efficiency of clay grouting to reduce the conductivity of fractured rock was verified. It was also confirmed that grouting into the EDZ by using grout holes oriented normal to the tunnel wall is difficult because many of the fractures in the EDZ are connected with the excavation surface and cannot be filled efficiently by pressurizing the grout slurry. A series of successively thicker bentonite slurries from 0.2 to 8.0% was used and it was concluded that the thin mud can fill the narrow areas of fractures and infiltrate through microcracks, potentially reducing the conductivity of the wide area. Successively thicker slurries can fill larger fracture apertures but infiltrate just around the injection boreholes.

We have seen in Chap. 3 that water saturation of the buffer clay can require a few years or a decade if it has unlimited access to groundwater, while it can take a hundred years or more to reach complete saturation in tighter rock. The negative consequences of such slow wetting can be a significant loss in expandability and tightness as discussed in Chap. 3. Rapid wetting, on the other hand, means that the inflow of water into the deposition holes in the buffer placement phase can be significant and generate problems by too quick hydration of the pellet fill. It can hence be heterogeneous and cause non-uniform wetting of the highly compacted buffer blocks, which in turn can lead to lateral displacement of the whole stack of buffer blocks.

Strong water inflow has been considered to be more critical than too little inflow and attempts have therefore been made to develop techniques for sealing the rock by injecting smectite clay grout. Realizing that the inflow in a 2 m diameter deposition hole can be one liter per minute it is obvious that the hole needs to be drained in the entire buffer placement phase, which lasts for at least one day. This is estimated to be difficult and impractical and grouting would hence be attractive. However, it seems questionable to do so since the resulting delay in wetting rate of the buffer blocks may cause permanent degradation of the buffer. The KBS-3 V concept and in fact all concepts presuming deposition in tight rock hence represent a dilemma.

Figure 4.8 shows an equipment, "Megapacker", developed in the Stripa project for grouting of large-diameter holes and tunnels. It is placed in the deposition hole to be grouted and the grout is injected into the narrow gap between the cylindrical packer and the rock from which it moves into water-bearing fractures. The injection is made by use of a pump that provides pressures of several MPa. Various tests have shown that a static pressure superimposed by oscillatory pressurizing can increase the fluidity of the grout very significantly. It has been successfully used in 0.76 m diameter cored holes for injecting clay and cement grouts and works

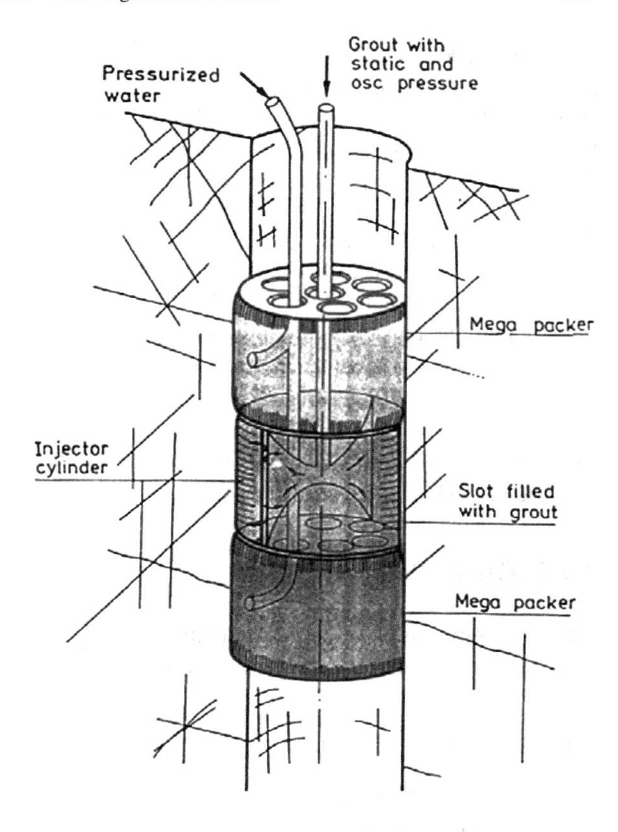

Fig. 4.8 The Megapacker developed for "dynamic" grouting [4]

where water-bearing fractures intersect vertical holes but is not expected to be of use where such holes are intersected by steep fractures in vertical deposition holes. The corresponding problem is even more pronounced in long horizontal tunnels, which most certainly will be intersected by numerous subhorizontal fractures.

4.3.6.2 Longevity Issues

Effectively sealing cement grout must have a water content (w/c ratio) lower than about 50% but this gives a high viscosity and poor penetrability in fractures. In practice, the latter property is improved by adding an organic superplasticizer, the role of which has been in focus for decades since it can possibly be carrier of radionuclides. Addition of silica in the form of amorphous "silica fume" is known to reduce formation of soluble cement components and this additive or some similar silica material is therefore given in grout recipies. The chemical stability of these modern cements has been in focus for some time and hydrothermal experiments have been performed and theoretical models derived for finding out if it can be determined. Unfortunately, there is no general consensus on this, partly because of the uncertainty respecting transition from amorphous to crystalline forms with associated syneresis effects and poor coherence. At present, cements are assumed to retain their properties for no more than 100 years [6, 7]. The very recently proposed "silica" cements of low pH type may turn out to remain intact for much longer periods of time but their sealing function is still probably not lasting long because of the thermally induced rock displacements in a repository that may disintegrate the brittle fracture fillings. The degradation is expected to be particularly strong of cement in narrow fractures because of the heterogeneous nature of the cement produced in the injection phase. Thus, cement in fractures with smaller aperture than about 50 μm contains voids and channels because of backward flow of water contained in the fractures [4,11]. In wider fractures than 50–100 μm the cement is homogeneous but it does not enter parts where the aperture is smaller than 10–15 μm.

4.4 Construction and Performance of Buffers and Backfills in Crystalline Rock

4.4.1 The KBS-3 V Concept

The most important barrier components, the buffer and the backfill, will be in focus here. We will see that the same viewpoints apply also to several other concepts.

4.4.1.1 Buffer

The selection and preparation of buffers are difficult and we will examine them in some detail, keeping in mind that canisters require remote handling and radiation protection. We will start by considering placement of the buffer in KBS-3 V holes, that has been planned in detail by SKB but that has not yet been made under realistic, radiating conditions. The issue is to decide how the buffer blocks and pellets, as well as the canisters, can be placed in the about 9 m deep holes with nearly 2 m diameter (Fig. 3.17). The problem is that the gap between blocks and rock needs to be sufficiently large to bring in the blocks and canisters into the deposition holes. By filling this space with pellets – a technique that has not yet been convincingly demonstrated in practice – the net density of the buffer clay can be kept high enough. Full-scale experiments have been made by placing blocks, one by one, in depositon holes to form columns into which canisters were then lowered (Figs. 4.9 and 4.10).

A problem with this concept is that the slightly more than 5 m long canisters can not be brought into the vertical holes in 5 m high tunnels unless the floor is shaped as indicated in Fig. 4.10. This will generate critical stress conditions in the floor and cause spalling if the primary rock stresses are just slightly higher than normal. Handling of the canisters and placing of the buffer blocks that have to be put on top of the canister, as well as placement of the uppermost part of the pellet filling, must be made remotely. The accuracy will not be high and variations in density are expected.

Fig. 4.9 Handling and placement of large buffer blocks. *Upper*: A 2 ton of KBS-3 V block lifted by vacuum technique for placement in a deposition hole. The block, which is blackish because of the lubricant molybdenum disulphide used for reducing the friction in the form at the uni-axial compression under 3E7 kg (300 MN) force. *Lower*: A prototype tool being used for inserting a model canister in its deposition hole [3]

Several problems are expected that may require stop of the planned operation and retrieval of buffer blocks and canisters that are half-way down in the holes: (1) unexpected strong inflow of water in the holes with difficulties in placing the clay blocks and canisters, (2) damage of clay blocks and canisters that fall or hit the walls, (3) fall of clay fragments that prevent blocks or canister to reach the intended position. We will see later that the risks can be minimized by applying the "supercontainer" concept .

4.4.1.2 Backfill

A not yet answered question is how tight the backfill in KBS-3 V tunnels has to be. The ideal case would be to backfill all excavated space so that it has the same

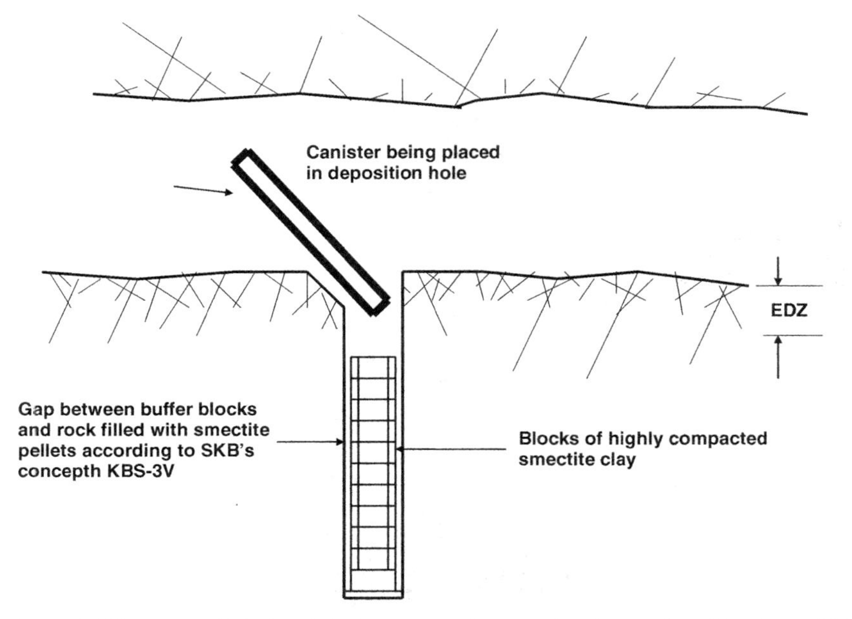

Fig. 4.10 Placement of blocks of highly compacted clay and canister in deposition holes (KBS-3 V concept)

water transmissivity as the rock before excavation and seal all major water-bearing fracture zones in the host rock. This would convert the permeable repository host rock into a big, tight body around which groundwater would flow, driven by regional and thermally induced hydraulic gradients. The cost would of course be too high but an ambition to backfill all rooms, tunnels and shafts so that they get the same transmissivity as the surrounding rock would be desirable and achievable. This principle is manifested by Fig. 4.2 indicating permeable but mechanically stable backfill where the tunnels are intersected by permeable fracture zones of 3rd order, and tight smectitic backfill in between. The matter is more difficult, however, because of the presence of the EDZ, which forms a continuous conductor throughout the repository rock. Strictly speaking the conductivity of the backfill in the tunnels between the intersected permeable zones would not have to be lower than the conductivity of the blast-induced EDZ. However, if the EDZ is cut off by constructing plugs that are keyed into the rock, thereby creating isolated hydraulic regimes, the conductivity of the backfill would have to be the same or lower as of the stress-disturbed EDZ surrounding the blast-disturbed one. It is estimated to be around 10 times higher than the average conductivity of the virgin rock, i.e. about E-9 m/s which is easily fulfilled.

As we saw in Chap. 2 the EDZ of TBM tunnels extends only to a few cm from the tunnel periphery in contrast to the several decimetre deep blast-disturbed EDZ. Sinze the EDZ of TBM-bored tunnels is also less permeable than that of blasted

Table 4.1 Composition of silica concrete [12]

Components	Kilograms per cubic meter of concrete
White cement (Aalborg Portland)	60
Water	150
Silica Fume	60
Fine ground α-quartz	200
Fine ground cristobalite quartz	150
Superplasticizer (Glenium 51)	4.38 (dry weight)
Aggregate 0–4 mm (Quartzite)	1679

tunnels the tightness of backfills in TBM tunnels would logically have to be better than for blasted tunnels. The matters are presently under discussion.

Where tunnels and shafts are intersected by permeable zones one should use physically stable material that is chemically compatible with the contacting tighter backfill in the tunnel that is located in low-permeable rock. A practical solution is to use silica concrete with little cement of low-pH type and ballast (aggregates) of quartz sand and quartzite. This type of concrete is largely inert with respect to chemical interaction with smectite clay and can be allowed to loose the cement component but still serve as permeable and rock-supporting fill if the size gradation of the ballast grains is suitable, i.e. selected so that the large majority of them are prevented to migrate into the rock fractures. The recipe of such a concrete is given in Table 4.1. It is easily pumpable and ultimately becomes very strong but hardens slowly. Backfilling of tunnels and shafts with clay materials where they are intersected by permeable fracture zones would be unsuitable since permeating water can erode and transport clay particles far out in the rock.

Figure 4.11 shows a method for filling and compacting backfills in slopes using vibratory plate compactors for densifying material close to the roof. It was applied and examined in SKB's underground laboratory at Äspö. The inclination of the backfilled and compacted layers was about 25 degrees. The evaluated median dry density using Friedland Ton was about 1450–1550 kg/m^3, yielding a density of up to 2000 kg/m^3 at complete water saturation. According to the lab test results with saturation and percolation of a significantly salt solution (3.5% $CaCl_2$) this would correspond to a hydraulic conductivity and swelling pressure of 5E-11 m/s and 500 kPa, respectively. Both are hence better than required assuming the average conductivity of the virgin rock to be E-10 m/s.

The way of preparing the slope to give a somewhat concave form gave good stability through arching by which it could be made rather steep up to the roof. Figure 4.12 illustrates that the density variations were considerable despite the careful work.

An alternative, more rational technique that has also been tested in the Äspö underground laboratory is to place packages of blocks of compacted clay powder for filling the major part of the tunnel and blow in pellets in the remaining space (Fig. 4.13). The blocks can occupy 75-80% of the tunnel if it has been excavated by contour blasting and less if the walls and roof are irregular. The density of the pellet fill determines the net density and thereby the hydraulic conductivity and swelling

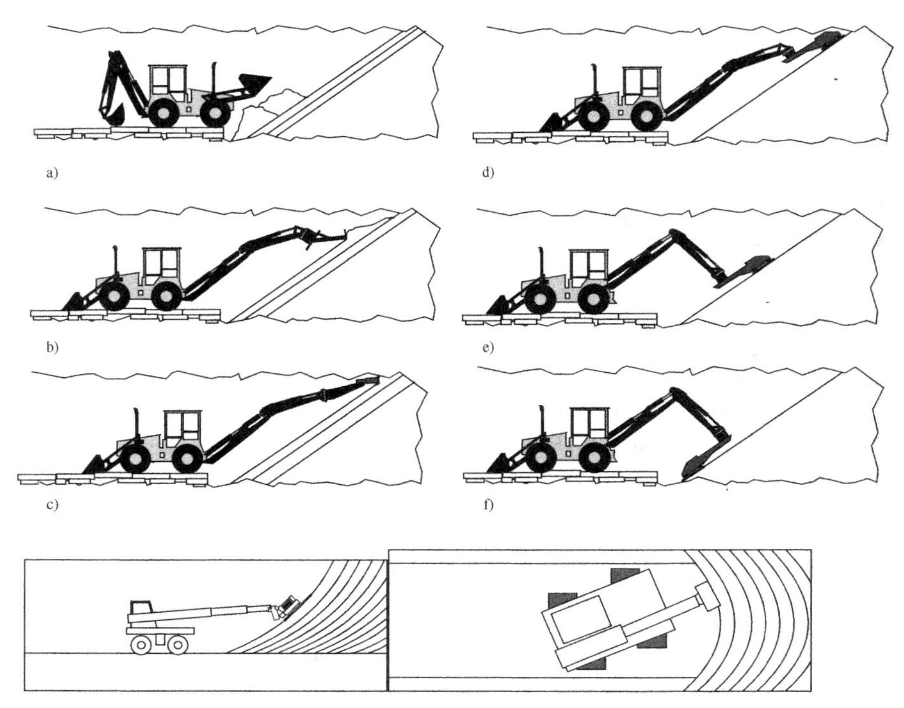

Fig. 4.11 Illustration of application (a, b, c) and compaction (d, e, f) of backfill according to the latest version of the "inclined layer principle" including use of a vibratory roof compactor for densifying material close to the roof. The use of the "roof" compactor is not shown. The lower picture shows the concave form of the layers that could be achieved by the mobility of the carrier and easy turn of the holder of the vibratory plate [13]

pressure of the peripheral part of the fill, which will remain softer than the rest despite the consolidation it will undergo by the expanding block assembly.

It appears that a high degree of block filling makes it difficult to bring in the pellets and the placement of blocks should ideally be so high that no pellets are required. The clay used for block preparation can then be a moderately expanding material, like Friedland Ton, while a filling degree of less than 75% will require pellets as additive for reaching conductivities of E-10 m/s and lower. Naturally, more smectite-rich blocks can be used but the cost would then be discouraging and the sensitivity to inflowing water in the placement phase stronger than for tightly fitting blocks of Friedland Ton or other clays consisting of mixed-layer minerals.

A way of effectively filling the space around the block masonry is to pump in a thick slurry of smectitic clay mud that also penetrates into the joints between the blocks and speeds up the maturation of the whole backfill. Smectitic muds with a density of 1200–1400 kg/m^3 are pumpable using dynamic injection technique or strong industrial pumps but these techniques require that temporary plugs have been constructed prior to the backfilling operation. The mud technique also requires that the space is drained until the mud is pumped in from below through perforated pipes

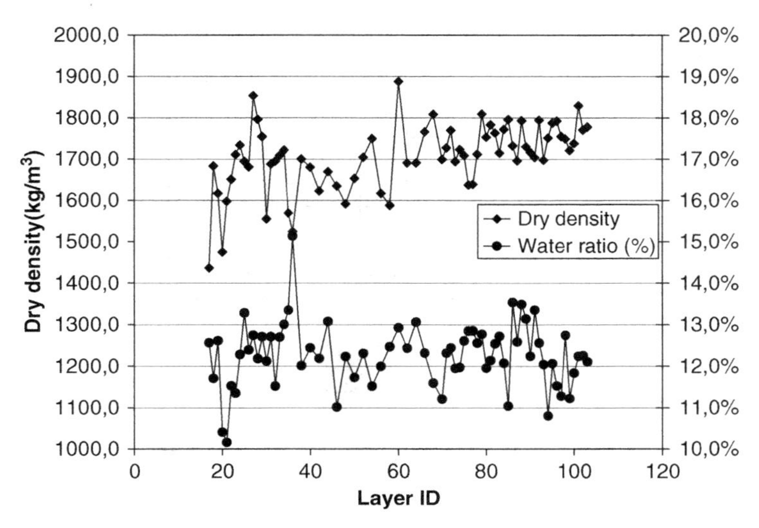

Fig. 4.12 Evaluated average density of compacted layers. The *upper* curve shows successive increase in density for the latest performed compaction events, which is explained by successively improved technique. The *lower* curve shows the water content (ratio), indicating that a rather small range can be obtained by careful preparation of the material [13]

and that pipes are placed at the crown of the tunnel for letting air out (Fig. 4.14). The pipes can easily be sealed, section-wise, by applying a suitable borehole plugging technique.

The problem with all these backfilling techniques is the water inflow from the rock. An example is a field experiment at 450 m depth in SKB's underground laboratory at Äspö where placement and compaction of a backfill with 30% bentonite and 70% crushed rock was made in a 5 m diameter TBM-bored tunnel in rock with the generalized rock structure shown in Fig. 4.15. The water pressure in the immediate vicinity of the tunnel was in the interval 100 kPa to 1.5 MPa. The average inflow of water in the drift exceeded 500 liter per day and meter length in the inner part of the drift (Sections 3585–3600) and 75 liter per day and meter length in the outer part (Sections 3560–3580). The conductivity of the rock matrix next to the backfill, corresponding to the EDZ of the bored drift, was E-10 to E-9 m/s. Backfilling was not possible in the wettest part, while it could be pursued without difficulties in the drier, 20 m long part.

In the 15 m long "wet" interval (Sections 3585–3600) there were approximately 10 fractures and the number of intersections of crossing fractures and the periphery of the drift, taken as inflow spots, were also about 10. The average inflow per spot was 0.5 l/min, which is hence too much to allow for problem-free backfilling. In the 20 m long "dry" interval (Sections 3560–3580) there were about 4 inflow spots yielding an inflow per spot of about 0.2 l/min, which did not cause problems at a backfilling rate of 6 m per day.

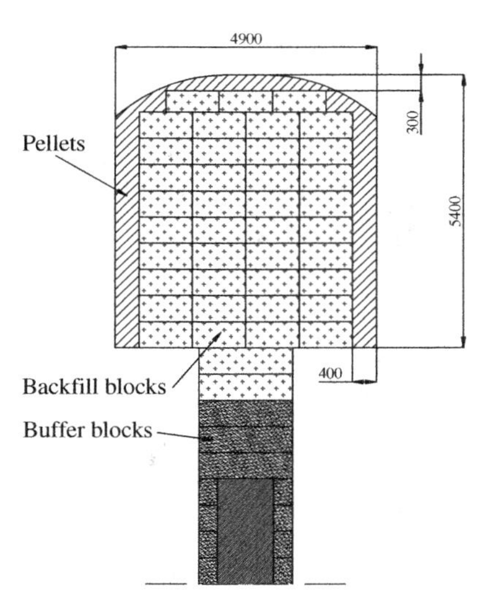

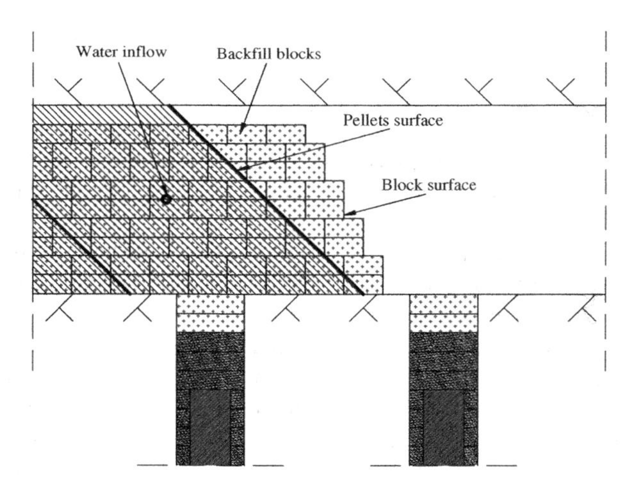

Fig. 4.13 Schematic sections of proposed backfilling of KBS-3 V deposition tunnels

Full-scale experiments using the installation shown in Fig. 4.16, i.e. with pellets filled around stacked blocks, and with controlled inflow of water in discrete spots ("water inflow") showed that about 0.2 l/minute per spot is near a critical state at a filling rate of around 6 m per day and that an inflow of 0.5 l/minute causes piping and softening in complete agreement with the earlier tests. The results from these two field experiments may not be directly comparable because of the difference in grain size of the backfills but they indicate the order of magnitude of spotwise inflow that can cause problems. Since hydraulic measurements have shown that the average hydraulic conductivity of the rock mass is about E-10 m/s and this is a very common value for crystalline rock, problems with unstable backfill states caused by

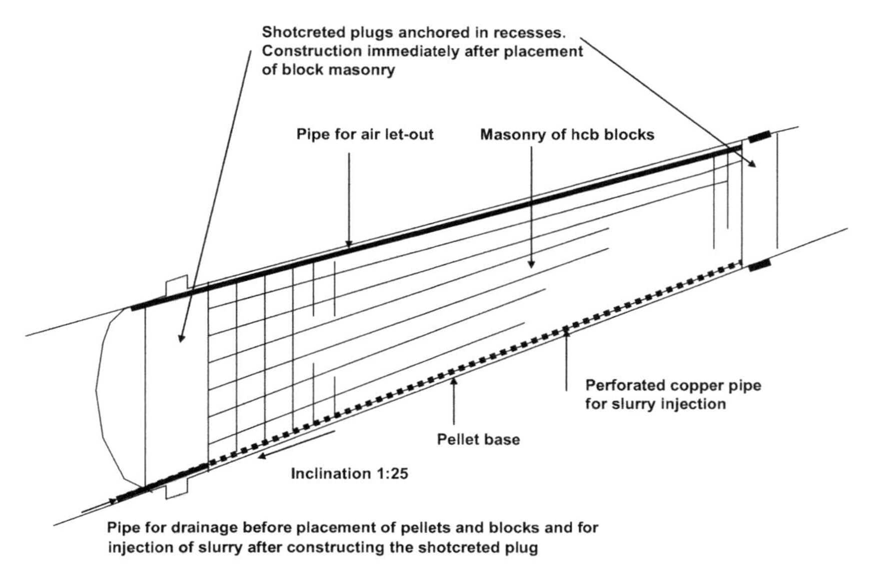

Fig. 4.14 Principle of placing a block masonry in a KBS-3 V tunnel segment under drained conditions and subsequent injection of clay mud. The depositions holes are not shown

water inflow are expected to be numerous. There is no doubt that one has to take steps to reduce the inflow, either by comprehensive grouting before excavation or by draining the rock in the backfilling stage.

4.4.1.3 General Strategy for Backfilling and Plugging

The various tunnels and rooms at the repository level will remain open for different periods of time: the deposition tunnels, of which some are planned to be constructed at the same time as waste is placed in others, are the first to be backfilled, while transport tunnels and the rooms in the central area will come next in the backfilling sequence. The ramp and the ventilation drifts will be backfilled and sealed subsequently, suitably at the same rate for avoiding generation of hydraulic gradients that can cause piping and erosion of clayey backfills. Certain rooms may have to be kept open for more than 100 years depending on possible extension of the repository or changes in regulations.

The fact that a number of the underground tunnels will be completely or partly backfilled while connecting transport tunnels are open makes it necessary to construct plugs at or near their points of junction. These plugs can be permanent or temporary depending on the chemical interaction of the plug materials and the backfills. It is therefore desired to have access to techniques for constructing lasting plugs as well as for placing lighter plugs that can serve as temporary supports or seals in the case of planned or unpredicted intermission of backfilling operations. The plugs can be constructed as indicated in Fig. 4.6 implying that the highly conductive part

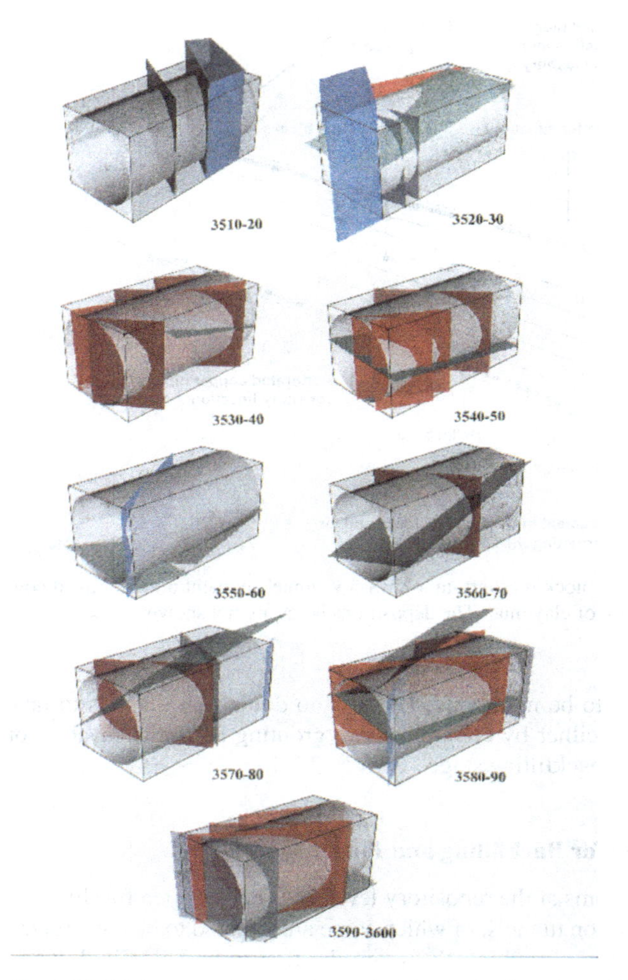

Fig. 4.15 Major water-bearing discontinuities in the Prototype drift. Plugs were constructed in Sections 3535.5–3538.5 and 3559.5–3562.5

of the EDZ of blasted tunnels can be cut off. Constructing of cement-stabilized silicate-rich fills is preferably made by pumping in the concrete between two plugs of which one is not completed until a late stage.

The uppermost parts of ramps and shafts have to be designed and constructed for protecting the tightly backfilled parts of the repository. Since the average hydraulic conductivity of the rock is higher closer to the ground surface than deeper down the backfills in the upper parts of the ramp and the shafts do not need to be very tight. A presently discussed principle is to seal the parts closer to the ground surface than 100 m with materials that are as mechanically and frost resistant as possible. Using rock for this purpose will also make human intrusion difficult but the material

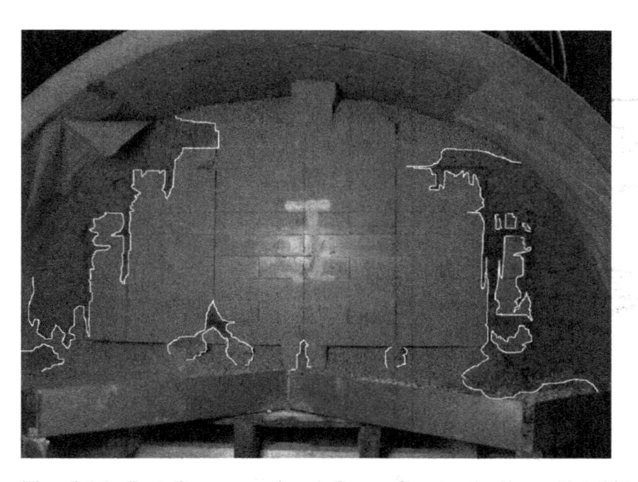

Fig. 4.16 Test for measuring inflow of water in the pellet filling surrounding the blocks in Fig. 4.13. Water was let in at a rate of 0.25 liter per minute in the right part through a nozzle located at the point where the uppermost white line starts at the plastic sheet, and 0.1 liter per minute in the separated left part through one hidden by the plastic sheet. In the right part water flowed largely downwards from start and in the left part it initially tended to flow upwards. The test was terminated after a few days for sampling

must be selected and placed so that it will not be attractive for use in road and dam construction or for preparing concrete. It is suggested here that special effort be made to make the uppermost parts of the backfill in the ramp and shafts as rock-like as possible. The uppermost 50 m of the ramp and shafts may preferably be backfilled with somewhat trimmed blocks from the stone industry to form a "jig-saw puzzle" with silica concrete poured in the joints. By this, the ramp and shafts will be sealed with material that resembles true rock with respect to appearance, porosity and resistance to excavation. Everyone experienced from drilling in natural rock with this consistency would confirm how difficult it is and that intrusion by people without access to advanced excavation techniques would be nearly impossible. Figure 4.17 shows how one can backfill the upper parts of a ramp and shaft.

4.4.1.4 Safety Issues

The KBS-3 V concept implies use of engineered barriers that have a potential of staying intact for at least 100 000 years in ordinary crystalline rock. This means that the rock does not need to serve as a tight barrier. It is required that it provides a stable confinement of the waste, which is offered by most unweathered crystalline rocks if the rock stress conditions are favourable as discussed in Chap. 3. The most valuable property of the concept is the great flexibility with respect to selection of canister positions. However, there are difficulties with safety implications as well, the major one being inflowing water in the canister and buffer application phases and in the tunnel backfilling stage of the construction of a KBS-3 V repository in crystalline rock with its numerous water-bearing fractures and fracture zones.

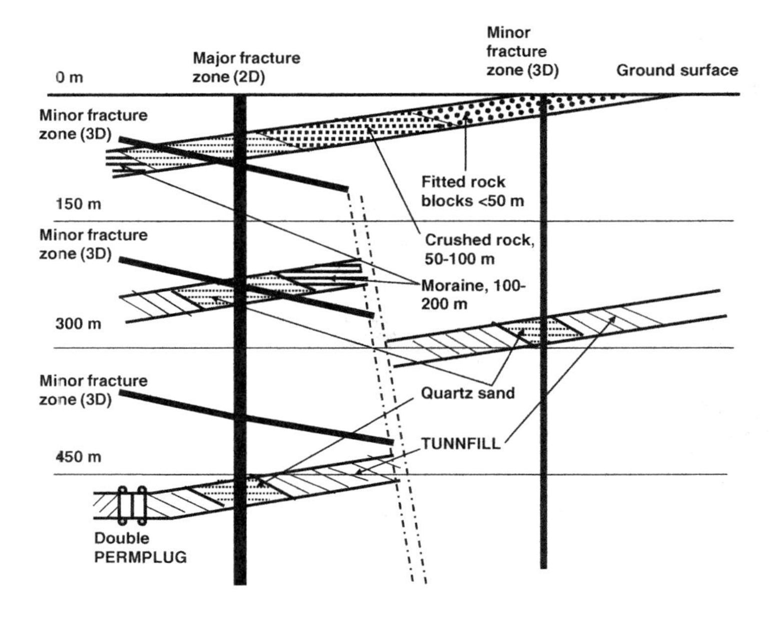

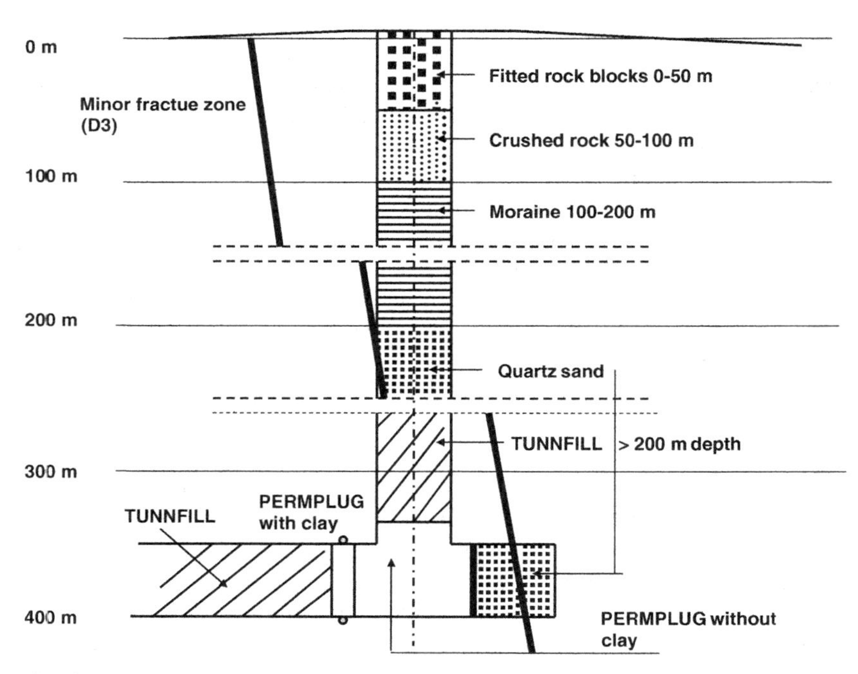

Fig. 4.17 Schematic pseudo-axial section of a ramp (*upper*) and a shaft. Dimensions not to scale. Cement-stabilized quartz sand is placed where the ramp is intersected by water-bearing fracture zones

4.4.2 KBS-3 V Type Concepts with More Than One Canisters

This concept deviates from KBS-3 V by hosting more than one canister in the deposition holes. This requires clay buffer blocks for separating the canisters. These buffer components will have to carry the load of the overlying canisters until the buffer clay surrounding the canisters has matured sufficiently much to assist in keeping the heavy metal bodies in position. A test with two genuine KBS-3 V canisters in has been conducted at about 400 m depth in the AEspoe underground laboratory. It has demonstrated that the function, although not very different from the one-canister case, is very complex and hard to model. The possibility of placing two or more canisters in the same vertical or steep hole is being considered by a few of the Organizations.

The performance of a series of buffer-embedded canisters, separated by clay blocks, will not deviate very significantly from that of the one- and two-canister cases but for a larger series of canisters the stress/strain conditions must be considered in detail. One hence needs to predict, for each canister/clay block and for the entire system of canisters and blocks, how large the compression or heave will be in the course of the hydration of the clay, which is not fully water saturated from start. Movements will develop very slowly and in a very complex way because of the transient hydraulic and mechanical interaction of clay and surrounding rock, and mechanical interaction of clay and canister. Early after placement, when the clay near the canister surface is not yet in tight contact with the canister surface because of the drying described in Chap. 3, the canister-separating clay will carry all the canister load, and it will fracture and undergo compression. This and several other processes were considered a number of years ago when the so-called Very Deep Hole ("VDH") concept was defined by SKB [14].

4.4.2.1 The VDH Concept

Disposal of hazardous waste in deep holes bored in crystalline rock is attractive because of the low hydraulic conductivity of rock at depth (Table 4.2). In combination with the high salt content that is expected at depth, up to 35000 ppm above 2000 m depth and up to 100 000 ppm at 2000–4000 m depth, groundwater flow at larger depth than 2000 m is very slow. Convection of groundwater caused by the temperature rise in the deployment part of the holes, which is below this depth, has been calculated and found to be insignificant [14]. It would not be measurable 50 m above the upper end of the deployment part.

In the late 80s SKB worked out a design implying that the HLW, confined in canisters, is placed in the lower part of tapered, very deep holes while their upper part is sealed with asphalt and concrete. SKB's earliest attempt some 20 years ago had the form in Fig. 4.18, with the "deployment" zone extending from 2000 to 4000 m depth where the hole has 800 mm diameter and with the "plugged" zone extending from the ground surface to 2000 m depth in which interval the hole is bored to 1300–1400 mm diameter. The uppermost part would be filled with asphalt

Table 4.2 Average hydraulic conductivity of granitic rock as a function of depth. The figures include contribution by minor facture zones, i.e. those of 3rd order [15]

Depth, m	Average conductivity, m/s
100	E-7
500	5E-10
700	E-10
1000	5E-11
1500	E-11
2000	5E-12
3000	E-12
4000	5E-13

and concrete down to 500 m depth and the part extending from 500 to 2000 m depth being "plugged" with dense clay.

The need for guaranteed stability was believed to require use of a permeable casing in both the plugged part and the deployment zone. It was intended to prevent rock fall and to guide canisters and highly compacted clay blocks in the placement phase. In a first version of the concept it was assumed that a boring mud should be used in the construction phase and that it would be replaced by a more viscous, thixotropic deployment mud before placing of canisters and clay blocks. They were planned to be put pair-wise in strong, permeable coarse-mesh metal cages, "supercontainers" (Fig. 4.19), that can be handled and placed remotely. This design principle was on the drawing tables of SKB in the early eighties when the method of plugging boreholes was first proposed and used. It appeared in the VDH concept a few years later [14].

The uppermost 500 m would be constructed by shaft sinking technique yielding an average diameter of a few meters [16]. The deployment part would contain dense clay buffer blocks separating canisters, while the part to be tightly plugged, i.e. from 500 to 2000 m depth, would contain only dense buffer blocks. SKB concluded from the study that the concept is feasible but involves a number of not yet solved problems. A major issue was that it would be difficult or impossible to retrieve canisters if required. We will look somewhat deeper into the matter here, paying attention to application of more recently developed techniques.

A suitable average spacing of the canisters was taken to be 1 m, which would hence also be the thickness of the buffer blocks. Figure 4.19 shows the originally proposed installation in the deployment part with canisters (W) separated by buffer blocks (B) and surrounded by a cage that serves as protection of a series of canisters and blocks when being placed in deployment mud. The drilling mud was proposed to be a Na smectite (bentonite) drilling mud. It serves as carrier of debris from the drilling head up to the ground surface and its thixotropy means that the rock fragments are "frozen in" at temporary stops and turns liquid again when the boring and pumping start. Petroleum and gas exploration and production companies often use heavy muds for preventing gas or oil to flow into the hole and for stabilizing the borehole walls. Smectite or palygorskite are main clay mineral components of

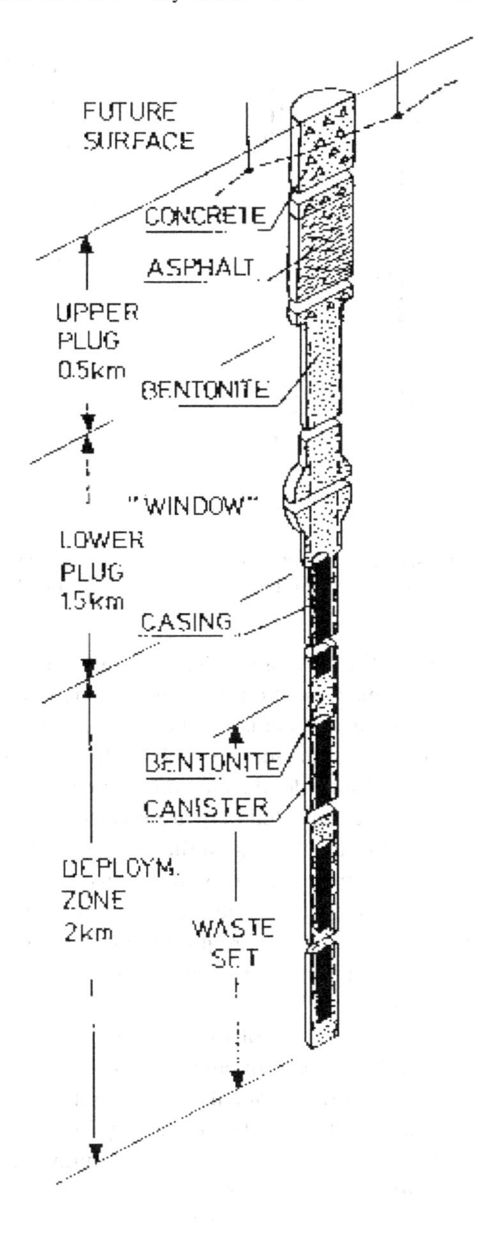

Fig. 4.18 The VDH concept [14,15]

the drill mud. In a later project phase it was suggested to use the same mud for drilling and deployment aiming at a mud of smectite-rich Ca bentonite and 10% $CaCl_2$ solution with a density of 1500 kg/m^3.

The concept implied placement of a cage of navy bronze with slightly smaller diameter than that of the hole (0.8 m) in the mud for supporting the rock all the way

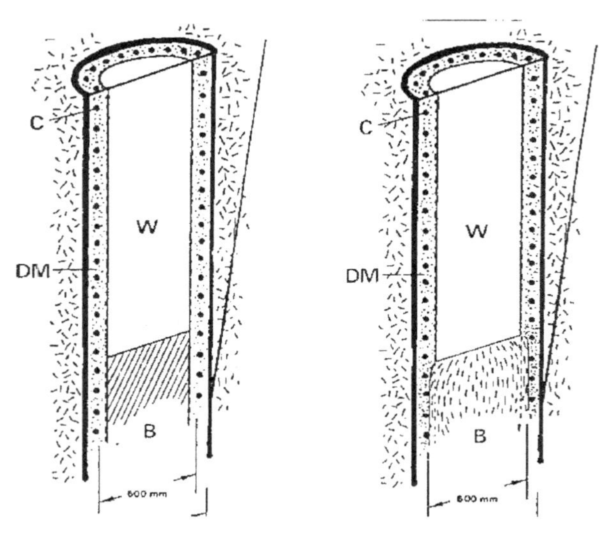

Fig. 4.19 Set of canisters and buffer blocks confined by a coarse-mesh metal cage. *Left*: Condition at placement. *Right*: Final stage after expansion of the buffer and consolidation of the mud

down to the base of the deployment part so that rock fall would be eliminated in the canister placement phase, in which cages with sets of one to five strongly compacted clay blocks and canisters were to be pushed down to form a continuous column of intermittent canisters from 4000 to 2000 m depth. For the clay-plugged zone, i.e. 2000 to 500m depth, the same principle would be applied. Here, cages with sets of highly compacted clay are pushed down inside the rock-supporting metal net in the mud in the 1.3–1.4 m diameter hole. The metal in the stabilizing net and in the cages was proposed to be navy bronze, which has about 90% copper zinc making up the rest. None of the metal components need to be operative after placement of the canisters but it is required that they do not cause mal-functioning and degradation of the clay components. This is believed to be guaranteed by their high copper content.

The net density of the ultimately matured system of expanded clay blocks and consolidated deployment mud was calculated to be 1890 kg/m^3 for the conservative case of slightly elliptical cross section, resulting from slight rock fall. The expected hydraulic conductivity of the clay in the plugged part would range between E-12 and E-11 m/s depending on the groundwater salinity. It would hence be tighter than the surrounding rock mass.

The requirement that the sets of clay blocks and of clay blocks separated by canisters must be placeable means that the viscosity and thixotropy must be suitably selected and hence also the content of clay minerals. Since the shear strength of the candidate Ca bentonite mud is estimated at 0.01–0.1 MPa the force needed for pushing the blocks down may be several tens of tons, requiring big drill rigs. Heavy pile vibrators would be suitable for reducing the shear resistance of the thixotropic depoyment mud.

A major question is the stability of deep boreholes, a matter that has been investigated in conjunction with drilling of the 12 km deep Kola hole near Murmansk in the former Soviet Union [16,17]. This hole, which was drilled by applying oil-drilling technique using a rotating borehead with bits, reamer and a bore mud that was probably clay-based, had a theoretical diameter of 214 mm, which was nearly the same as the true measure down to about 500 m depth, while the actual maximum/minimum diameters were 290/225 mm at 2400 m depth, 390/250 mm at 5000 m depth, 570/225 mm at 8450 m depth and 370/240 mm at 12000 m depth. One finds that the borehole diameter is significantly larger at 5000 m depth than at about 1000 m depth and that the variations and lack of axial symmetry has implications. Thus, where the hole is wider than planned the clay buffer has to expand more by which its density and bearing capacity are reduced and its hydraulic conductivity increased.

A number of objections to the VDH projects were initially raised but they have later been ruled out as described below:

- The canisters are not retrievable – The canisters are retrievable by reboring although this will be very difficult and expensive. The same is valid for most other concepts.
- The clay components can be eroded and lost where the holes are intersected by water-bearing fracture zones – This is a problem but it can be totally eliminated by filling the holes with silica concrete where they are intersected by such zones. This principle has been worked out by SKB for sealing deep boreholes [12].
- The canisters will be placed where there are no major fracture zones with a potential to undego shearing but rock displacement may take place also in other parts and cause failure of the canisters – With canisters of HIPOW type the problem is eliminated since they can sustain large strain without undergoing breakage.
- Tectonically induced shearing of the rock may break the canisters – Such shearing will take place in major fracture zones only and not in the segments where the canisters are located.
- The concept implies that there is only one barrier – The operative lifetime of the clay components is as long as that of the KBS-3 V concept – The overall performance of the VDH concept is to provide effective tigthness in the axial direction of the holes in the parts where the canisters are located and accept groundwater flow in fracture zones that intersect the holes.
- Drilling of holes with 1.3–1.4 m diameter to 2000 m depth is not considered to be a problem but it is questionable whether 0.8 m diameter holes can be bored to 4000 m depth – British and German drilling companies expect this to be possible but propose practical testing for finding optimum muds for drilling and deployment.

A detailed study of the VDH concept with somewhat different dimension was made in year 2000 [16]. The diameter of the proposed well design was 1170 mm to 500 m depth using a 1067 mm casing cemented in the hole. The diameter between 500 and 2000 m would be 1016 mm and a 914 mm casing placed with bentonite used for sealing the annulus except for the lowest 100 m which would be cemented. Between 2000 and 4000 m the diameter would be 838 mm with a 762 mm slotted casing. Drilling would be made by a down-hole hammer with foam as drilling fluid. The time to drill and case such a hole would be 137 days and cost about 35 million BP as per May, 2000. This is of course attractive.

4.4.3 Wide Rooms with Arrays of Canisters

The method of storing many containers in big rooms is common in disposal of hazardous chemical waste in rock. A recently completed EC project dealing with disposal of such waste in abandoned mines has shown that it is rational and safe provided that the waste is embedded in "buffer" of smectite-rich clay [18]. It is believed that the method would work excellently for low-level radioactive waste and intermediate-level waste that generates little heat but HLW stored in this fashion causes a considerable temperature rise and can also involve a risk of criticality.

One should distinguish between the "dry" case represented by the US concept of storing HLW at depth in rock that is virtually dry (Yucca Mountain), i.e. above the groundwater level, and the case with the repository located below this level. The firstmentioned case offers advantages but can not offer dry conditions over 100 000 years because of climatic uncertainties – deserts may be wetlands and wetlands turn to deserts. Location at a few hundred meters depth below the groundwater level is believed to be less sensitive to changes in climate and is the principle followed here.

The wetting rate of the buffer resulting from its sorption of water that migrates from the rock will be extremely slow and will not compensate for the heat-generated desiccation of the buffer. Hence, all the problems with drying and salt enrichment that were discussed in Chap. 3 will be magnified in such a repository. However, using HIPOW canisters the case would be quite different, their very low corrosion rate puts less demand on the buffer and it may be allowed to undergo significant changes. One may in fact consider quite other materials like crushed basalt that will be successively altered to smectite under the hydrothermal conditions that will prevail for some thousand years. Quite comprehensive research would be required, howeverless, for finding out whether the concept is feasible. If so it would be very cost effective.

4.4.4 Inclined Deposition Holes with Single Canisters

The impractical handling of canisters and buffer blocks implied by the KBS-3 V concept has led to somewhat altered versions and one of them is believed to have a number of advantages. It is basically the same as the KBS-3 V but has the deposition holes inclined and oriented in the fashion shown in Fig. 4.20. The advantages of this KBS-3i concept compared to the original are:

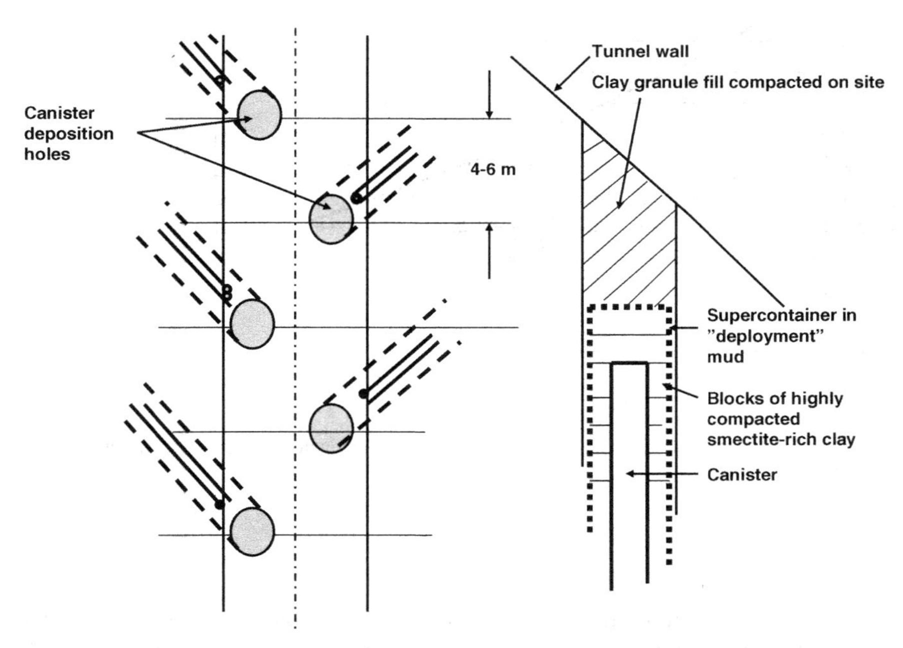

Fig. 4.20 The KBS-3i concept with 45° dip and +/–45° deviation from the tunnel axis

- The larger distance between the canisters causes less overlap of the local temperature fields and hence lower buffer temperature.
- The supercontainer that contains the canister and tightly fitting clay blocks is prepared in a specially equipped facility on the ground level and transported with simple radiation shielding to the hole. The hole is partly filled with "deployment" mud in which the supercontainer is submerged (cf. Fig. 4.22).
- On top of the supercontainer granulated expandable clay of Friedland Ton type, or similar, is filled and compacted. Its uppermost surface is covered by a removable metal lid anchored to the rock.

The significantly larger spacing of the canisters than for the KBS-3 V concept means that the overlap of individual temperature fields is smaller and the maximum temperature of the buffer somewhat lower than 90°C despite the more effective utilization of the rock (center-to center 4–6 m). The real advantage of the KBS-3i concept is the ease with which the canister and clay blocks can be placed by using the supercontainer principle. For reaching the required net density after maturation, i.e. no less than 1900 kg/m^3, the "deployment mud" – borrowing the term from the VDH concept – must have a sufficient density. For VDH the proposed mud of smectite-rich Ca bentonite and 10% $CaCl_2$ solution was concluded to have a minimum value with of 1500 kg/m^3, but it can be considerably lower for a KBS-3 hole if the clay blocks can be compressed to a very high density. It has been demonstrated that

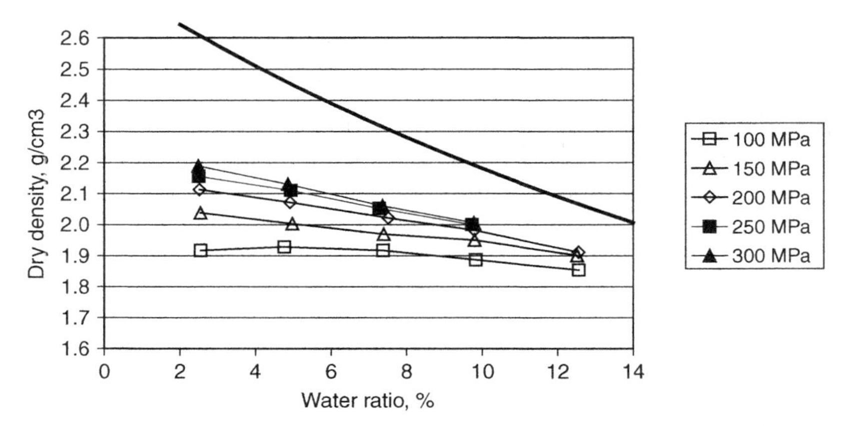

Fig. 4.21 Dry densities obtained by compacting MX-80 clay powder with different water contents (ratio) by using compaction pressures of 100 to 300 MPa [12]

compaction of smectite-rich clay powder with a water content of about 6% under 250 MPa pressure can yield a dry density of more than 2100 kg/m^3 (Fig. 4.21). A suitable grain size distribution of the clay powder for achieving such high densities is 20% of grains with 2–8 mm diameter, 20% with 1–2 mm, 42% with 0.1–1 mm diameter, and 17% finer than 0.1 mm. For a dry density of this order of magnitude the "deployment mud" does not need to be denser than 1100 kg/m^3.

There can be a problem, however, if clay expands from the dense buffer into fractures that are open or can be opened by high swelling pressures. This matter has been investigated experimentally and theoretically assuming expansion of buffer into fractures. The "wall" friction will limit the penetration of the clay, which is driven by the swelling pressure gradient that is several megapascals at the periphery of the deposition holes and nil at the front of the penetrating clay. It has been feared, however, that if the groundwater flow in the fractures is too high the soft front may be eroded and brought away from the canister deposition holes, making more clay move away and ultimately cause a significant loss of canister-embedding clay. This fear is not justified as far as the canister deposition holes are concerned since fractures with significant aperture will not be accepted in the holes. However, other concepts like the VDH will imply intersection of fractures with apertures that are large enough to cause practically important loss of buffer clay. For these cases the mud should have a filtering function so that it blocks the fractures for which one can think of several solutions. One is to use a "deployment" mud of smectite clay mixed with suitable graded quartz particles or consist of the smectite species palygorskite (attapulgite), which can cause arching through the needle-shape of the particles and prevent buffer smectite particle aggregates from being moved into the fractures. The principle is shown in Fig. 4.22. Figure 4.23 illustrates the shape and size of palygorskite particles.

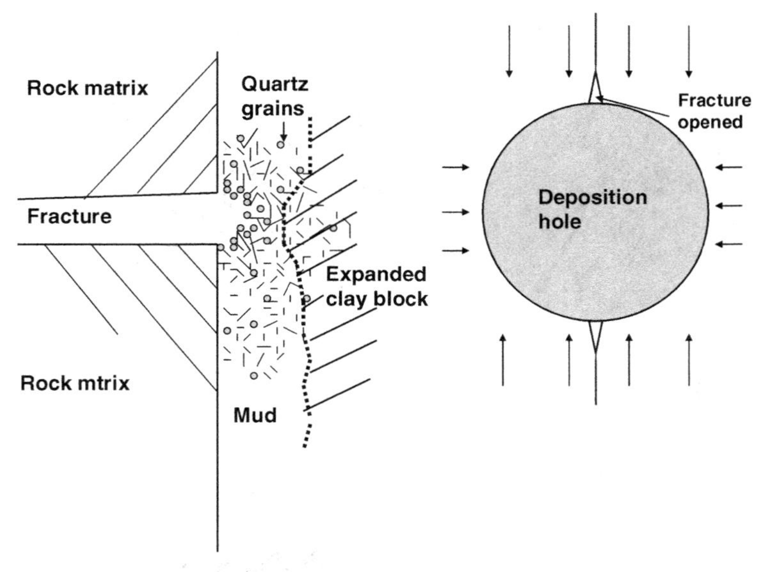

Fig. 4.22 Clogging of fractures by quartz grains forming arches. Right: Deposition hole in rock with strongly anisotropic stress field that causes widening of a steep intersecting fracture into which the buffer clay can move

4.4.5 Very Long Holes (KBS-3H)

Some decades ago NAGRA and ENRESA proposed concepts with very long holes, and performed field experiments in TBM-bored tunnels with about 2 m diameter in crystalline rock. This early concept implied that the HLW canisters be surrounded by sector-shaped, densely compacted large blocks of smectite-rich clay. Placement of such blocks in rock that gives off only little water in the large holes is possible, which means that the concept is feasible for argillaceous rock, while the commonly stronger inflow in crystalline rock can cause great difficulties with dispersion of the clay blocks, slippery floor etc. For minimizing these problems and making the handling of buffer and canisters more rationally the idea of using supercontainers was born.

4.4.5.1 The Supercontainer Principle

Preparation of perforated tubes confining tightly fitting sector-shaped buffer blocks around centrally placed canisters has been studied by several of the Organizations and it is a key component of the Belgian concept adaptated to clastic clay for hosting a HLW repository. The supercontainers can be prepared elsewhere in the repository area and transported to the intended site and put there by pushing it into the deposition hole using air cushions for minimizing friction. Figure 4.24 shows the

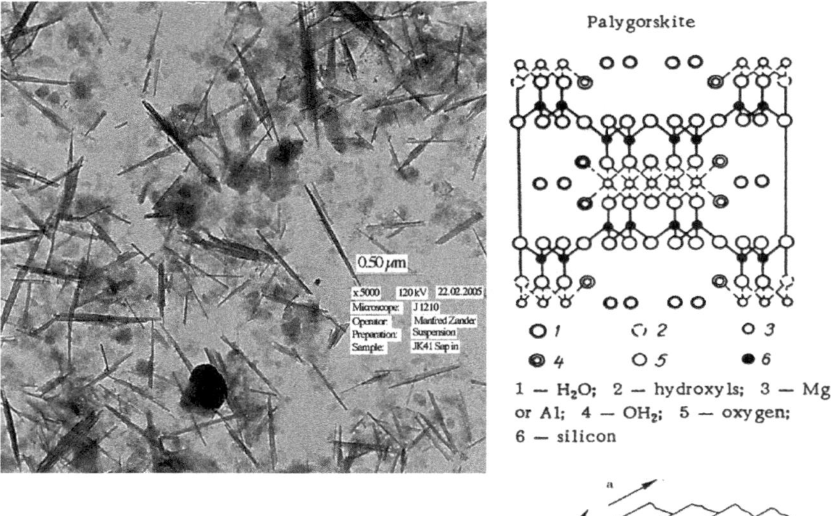

Fig. 4.23 Palygorskite crystal lattice and morphology

components of SKB's supercontainer, which is based on the same principle as a borehole plug concept that we will consider later in this chapter.

For eliminating the risk of criticality and for limiting the temperature the supercontainers must be separated by placing large buffer blocks between them corresponding to those of the VDH concept. The fitting of these "distance blocks" and the rock wall has to be very tight, i.e. with clearings of less than 15 mm, for minimizing leakage along the rock wall from the placed units.

The supercontainers are proposed to be made of iron for the reason that it provides high strength and hence allows for small dimensions and low weight. However, this choice is associated with a number of problems that we will consider here.

4.4.5.2 Impact on the Chemical Integrity of the Buffer Clay

A major problem with supercontainers of iron is caused by the presence of the precious metal copper in the canisters. Copper will survive while the iron will be sacrificed and converted to magnetite in conjunction with production of hydrogen gas. Magnetite has a significantly lower density than iron and occupies a larger space than iron in uncorroded form. This means that a pressure would be exerted on the clay blocks and on the clay formed around the containers, which will tend

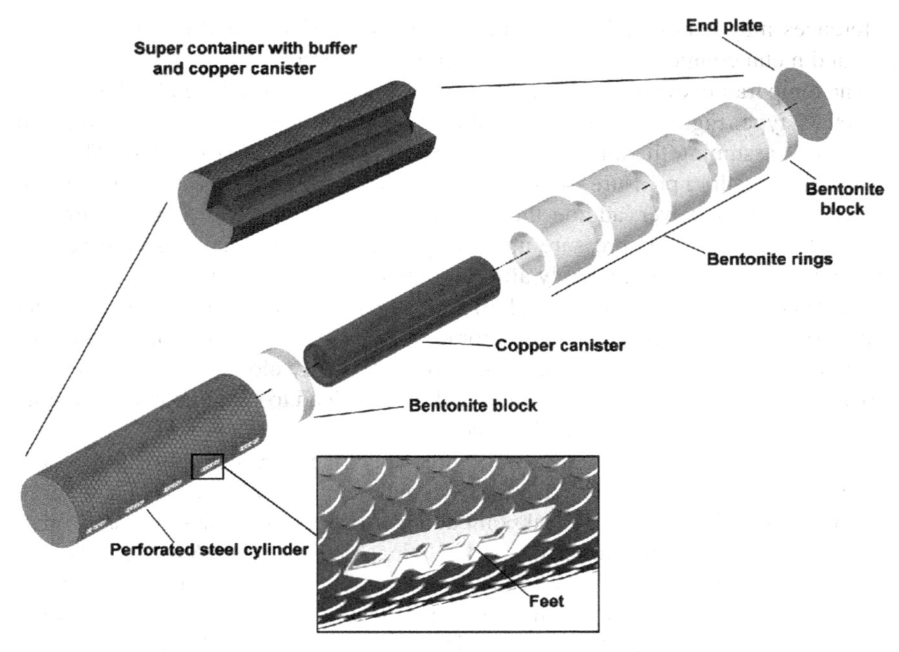

Fig. 4.24 The components of a supercontainer of KBS-3H type [1]

to compress both. It may be that the pressure induced by the iron corrosion and the swelling pressure of the clay blocks combine to give such a high pressure on the rock that steep fractures that intersect the deposition holes open and let clay out, similar to what can happen by corrosion of the iron canisters and the insert within KBS-3 V canisters (Fig. 3.3). The role of iron is twofold: (1) iron ions will replace originally sorbed sodium and cause reduction of the spacing of the smectite lamellae, which reduces the self-healing potential and increases the hydraulic conductivity, and (2) iron compounds are formed, serving as cement that binds stacks of smectite lamellae together. Fe reduction in smectite clay appears to cause reduction in swelling pressure by such cementation and increase in hydraulic conductivity.

The other effect, gas formation, means that channels can be formed and ruin the homogeneity of the clay that is formed outside the supercontainers. The problem is that anaerobic corrosion with production of hydrogen gas can be very extensive. We will come back to it later in this chapter.

4.4.5.3 Evolution of Buffer Clay

The design density and porosity of the buffer components following saturation and swelling into the gap around the supercontainer are 2000 kg/m^3 and 0.44, respectively. Some differences in buffer density are, however, likely to occur, especially between the saturated buffer inside and outside the supercontainer. These differences may persist indefinitely because internal friction and friction between clay and metal components and rock may limit homogenization.

The temperature evolution naturally depends on the heat produced by the respective HLW type. Figure 4.25 illustrates the predicted temperature evolution for two geometries assuming the ultimate density of the buffer to be 2000 kg/m^3. The corresponding swelling pressure for a porewater salinity of 1 M NaCl solution will not exceed 2 MPa. However, adding the pressure caused by the conversion of iron to magnetite can bring this pressure up to very high and even critically high values. The temperature distribution early after placement is exemplified in Fig. 4.26.

There is also a chemo/mechanical effect that needs to be considered, namely the impact of the change in shape, i.e. perforation etc, of the corroding supercontainer on the expanding clay. The corrosion is expected to vary along the supercontainer depending on the clay's access to water and this can lead to considerable variations of the density of the clay outside the supercontainer.

The diagram in Fig. 4.27 shows the predicted rate of water saturation of the clay buffer as a function of the average hydraulic conductivity of the surrounding rock. One finds that there is a breakpoint at the conductivity value E-11 m/s, which is in agreement with the earlier derived relationship for KBS-3 V (Chap. 3). For conductivities lower than E-11 m/s the initial water migration in fact takes place from the buffer to the rock, which causes significant desiccation of the entire clay mass leading to formation of fractures and possible permanent collapse of the stacks

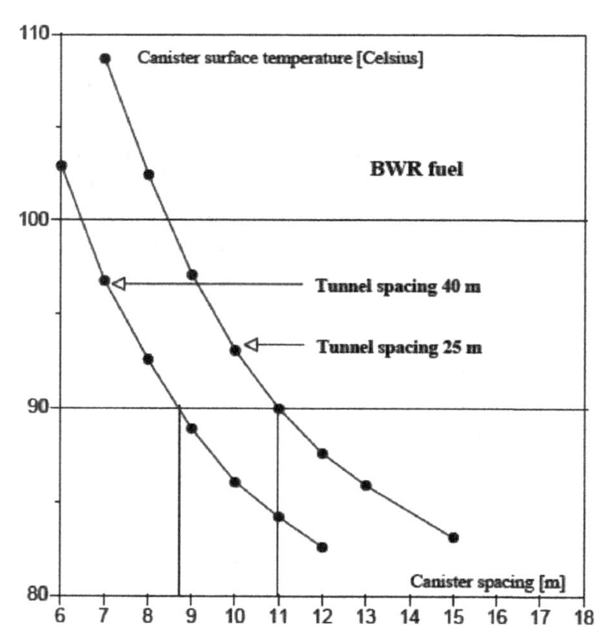

Fig. 4.25 Temperature in the center of a one-level repository containing 1500 BWR canisters (1700 W) as a function of the spacing of supercontainers and deposition drifts [19]

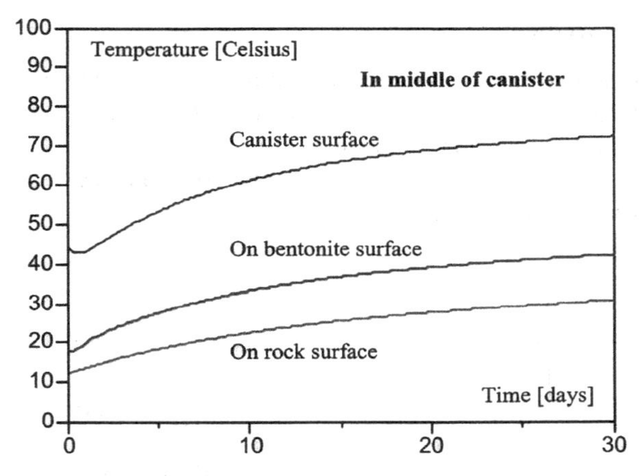

Fig. 4.26 The temperature evolution (vertical axis) over the initial 30-day period after placement of supercontainer with buffer and canister [20]

of smectite lamellae. This deteriorating effect is small for the KBS-3 V concept since water for hydration is supplied by the overlying backfill but it can be strong for the KBS-3H version and particularly strong for a repository of this type in argillaceous rock as we will see later in this chapter.

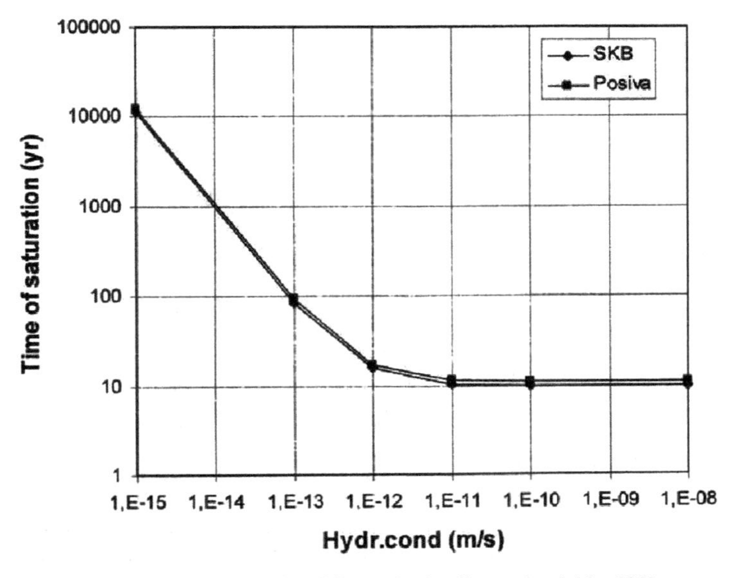

Fig. 4.27 Time to saturation for different hydraulic conductivities [20]

4.4.5.4 Impact of Tectonic Movements

A major difference between the two KBS-3 concepts is that a smaller portion of a completed deposition tunnel can be used for placing canisters for KBS-3H than for KBS-3 V because of the possibility to better adaptate to the rock structure in the latter case. Figure 4.28 shows the planned isolation of a fracture zone intersecting a KBS-3H tunnel by constructing plugs on both sides of the zone.

4.4.5.5 Practicality

The technique described in Fig. 4.22 for minimizing or eliminating loss of clay to the surrounding rock, i.e. by filling the space between buffer blocks and rock with an erosion-resisting mud, is of course not possible for the horizontally oriented deposition tunnels. This is a serious drawback of the concept.

Turning finally to practical issues, one needs to assess the risk that the heavy supercontainers can become stuck in the placement phase because of failure of the sophisticated machine, rock fall or that the operation is delayed by unexpected strong water inflow. Adding the risk of loosing unacceptable amounts of buffer clay to the other difficulties makes the KBS-3H concept less attractive than the other ones.

4.4.6 Assessment of the Concepts Applied to Crystalline Rock

Comparison of the four concepts in Fig. 4.3 can be made with respect to the following properties:

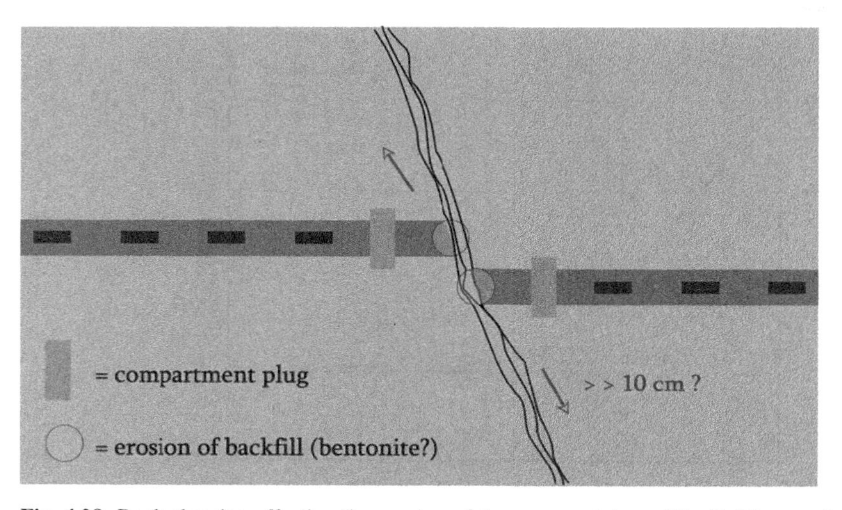

Fig. 4.28 Rock shearing affecting the spacing of the supercontainers (*black*). The questionmark indicates the unknown actual shear strain that the "compartment plug", i.e. distance block fill, may undergo

- Uncertainties.
- Practicalities.
- Cost.

4.4.6.1 KBS-3 V – Reference Case

In its present form there are some major uncertainties and possible deviations from the expected performance: (1) The time to reach complete water saturation can be very long, which can cause permanent reduction of the expandability of the buffer and increase in hydraulic conductivity. (2) In the early wetting phase the canisters will be exposed to tension since the upward expansion of the upper part of the buffer will tend to pull up the canisters while their lower ends are fixed in the buffer. (3) After complete hydration the canisters will settle but it is not known with certainty what the settlement will be and how it can affect the buffer clay.

These matters bring us back to the question raised in Chap. 3 of whether conventional "effective stress" soil mechanics applies also to smectite-rich clays and hence what the stress/strain properties of the interparticle contacts are. As to the settlement of the canisters this theory implies that it would stop after dissipation of the porewater overpressure induced by the canister load. However, it is a well-known fact that creep under constant volume conditions causes additional strain, the mechanisms being described in Chap. 3, and it is particularly important for smectite-rich clays with their viscous character. Thus, the mechanical interaction of adjacent particles is via the water hulls with dissolved cations, a medium that can be understood as being poorly ordered and hence apt to undergo time-dependent strain of viscous type. As long as shear stresses prevail in the buffer due to the canister load, creep will go on for ever at a retarding or constant rate depending on the magnitude of the generated shear stresses and possible geometrical constraints.

Placement of clay blocks and canisters under realistic conditions has not yet been made and it may turn out to be impossible to place the pellet fill around the dense clay blocks with required homogeneity, or the blocks without damaging them. Use of supercontainers can solve these problems but would require that the height of the tunnels is increased to 8–9 m.

A valuable property of the concept is that the position of the deponition holes can be adaptated to the rock structure and that the canisters are retrievable, although with difficulty. The cost for construction, waste application and sealing is believed to be substantial. A rough estimate is that the cost for constructing a repository hosting 5000 canisters of the presently considered SKB type is on the order of 2 billion Euros excluding site investigations and design.

4.4.6.2 Wide Rooms with Deposition Holes with Single Canisters

The temperature of the buffer will be very much higher than for KBS-3 V and may cause strong degradation of the buffer clay. Water saturation will take a very long time, the central parts of backfilled rooms may in fact require tens of thousands of years or even more [18]. This is a valuable property of the concept as long as the major part of the buffer stags intact, and so is the fact that the canisters will be easily retrievable. The cost for constructing a repository hosting 5000 canisters of the presently considered SKB type is expected to be on the order 1 billion Euros excluding site investigations and design.

4.4.6.3 Steep Holes with Two or More Canisters

The performance of the buffer and canisters is similar to that of a KBS-3 V repository, the difference being that the weight of two or more canisters may cause larger settlement. For the VDH concept the deep location of the buffer will make it totally saturated in less than one year, thereby causing less degradation than for the earlier mentioned concepts. Another positive feature is that the transport paths for possibly released radionuclides is at least 5 times longer than that of the KBS-3 V concept. The pressure conditions in the "deployment" zone make it necessary to use HIPOW canisters, which has the advantage of keeping the waste perfectly confined in very rigid units. Careful identification of unsuitable canister positions for avoiding them is therefore not at all as important as for canisters that are presently favoured by SKB but also for VDH it is required to avoid location of the canisters where significant fracture zones intersect the holes. These parts should be filled with physically stable concrete that is cast on site. The experience from applying the technique for plugging of boreholes with 500 m depth indicates that it may be difficult at larger depths.

A most important issue is that the large number of canisters and separating expanding/contracting clay blocks may cause substantial internal movements in axial direction. One has to make sure that there is force equilibrium within and between each set of canister/blocks, not only under the initial conditions when the buffer clay is largely intact and can provide "self-sealing" but also at later stages when its stress/strain properties will change as discussed in Chap. 3.

Tectonic impact on VDH is not believed to be critical. Thus, shear strain generated by large-scale stress changes and displacements of block units will take place along major discontinuities, at the passage of which, the holes will be backfilled with coarse material that is not required to be low-permeable and that can undergo large shear strain without loosing its physical stability, i.e. silica concrete.

The canisters are retrievable but only with great difficulties. A rough estimate of the cost for constructing a repository hosting 5000 canisters of the HIPOW type gives a figure of about 0.5 billion Euros excluding site investigations and design. It is the cheapest repository concept since no deep underground rooms have to be constructed but it requires development of technique for production of HIPOW canisters. From the point of economy it would certainly be attractive to the Organizations that can think of depositing reprocessed reactor fuel confined in very strong canisters.

4.4.6.4 Inclined Deposition Holes with Single Canisters

It is a bit remarkable that this concept has not been formally proposed as a candidate by the Organizations despite its advantages. As to the long term performance it is almost the same as for KBS-3 V but for the placement phase it would solve the problem with KBS-3 V requiring a high deposition tunnel or cutting "ramps" at the upper end of the deposition holes. It would also be more economic because more canisters can be hosted per deposition tunnel, fulfilling the canister surface temperature criterion. Furthermore, it offers safer placement conditions since the canister cannot fall down in the holes, the supercontainer with the canister surrounded by clay blocks will simply slide down along the wall if gripping would fail. We will be right back to the "optimal concept" in a few pages.

4.4.6.5 Very Long Holes

Despite its apparent simplicity the concept is not very practical and the difficulties met with in placement of the supercontainer units are discouraging. However, the most critical factor is the evolution of the container/buffer/canister system, for which one foresees the problems described earlier. Thus, the proposed metal for the supercontainers, iron, will undergo dissolution and neoformation of magnetite, which will cause irregular deformation and displacement of the metal cage and thereby heterogeneity, along the supercontainers, of the clay that has migrated through their openings. It is expected that the heterogenity of the clay surrounding the metal cage, resulting from variations in penetration rate of the clay expanding out from the dense blocks, can cause stress concentrations that may break and deform the cage such that it becomes difficult or impossible for the clay both outside and inside the containers to become homogeneous.

The combined pressures of the buffer clay and magnetite is another major problem because it can exceed the hoop stress in the deposition holes and widen steep fractures that intersect the hole in axial direction so that buffer clay can migrate out into the surrounding rock. A third issue is the impact of iron in ionic form, which can replace the originally sorbed sodium by cation exchange causing an increase in hydraulic conductivity, and form cement that reduces the self-sealing potential of the clay and make channels caused by gas penetration permanent. Production of hydrogen gas in large amounts by anaerobic corrosion of iron supercontainers is really a major threat to the required function of the clay. All these chemical and physico/chemical effects may be slow and have only a small impact on the buffer properties before it is largely water saturated, but they will all appear in due time even if the rock is very tight and provides the buffer clay with only little or no water in the first century. This would, on the other hand, cause considerable degradation of the buffer clay by desiccation: the smectite particles consisting of stacks of lamellae will contract and stay so by being prevented to expand by precipitated silica or iron as outlined in Chap. 3.

As to practicality, there are considerable problems in crystalline rock because of local strong inflow of water into the working area. It may be so high that placement of the containers has to be stopped and temporary plugs constructed in a hurry, all under very difficult conditions with the floor covered by sticky and slippery clay emanating from wetted supercontainers. This problem is expected to appear in many long deposition holes in this type of rock and can only be solved by very effective pregrouting or by tightening the rock by freezing before the whole operation with placement of containers starts. The most obvious difficulty is the risk that the heavy supercontainers become stuck in the placement phase because of failure of the sophisticated machine, rock fall or delay caused by unexpected strong water inflow. In summary, the KBS-3H, and other "very long hole" concepts in crystalline rock, are not attractive. They are applicable in clastic clay if a strong liner has been constructed like in the HADES URL, which would also be required in salt rock because of the quick creep and unsatisfactory stability. The concept of very long holes, lined or unlined, is applicable to argillaceous rock, however, provided that the holes are kept effectively ventilated until the containers are placed in the unlined case. The risk of rock failure in the waste placement still suggests that the holes should be short.

As to cost, the concept would give cheaper disposal than the competing ones where it can be realized but for crystalline rock the real cost may be at least the same as for KBS-3 V or higher if sealing of the rock requires comprehensive grouting or freezing.

4.4.7 The "Optimal" Concept

The search for a simpler and safer buffer placement technique than for the KBS-3 V concept has led to the KBS-3i version, keeping in mind that canisters require remote handling and radiation protection. In fact, placement of true 25 t canisters with HLW under realistic conditions in 9 m deep holes with nearly 2 m diameter has not yet been made, nor has filling of pellets according to the KBS-3 V concept been made under realistic conditions. The pellets are required for reaching a sufficiently high ultimate density of the buffer but would preferably be replaced by using a clay mud with sufficient density.

Problems and risks are manifold in realizing the KBS-3 V concept. In addition to the recently identified need of cutting 45° ramps at the floor for getting the canisters down in the holes, which changes the stress conditions with not yet fully realized rock mechanical consequences, one identifies the following potential troubles: (1) unexpected strong inflow of water in the holes with difficulties in placing clay blocks and canisters, (2) fall of clay blocks and canisters, (3) fall of clay fragments that prevent the clay blocks to come in contact, (4) fall of clay fragments that make blocks stuck before they are down. The risks can be minimized by applying the "supercontainer" concept and inclining the deposition holes according to the KBS-3i concept by providing mechanically tight confinement of canister and clay blocks, and by supporting the rock by using mud instead of pellet fill.

We have already seen the basic features of the KBS-3i concept, and will examine it a little more from the point of short- and long-term performance, paying particular attention to practicalities. The basis of the assessment is the physical model shown in Fig. 4.20 and the general description provided earlier in the text. As said, the concept is very much alike the KBS-3 V concept and merely includes improvements that are required.

4.4.7.1 Design Principle

The basic principle of the KBS-3 V concept is technically very sound and is applied also in the KBS-3i concept. However, the latter is superior because of the use of supercontainers, which makes handling and transport of canister and clay rational and safe. It is of course required that the clay can migrate through the metal cage of the supercontainer and embed it in homogeneous form, the driving force being the suction potential of the dense clay blocks inside the cage. The expansion of the buffer clay is associated with migration of water in the opposite direction, i.e. through the expanding clay. The process is complex but the major involved processes have been identified and theoretical models developed for predicting the maturation rate (Ch. 3).

The problem with KBS-3 V is that the criteria for selecting acceptable positions for the deposition holes requires insignificant inflow of water in the holes, which will cause variation in maturation along the column of clay blocks tangentially. The difficulty with little inflow of water in the holes, it may be less than one liter per day, is that desiccation and evaporation can delay saturation of the buffer associated with permanent loss in expandability. The KBS-3i concept largely reduces this problem by providing a clay mud in the deposition holes. The mud gives off water uniformly to the dense clay blocks, which should be made with the same rigorous criteria as for KBS-3 V concerning accessory minerals (pyrite, carbonates etc). The mud also has the purpose of improving the heat conductivity of the system and of providing the rock with mechanical support in the placement phase. Naturally it also supplies the buffer with clay substance to reach the required density.

Early maturation of the buffer in deposition holes of KBS-3 type causes the water pressure to rise in the lower part of the holes and the whole package of blocks and canister can be lifted and begin to move up into the tunnel backfill. The pressure is released by flow of water from the bottom through the pellet fill in which channels are formed that can be permanent. If the buffer becomes tight enough to resist piping before the tunnel backfill has become water saturated and has its porewater pressure approaching that in the rock, the force on the lower end of the clay blocks will exceed the weight of the canister and blocks by orders of magnitude and result in large deformation of the clay blocks, ruining the coherence of the whole EBS. The problem exists also for the KBS-3i concept but the supercanister will confine the clay and canister and minimize problems with possible lifting. For totally avoiding the risk of significant upward displacement of the EBS in the deposition holes they must be secured by casting concrete plugs that extend into reamed recesses in the rock, applying a technique worked out for borehole sealing that we will look at later in the chapter. Under all circumstances it is required to backfill the tunnel immediately after having brought down each individual supercanister.

4.4.7.2 The Canister

It is no demand that the canisters should be of the HIPOW type, which would eliminate all possible risks of leakage of radionuclides from the canisters at any time, so the ordinary presently proposed KBS-3 copper/iron canisters, surrounded by tightly fitting clay blocks in supercontainers, may well be used in a KBS-3i repository.

4.4.7.3 The Supercontainer

It is essential to realize that if the supercontainer units are prepared so that there is very good fitting of the dense blocks in the supercontainer as well as mutually, the units should become very rigid. Copper is naturally a possible candidate material for manufacturing supercontainers and it may well be possible to make them strong and rigid enough for being placed in the inclined deposition holes with suitable support and guidance. Titanium is an expensive alternative which is significantly lighter than copper (4500 kg/m^3), and as chemically stable as copper, having a tensile strength that significantly exceeds that of copper. However, Navy Bronze is proposed to be used. It is an alloy weighing 8930 kg/m^3 with 90–95% Cu and 5–10% Zn and has a chemical stability that is believed to be similar to that of copper. Its tensile strength is close to that of steel [12].

The perforation ratio of the cage shall be at least 50% for giving the clay inside and outside the supercontainer ultimately a similar, high degree of homogeneity and density. It is probable that the use of round holes will lead to some permanent differences in density of the clay surrounding the containers, and a cage with coarse bars with square-shaped openings and a 50–100 mm aperture would be ideal.

4.4.7.4 The Buffer

A number of large-scale field tests simulating the hydrothermal conditions in a deep repository in crystalline rock have indicated that montmorillonite, which is the major smectitic component in several of the buffer materials that are presently considered by the Organizations, remains chemically unchanged in experiments with lower temperature than about 90°C when the groundwater clay is of low-electrolyte type. Both theoretical degradation models and natural analogues show, however, that such buffer clay will loose a lot of its isolation capacity with time and it is not known whether it performs satisfactorily even for a very limited period of time. For widening the margin to critical conditions and for economical reasons one can think of using other buffer materials that can sustain higher temperatures and there are at least three common species available in sufficiently large amounts to be used commercially: saponite, palygorskite and mixed-layer clays like Friedland Ton. We will consider them here briefly and take as a common basis the data in Table 4.3, which shows laboratory data of the three clays for two different electrolyte contents in the porewater and two densities, 1800 and 2000 kg/m^3 in water saturated form, of

Table 4.3 Typical properties of some candidate buffer materials

Clay	Density at water saturation kg/m^3	Hydraulic conductivity, m/s*	Swelling pressure, MPa*	Cation exchange capacity, meq/100 g
Montmorillonite-rich (MX-80)	1800 2000	9E-13/2E-11 4E-13/2E-13	0.60/0.20 7.30/4.70	120
Saponite GeoHellas SA	1800 2000	E-12/5E-12 5E-13/E-12	2.50/2.00 8.80/5.00	110
Mixed-layer (Friedland Ton), Frieton GmbH	1800 2000	5E-11/E-10 4E-12/2E-11	0.15/0.10 1.00/0.65	60

* First figure represents saturation with distilled water, lower saturation with 3.5% $CaCl_2$.

which the higher is representative of buffer clay in HLW repositories and the lower representing the density of clay component in mixtures of clay and ballast.

The MX-80 clay has been selected as reference buffer material by several of the Organizations for two major reasons: clays of this type have an extremely low hydraulic conductivity for the density required for buffer clay, and its strong expandability provides a high degree of sealf-healing and tight contacts with the rock. A further reason is that montmorillonitic clays are commercially available at a reasonable price in nearly all parts of the world. In all these respects montmorillonite-rich material is superior to the two other candidates buffer types but saponite performs almost as well. The mixed-layer clay is clearly more permeable and less expandable but has the advantage of being less sensitive to salt water respecting hydraulic conductivity and expandability, and be less expensive. For use as buffer the density at water saturation should be higher than 2000 kg/m^3. For montmorillonite and saponite used as buffer the swelling pressure may be critically high if its density happens to be higher than planned and if expanded canister corrosion products contribute to the net pressure. In this respect the mixed-layer clay represents much less risk. The most important issue is, however, the chemical stability of buffer clays and since saponite is considered to be more stable than montmorillonite it should be ranked higher than the lastmentioned. Mixed-layer clay of type Friedland Ton has less content of vulnerable components than pure smectites and is therefore judged to be more stable than these. It should therefore be ranked high with respect to chemical stability and is a top candidate if its physical properties are acceptable.

4.4.7.5 The Clay Mud

The mud in which KBS-3i supercontainers are proposed to be submerged has four major missions: (1) to serve as lubricant for minimizing friction in the canister placement phase, (2) to contribute to the amount of clay in the buffer, and (3) to prevent erosion and loss of clay material from the buffer expanded into fractures in the rock surrounding the deposition holes (cf. Fig. 4.22), and (4) to provide support to the rock for minimizing minor rock fall. As for the "deep" hole concept the main idea is to use a mud with highest possible density and a sufficiently high smectite content for reaching a net hydraulic conductivity of the buffer that can be accepted, and also with a markedly thixotropic behaviour with low shear strength at agitation. The oil and gas companies have huge experience in using clay mud for boring and sealing holes and they have also extensive knowledge in composing muds that have the required properties.

A number of muds have been tested in the laboratory as briefly described here. The idea was to mix highly compacted clay pellets in a mud with relatively large or strongly anisotropic particles for making them form a filter-cake at the aperture for retarding or halting migration of smectite particles from the dense pellets. A pellet fill of montmorillonite-rich, dense pellets (MX-80) has a dry density of 1000 kg/m^3, i.e. about 1650 kg/m^3 at water saturation, which corresponds to a hydraulic conductivity and swelling pressure of E-11 m/s and 100 kPa, respectively, for brackish groundwater. With the voids in the pellet mass filled with a mud with a density of about 1200 kg/m^3 the average density will be raised to 1750 kg/m^3 and it will be fairly homogeneous since the mud is expected to be consolidated by the expanding pellets. Three muds were used in a representative test series:

1. Mixture of 50% palygorskite and 50% silica flour. Tap water added to 1260 kg/m^3.
2. Mixture of 50% MX-80 clay powder and 50% silica flour (particle size smaller than 25 μm). Tap water added to 1200 kg/m^3 density at water saturation.
3. Mixture of 40% Na montmorillonite-rich bentonite and 60% crushed granite with Fuller-type grain size distribution ranging between 0.02 and 2 mm. Tap water added to 1300 kg/m^3 density at water saturation.

The palygorskite clay contained 15% saponite and 5% quartz. The average hydraulic conductivity of palygorskite clay with a density at saturation of 1550 kg/m^3 is less than 2E-11 m/s and this density was expected to be reached by consolidation under the swelling pressure exerted by the expanding MX-80 pellets. Figure 4.23 shows the tubular shape of the particles, which, like zeolites, contains free water that flows more readily than in the interlamellar water of montmorillonite. The hydraulic conductivity of palygorskite is therefore higher than that of clay consisting of the lastmentioned mineral. In contrast, however, saponite exerts a higher swelling pressure on the confinement and is believed to be more chemically stable.

The tests involved filling the muds in plastic tubes with rows of 2 mm slots and submerging the tubes in containers in which freshwater was pumped around continuously for several weeks. Immediately after starting the respective test it was obvious that clay material was at once released from the muds and eroded, forming particles with a size ranging from a few hundred micrometers to a few millimetres that migrated through the slots. This was obvious even for an experiment with MX-80 pellets in tap water, which hence indicates that the risk of dispersion and significant loss of clay material by erosion into fractures in the surrounding rock may not be high. Figure 4.29 shows a sectioned tube with MX-80 pellets immersed

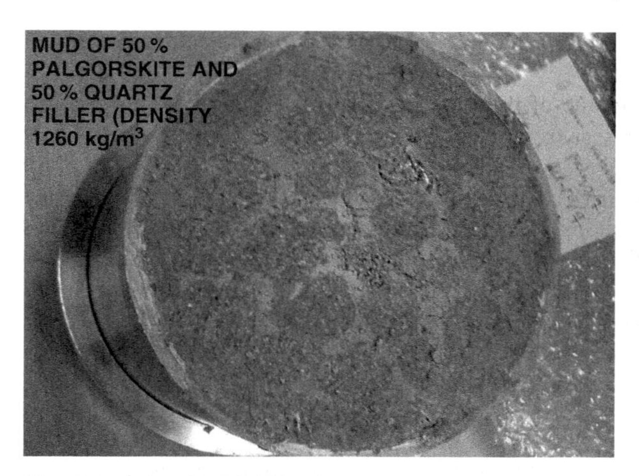

Fig. 4.29 Tube with MX-80 pellets embedded in mud of 50/50 palygorskite and silica flour sectioned after 10 days of exposure to tap water via 100 slots with 2 mm height and 30 mm length in the tube. The water was pumped to flow at a rate of 50 mm/s around the tube. In the central part the pellets have sucked water from the mud causing desiccation of the mud and expansion of the pellets

in mud of 50% by weight of palygorskite and 50% silica flour. As predicted by applying current theories for calculating the rate of maturation of the mud-saturated pellet fills in the tubes they were far from complete after 10 days in their central parts where the pellets had sucked water from the mud causing desiccation of the mud and expansion of the pellets. Figure 4.30 shows the excellent degree of homogeneity of pellets in a mud of 50% by weight of MX-80 and 50% silica flour. As expected, the

Fig. 4.30 Tube with MX-80 pellets embedded in mud of 50/50 MX-80 and silica flour sectioned after 40 days of exposure to tap water flowing around the tube surface at a rate of 50 mm/s. The plug appears to be very homogeneous and the pellets can hardly be discerned

plug appeared to be largely water saturated and homogeneous when the maturation periods were long (40 days).

The most important outcome of the tests was that the erodablity, expressed as the average rate of material loss from the pellets immersed in mud, was three times higher for the case with pellets in tap water, i.e. with no mud, than for the case with pellets in mud of 50/50 MX-80/silica flour. This confirms the hypothesis that coarser, "inert" particles, making up a considerable part of the mud, minimizes loss of clay from dense smectite-rich objects, like pellets in the described experiments and compacted clay blocks in deposition holes of KBS-3 type. The filter-cake principle was hence documented by all the tests.

4.4.7.6 The Backfill

As outlined earlier in this chapter different techniques are preferably used in deposition and transport tunnels, and in ramps and shafts. For the optimal repository one can think of a special backfilling of the deposition tunnels, namely placement of a masonry of highly compacted blocks of a clay of moderate expandability, like the mixed-layer Friedland Ton (Figs. 4.16, 4.31), and filling the space between the masonry and the rock with a mud with fairly high density and capability for easy pumping. It is termed grout when used in tunnels and can be similar to the muds described above although with cheaper materials, like mixed-layer S/I clay minerals. The thixotropic strength regain should be high for quick stiffening and grouts based on Friedland Ton mixed with 5% slag cement and strongly compacted MX-80 pellets and with a water content of about 120% seem to be optimal. They exhibit an increase in shear strength from 1 kPa to 10 kPa in one day and to 100 kPa in one week. Easily established "temporary" plugs of steel in prepared positions with keyed-in concrete frames for supporting the steel plates that are equipped with peripheral rubber bladders are required for creating confinement of 25–50 m long tunnel segments that are backfilled at a time.

When applied on site, the grout in the gap between rock and compacted blocks will consolidate and stiffen quickly. It is expected to resist piping if the pressure of

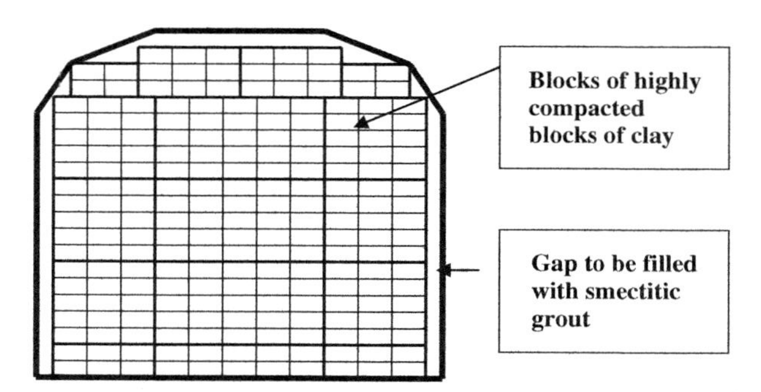

Fig. 4.31 Clay block masonry. The gap between rock and clay blocks can be filled with clay pellets following SKB's present concept, or with grout

water flowing in from rock fractures is not too high and spot-wise inflow rates do not exceed about 0.2 l/minute. Systematic tests with grouts of different compositions are required to assess the potential value of the method. Tests with Friedland clay in contact with saturated cement water have indicated considerable resistance to chemical changes of the clay as reported in Chap. 3.

4.4.7.7 Backfilling of Other Rooms Than Deposition Tunnels and Holes

The following criteria for selection of suitable backfills have been defined by SKB:

- The hydraulic conductivity of backfills in tunnels and rooms connected to deposition tunnels should not exceed that of the surrounding rock, interpreted as the conductivity of the EDZ.
- The backfill must be able to support the roof and walls of the room. Considering common rock structural features the weight of potentially unstable blocks requires that the backfill exerts a pressure on the rock of at least 100 kPa.
- The backfill must not contain minerals or other constituents that can have a negative influence on buffer or cementitious components like concrete and steel plugs.

The block-and-mud backfill with the blocks made of mixed-layer clay like Friedland Ton, and filling the space between rock and blocks with highly compacted pellets, as implied by SKB's concept KBS-3 V or with mud as proposed for KBS-3i, will fulfil the criteria. For other rooms, simpler techniques like the one illustrated in Fig. 4.11, can be employed and a cheap and technically good solution can be to use granulated Friedland Ton compacted on site to a dry density of about 1425–1475 kg/m^3 corresponding to a density at saturation of slightly more than 1900 kg/m^3. It would also fulfil the requirements when used for backfilling of blasted tunnels since its hydraulic conductivity is lower than that of the EDZ of the surrounding rock and the swelling pressure at least 100 kPa. Field tests indicate that, with properly selected water content and granulometry, this density can be achieved using 400 kg vibratory plates (Fig. 4.32) but more effective equipment like sheep- or padfoot vibratory rollers (Fig. 4.33), or heavy stamping tools of the type once used long ago for the construction of German Autobahns are recommended. Table 4.4 serves to form a basis of selection of useful clay barriers including buffers. Materials suitable as backfills in other rooms than those hosting HLW canisters have bold figures.

The Lithuanian clay, which is of Triassic age and has some 25% montmorillonite content, was identified in the course of development of the "Near-Surface-Repository" for Lithuanian LLW and MLW. It is being mined for cement production (Fig. 4.34) and has been investigated in the lab and in the field. Compaction of 10–12 cm thick layers by 10 runs of 400 kg vibrating plate gave a dry density for unprocessed "raw" material with about 20% water content of 1680 kg/m^3, corresponding to more than 2000 kg/m^3 at water saturation.

Fig. 4.32 Vibratory plate held and manouvered by a tractor in SKB's slope compaction tests [13]

4.4.7.8 Detailed Strategy for Optimal Backfilling of Tunnels, Rooms and Shafts

A very important matter is how backfilling should be made of the parts of tunnels and shafts that intersect permeated fracture zones. They offer difficulties in the construction and waste placement phases by bringing in water into these parts and further to adjacent rooms, and they serve as major transport paths for groundwater after closure by which the geochemical conditions can be quickly altered, and for possibly released radionuclides.

The basic principle is to backfill all rooms, tunnels and shafts such that they get the same transmissivity as the original rock. This ambition is followed in the current development of techniques for borehole plugging and would imply that tight backfill is placed in the parts of tunnels and rooms that are not intersected by water-bearing fracture zones, and that the space where such zones occur is backfilled with material

Layer	Dry density, kg/m^3	Density at water saturation, kg/m^3	Water content, %	Hydraulic conductivity, m/s
1st layer	1500	1945	29	1.9E-10
2nd layer	1550	1975	26	2.1E-10
3rd layer	1510	1950	29	1.1E-10

Fig. 4.33 Compaction of suitably wetted Friedland "Clay noodles". *Upper*: Placed before compaction, *Center*: Compaction using a 5 ton vibratory pad-foot BOMAG roller. *Lower*: Homogeneous state reached after 10 runs. The table shows the results of the compaction efforts [18]

Table 4.4 Hydraulic conductivity and swelling pressure of proposed commercially available candidate clay-based materials saturated with 3.5% $CaCl_2$ solution. K is the hydraulic conductivity, and p_s the swelling pressure

Material	Dry density, kg/m^3	Density at saturation, kg/m^3	K-Lab, m/s	K-Field, m/s	p_s–Lab, kPa	p_s–Field, kPa
MX-80-type	Content of expandable clay minerals (montmomrillonite) >70 %					
	1110	1700	E-10	5E-10	100	50
	1270	**1800**	**3E-12**	**6E-11**	**600**	**300**
	1430	1900	E-12	5E-12	1500	750
	1590	2000	3E-13	E-12	4000	2000
	1750	2100	5E-14	2E-13	10000	5000
Friedland	Content of expandable (mixed layer + montm.) clay minerals 40%					
Ton	1270	1800	E-9	5E-9	75	35
	1430	**1900**	**2E-10**	**E-9**	**200**	**100**
	1590	2000	5E-11	3E-10	600	300
	1750	2100	<E-12	5E-11	2000	1000
Lithuanian	Content of expandable clay minerals 25%					
Triassic clay*	1200	1800	1.2E-9	6E-9	<50	<10
	1400	1900	E-10	5E-10	<100	<50
	1550	2000	E-11	5E-11	100	50
	1720	**2100**	**E-12**	**5E-12**	**200**	**100**

that can be more permeable than the surrounding rock but remains physically stable. The principle is illustrated in Figs. 4.2 and 4.17 for the entire repository and is justified by the fact that groundwater will anyhow flow through the fracture zones around the tunnels and rooms making tightening of the permeable parts of the rock unnecessary. The material considered is the so-called silica concrete with the recipe given in Table 4.1. Experience has shown that addition of 10–20% of suitable graded crushed quartzite improves the mechanical strength and stiffness. The aim is to

Fig. 4.34 Stockpiled smectitic Lithuanian clay

minimize the cement content for limiting pH and for maintaining physical stability of the fill even if complete dissolution and loss of the cement component would take place. Ordinary concrete mixers can be used for the preparation.

There appears to be four possibilities to minimize or eliminate critically strong inflow of water in the backfilling phase:

1. *Reduction of the water pressure in fractures intersecting the tunnels*. This can be made by drilling nearly axial ("small angle") holes around the tunnel periphery and keeping them drained until the backfill has proceeded to their outer ends after which they can be plugged by use of any of the techniques that are available and described later in this chapter. In this context it is important to know how to select the most suitable excavation technique: TBM technique largely preserves the rock structure and the water pressure and flow capacity of fractures that are more or less perpendicular to the tunnel, while drill-and-blast causes an EDZ that has a much higher hydraulic conductivity than the undisturbed rock and hence yields a much lower water pressure in the high number of interacting fractures.
2. *Sealing of the rock by grouting*. This can be made by injecting suitable cementitious or cement/clay grouts in long boreholes extending from the front of the advancing tunnel in the construction phase, and/or around the periphery for reducing the water pressure in a first phase and grouting them in concjunction with the backfilling.
3. *Sealing of the rock by freezing*. The most radical, effective – and expensive – method of sealing pervious rock is to apply freezing. This can be made by inserting freezing tubes in long boreholes around the periphery as for reducing the water pressure.
4. *Increasing the density of the backfill*. The possibility to effectively increase the bulk density of the backfill and thereby to effectively reduce the void sizes and the risk of piping and erosion depends on what field compaction technique and equipment that can be chosen and that is more effective than ordinary vibrating rollers. For this purpose sheep- or pad-foot vibrating rollers can be considered and also heavy dynamic compaction machines. The problem to get sufficient density at the roof will still probably remain and blocks of highly compacted smectite-rich clay will therefore have to be placed there.

Backfilling of deposition tunnels of KBS-3 type with clay blocks and injection of smectitic grout between pre-constructed plugs is not expected to cause severe problems, while backfilling of granulated smectitic material, like Friedland Ton, forming a moving slope may be very difficult because of too quickly inflowing water that can transform the backfill to a slurry. The question is whether piping and subsequent erosion is caused by a critically high pressure of the water that penetrates the backfill, or whether a sufficiently high hydraulic gradient can cause piping. Figure 4.35 illustrates the conditions in the backfilling operation: Case I represents the phase when the front of the filling is temporarily in tight rock, while in Case II it has just passed a water-bearing fracture in the rock. In the first case physical stability is not jeopardized by water entering the backfill from contacting rock but water dripping from the roof on the slope under construction will "liquify"

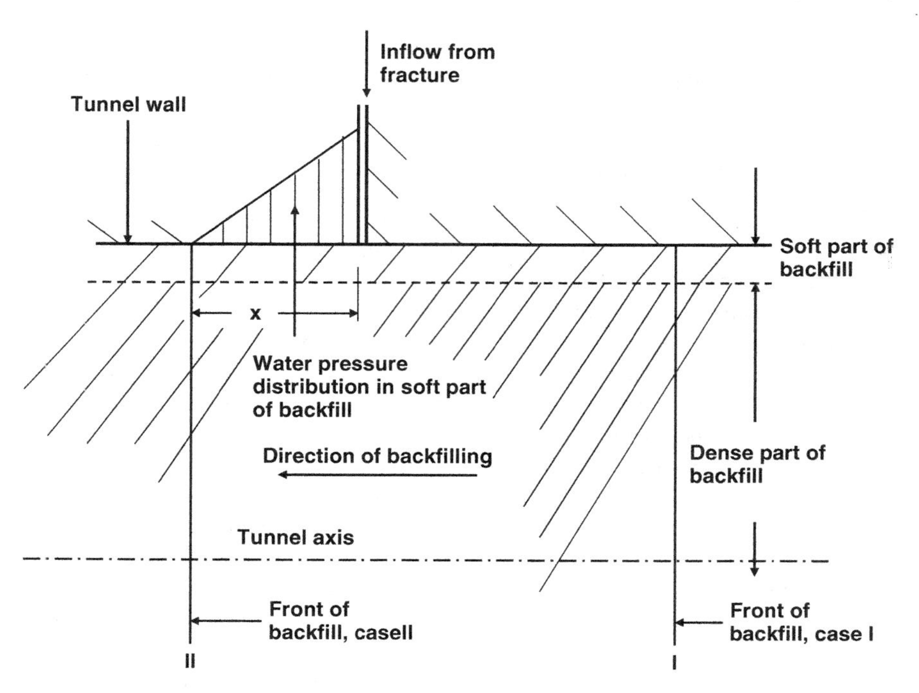

Fig. 4.35 Conceptual model of inflow of water from fracture into tunnel backfill

it to a few centimeters depth. In the second case water enters the backfill through a fracture in the rock and can displace the backfill material by forming a channel that appears at the surface of the slope. If the backfill has an initial high degree of water saturation piping can be assumed to take place when a critical gradient is formed.

Applying the conclusion from laboratory investigations on saturated smectitic clay that piping and erosion is initiated at a hydraulic gradient of 30–100 depending on the density [4], one can estimate the risk of such distortion as a function of the distance x between the free end of the grout and the nearest, passed water-bearing fracture. The obvious problem with predicting the distribution and extent of water inflow into tunnels for selecting suitable backfilling rates and for handling unexpected flooding and comprehensive erosion speaks in favour of using thixoptropic clay mud for filling the space between block masonries and the surrounding rock in tunnel segments confined between plugs. The big advantage of this procedure is that inflow of water from the rock causes only small, local hydraulic gradients, that displacement and erosion of the mud can not be very significant, and that the potential for self-healing is good if the density is not too low.

4.4.7.9 Plugs

Regardless of the emplacement strategy, some access and transport tunnels have to remain open for a long time. This period can last for many years and possible need for monitoring of various processes in the repository before final closure may further extend it. The intersection of waste emplacement rooms/drifts and the access tunnels and rooms need to be physically closed in order to ensure that the canisters remain undisturbed and that no undesirable hydraulic conditions are allowed to develop within the backfilled volume. However, they still need to be accessible and it is therefore essential to apply a plug design and construction technique that make both permanent sealing and access to these rooms possible. As a result of these requirements, the generic guidelines specified by most national programs considering deep geologic repositories call for the installation of plugs at the junction of the emplacement room/drift and the access/transport tunnels, leaving open what design shall be selected.

In order to distinguish between constructions that are intended to provide short- or long-term performance, the terms "plug" and "seal" are used very specifically in this report. "Plugs" are those constructions that are intended to provide mechanical restraint, physical security and hydraulic control functions over the short-term, i.e. repository operational and pre-closure monitoring periods. Constructions that are to prove longer-term performance (post-closure) are termed "Seals". Seals are therefore permanent installations while plugs may or may not have a sealing function over the longer term. Plugs may become Seals at the time of repository closure. The distinction between these two functions may seem subtle but is important when it comes to defining their performance requirements.

This chapter focuses on the role and requirements of the plugs that are to be installed at emplacement room/tunnel/drift entrances or in other locations within the repository that may require temporary mechanical or hydraulic control structures. An example is a section of an emplacement excavation that is unsuitable hydraulically and requires remedial action to prevent unacceptably high water influx to occur into adjacent areas. These plugs are not necessarily a permanent feature of the repository and may, if required, be removed for later installation of a permanent seal. Room/Drift plugs are also by their defined function, physically accessible during repository operation so their performance can be monitored and remedial actions taken if necessary, e.g. increased seepage past the plug.

The construction principles, preferably shotcreting in several layers for temporary plugs and on-site casting of permanent concrete plugs, are known and tested through full-scale tests in the ESDRED project [21] among others, but what remains to be considered is their long-term stability and chemical interaction with adjacent smectitic backfills. The matter has been in focus for a long time since the high pH of cement materials is known to be a threat to smectite minerals. The matter was considered in Chap. 3, from which it appears that concrete with low-pH cement in contact with clays of Friedland Ton type is not expected to lead to practically important degradation of either of the components even in a long-term perspective. For concrete based on Portland cement and backfills of montmorillonite, the long-term chemical stability remains to be demonstrated.

4.5 Construction and Performance of Buffers and Backfills in Salt Rock

4.5.1 General

In principle, repositories in salt rock are similar to those in crystalline rock, they comprise transport tunnels, shafts and ramps and the most important part – deposition tunnels or holes. Only one repository has in fact been licensed and put in operation, the WIPP repository in New Mexico, USA, for LLW and ILW, and we will examine it in some detail in this chapter [22]. We will also consider design principles proposed for HLW.

One distinguishes between two types of rock salt, bedded and domal salt as described earlier in the book, the main differences being that bedded salt commonly has limited thickness and is interfoliated by layers of clay and silt or sand, which causes local high hydraulic conductivity and poor mechanical stability.

In New Mexico, United States, a national repository, "WIPP[1]" (Fig. 4.36) has been constructed and licensed for long-lived radioactive transuranic radioactive waste.[2] Repository concepts have been worked out in Germany for HLW but not substantiated in Germany (Fig. 4.37). The main difference between the waste types is that the waste in WIPP does not produce heat in contrast with the German HLW. While the main focus has been on domal salt in Germany during the past decades it has been on bedded salt in the US. The German concept is not yet completely defined and several versions have been investigated, the following principles being considered:

- Borehole emplacement of vitrified HLW
- Drift emplacement of spent fuel
- Combined drift and borehole emplacement of spent fuel and vitrified HLW

4.5.2 Description of Disposal Concepts

4.5.2.1 Bedded Salt

Following applicable US law and related regulations the post-closure safety of the WIPP repository is determined by the radionuclide containment and the isolation provided by the geology and the engineered barriers (EBS). The fundamental premise is that virtually all the radionuclide containment and isolation required to meet the various performance criteria are provided by the geological media, while

[1] Waste Isolation Pilot Plant.

[2] TRUW.

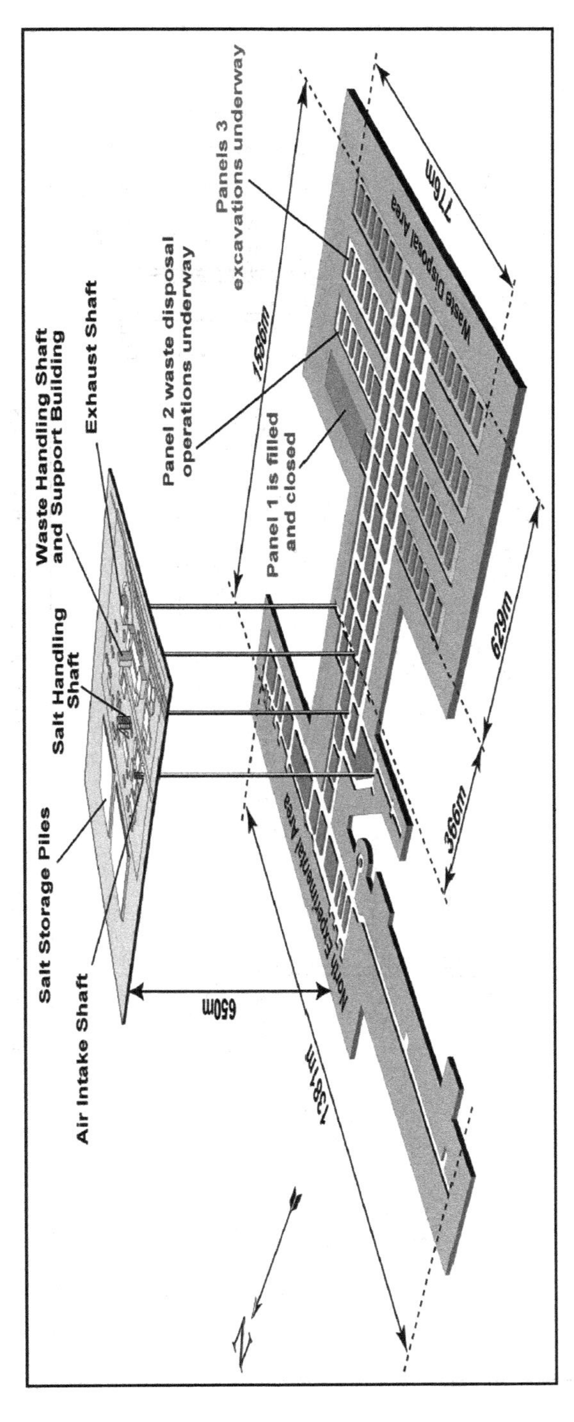

Fig. 4.36 Schematic illustration of surface and subsurface facilities at the WIPP site [22] (By Courtesy of US EPA)

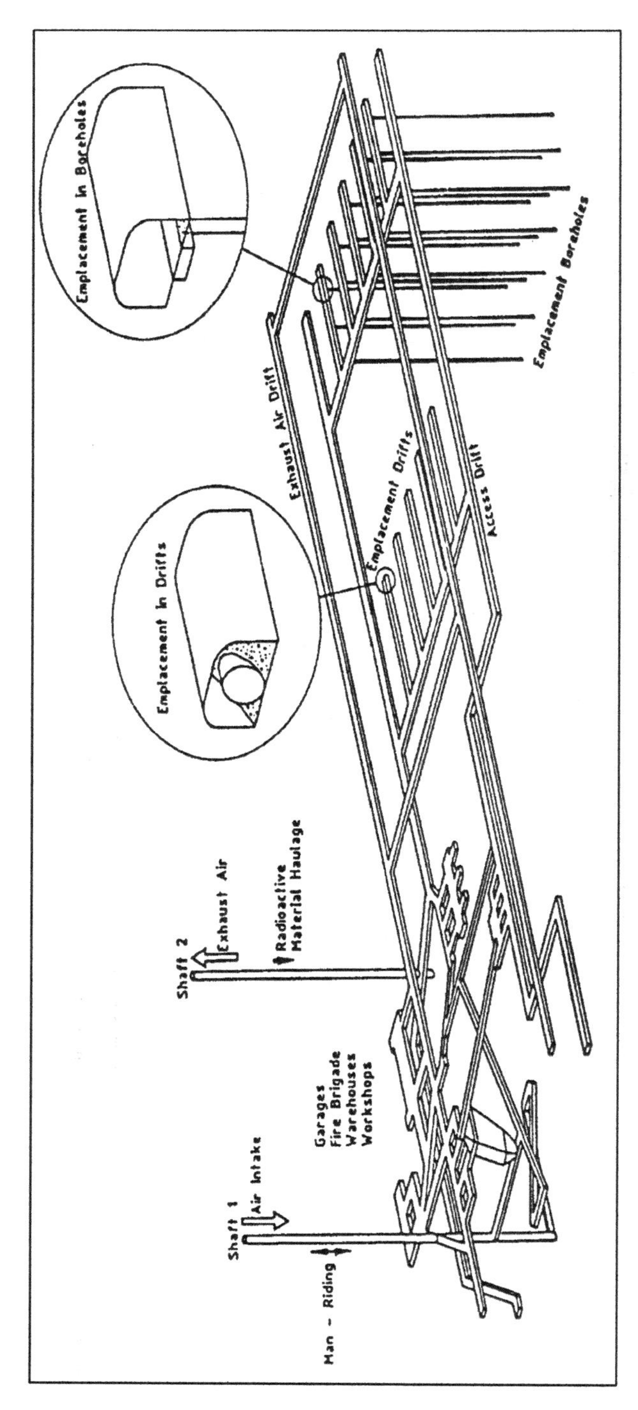

Fig. 4.37 Typical layout of a HLW repository in a domal salt formation in Germany [22] (By Courtesy of GRS)

the post-closure safety and performance of the WIPP repository is governed by man-made barriers, of which only the magnesium oxide (Mg) buffer/backfill is the most important [22].

Similar to the domal salt repository concepts, a particularly important component of the WIPP repository concept is the inherent rheological (creep) characteristics of the salt rock hosting the repository. The closure rates during the first five years after closure are predicted to be ~10 centimeters per year vertically, and ~7 cm per year horizontally, followed by lower rates, i.e. ~6.5 cm per year vertically, and ~3.5 cm per year horizontally. These closure rates will gradually encapsulate the emplaced waste and turn the WIPP repository into a virtually impermeable monolith within 300 years after closure.

4.5.2.2 Domal Salt

German HLW remaining from reprocessing of spent fuel will be vitrified in steel canisters for disposal in 0.6 m diameter vertical boreholes extending down from a disposal drift to about 900 m below the ground surface (Fig. 4.37, 4.38). It is planned to place about 200 canisters in each borehole. The gap between the canisters and the borehole wall will be backfilled with crushed salt for ultimate transfer of the weight of the canister to the surrounding rock mass. Since salt has a very significant creep potential, as was discussed in Chap. 3, the deposition holes will converge and gradually compact the crushed salt, by which complete encapsulation of the HLW canisters will take place within a very short period of time [22].

An alternative German repository concept has been developed for direct disposal of spent fuel according to which large, self-shielding Pollux casks containing LWR elements will be emplaced in horizontal drifts with about 200 m length, 4.5 m width and 3.5 m height (Fig. 4.38). Backfilling is planned to be made with crushed salt.

The permeability of the backfill material is of special importance to the long-term safety of a repository in the event of a brine influx to the repository from the overburden or from an undetected brine pocket in the rock mass. In addition to the

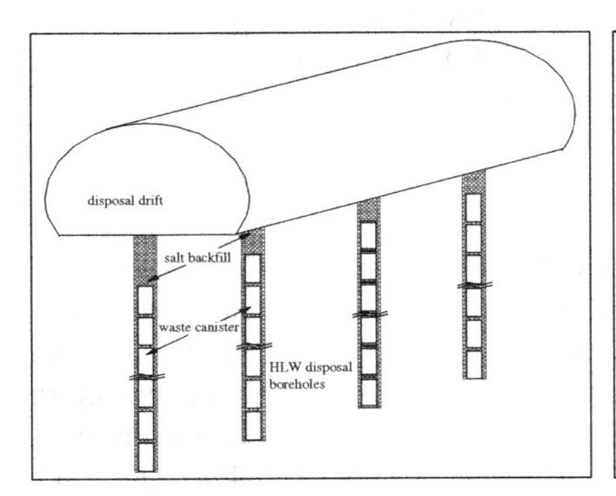

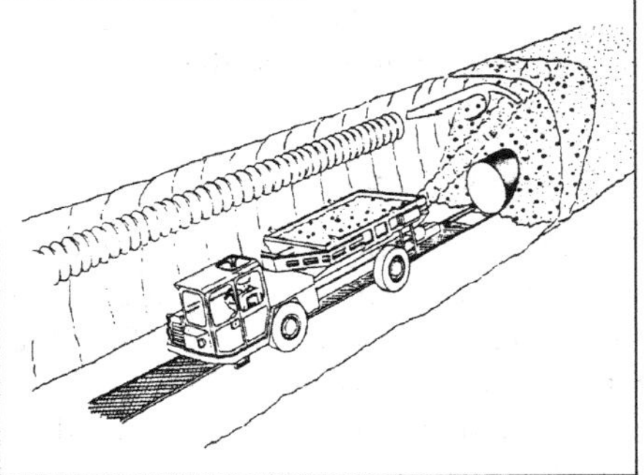

Fig. 4.38 Disposal technique in Germany. *Left*: Vertically placed small canisters in deep boreholes. *Right*: Horizontally placed large Pollux casks [22] (By Courtesy of GRS)

natural compaction of the backfill material caused by the creeping rock mass, the heat released from the emplaced waste will increase the temperature in both the backfill and the surrounding rock mass, which, in turn, will accelerate both drift-closure and backfill-compaction rates. The maximum salt temperature is kept at about 200°C.

4.5.3 Required Function of the Repository

4.5.3.1 Bedded Salt

The WIPP disposal system occupies a volume of 76 km^3 and isolates up to 175 584 m^3 of TRUW for at least 10 000 years, also referred to as the regulatory period. The conditions are as follows:

- If the repository is only affected by likely natural events, also referred to as the "undisturbed case", the maximum annual effective dose to a member of the general public living at the WIPP site boundary may not exceed 0.15 mSv/a.
- If the repository is affected by inadvertent human intrusions with a probability of occurrence equal to or greater than E-8, also referred to as the "disturbed case", the amount of cumulative releases of radionuclides to the accessible environment, i.e., anywhere outside the boundaries of the WIPP disposal system, may not exceed two very stringent, waste-inventory-related limits, i.e., there is not a set radionuclide release limit value for all repository inventories.

As illustrated by the mean and overall complementary cumulative distribution functions, even in the event of multiple borehole intrusions and all other natural and human-induced events with a probability of occurrence equal to or greater than E-8, the predicted mean and overall mean cumulative releases of radionuclides from the WIPP repository during the 10 000 year regulatory period are less 10% of the very strict limits defined by the regulatory authorities.

The environmental conditions of the WIPP site have been monitored since 1974 and the performance of the WIPP repository will be monitored throughout the 25–35 year operational, the 10 year closure, and the first 100 years of the post-closure periods, also referred to as the 100 year active institutional controls.

4.5.3.2 Domal Salt

According to paragraph 45 of the German Radiation Protection Ordinance, the individual dose to humans caused by radionuclides released from a repository must remain below 0.3 mSv/a. To achieve this objective, the required long-term containment and isolation of all kinds of radioactive waste emplaced in a repository have to be provided by the rock and the EBS. In all the domal salt repository concepts for HLW, the host rock represents the main barrier to radionuclide migration. However, as described further below, EBSs are also required to ensure adequate containment and isolation of the disposed HLW. Retrievability is not foreseen in any of the present domal salt repository concepts. Measures to maintain access to the disposal areas are considered contradictory to the long-term safety of a repository in salt rock [22].

4.5.4 Current Repository Design Principles

4.5.4.1 Bedded Salt

The WIPP-repository design principle implies placement of the containers (drums) in tight spacing in fairly large rooms with crushed MgO filled between them [22].

In WIPP a 400 m wide rock volume ("shaft pillar") hosts four shafts with and without lifts. The baseline room-and-pillar repository design of the repository comprises eight panels each containing seven disposal rooms and separated laterally from the adjacent panel by a 61 m wide salt rock pillar. Each disposal room is 4 m in height, 10 m in width, and 91 m in length. It is separated laterally from adjacent disposal rooms by a 30 m wide salt rock pillar. Essentially all underground openings are excavated by means of mechanical mining methods to minimize the extent of the EDZ. A baseline design assumption is that underground openings will be closed by self-sealing through the inherent creep of the repository host rock. Figure 4.39 shows disposal of RH-TRUW canisters for ILW in WIPP in 0.75 m diameter, 4 m deep, horizontal boreholes located 1.2 m above the invert/floor and spaced 2.4 m apart in the walls of the disposal rooms.

The very effective isolation of the waste in bedded salt takes place at a depth of about 660 m and the only pathways for possibly released radionuclides is through the four shafts, which hence are planned to be tightly sealed.

Fig. 4.39 Simulated disposal of RH-TRUW canisters in WIPP [22] (By Courtesy of US EPA)

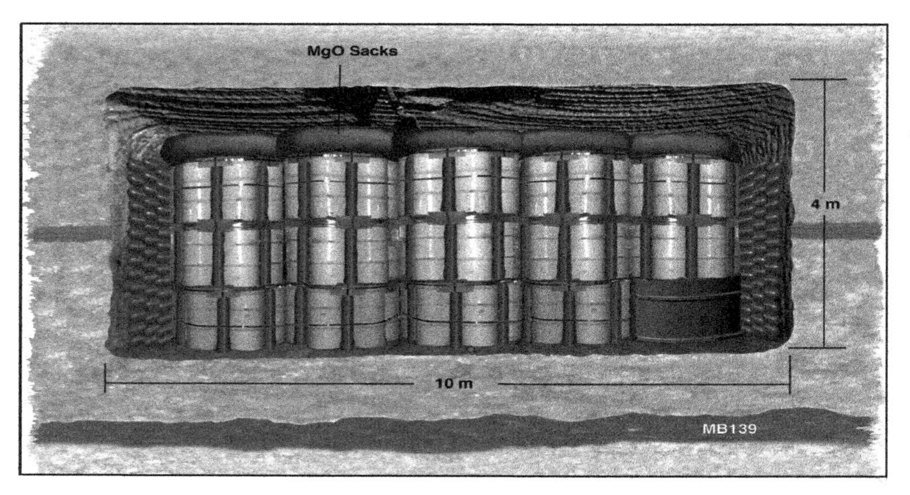

Fig. 4.40 Schematic illustration of the emplacement configuration for CH-TRUW containers and MgO sacks in the WIPP repository [22] (By Courtesy of US EPA)

As illustrated in Fig. 4.40, CH-TRUW (LLW) is emplaced in the disposal rooms. Granulated MgO buffer/backfill is emplaced in portions of the disposal rooms in step with the gradual disposal of CH-TRUW. The MgO buffer/backfill is the only man-made barrier considered an EBS by the US Environmental Protection Agency, EPA.

4.5.4.2 Domal Salt

Criteria used in the conceptual studies for designing a HLW repository in a salt dome in Germany include the following major items:

- Retrievability is not to be considered.
- Two shafts are sufficient for constructing/developing and operating the repository, including all kinds of transport and the ventilation system.
- The mine/underground workings shall have a lateral safety distance to the flanks of the salt dome of at least 200 m and a vertical safety zone beneath the top of the salt dome and the cap rock of at least 300 m. Additionally, a safety area that should not be used for disposal purposes, shall remain to 300 m distance around each shaft.
- All disposal areas are located at the same level.
- Rock material (salt) and waste are transported in different drifts, preferably in a one-way system. However, the number of drifts and other excavations should be kept as low as possible to minimize the required number of drift seals (plugs).

One of the two shafts will serve for bringing fresh air and for personnel transportation, while the other will serve for exhaust air and waste transportation. The final layout of the disposal areas and the drifts depends on the geology at the

Table 4.5 Possible geometrical data of the Gorleben repository concept [22]

Drift disposal of Pollux casks	Interim storage period, years	Cask spacing, m	Drift spacing, m
	15	6	28
	30	1	36
Borehole disposal of HLW canisters	Interim storage period (years)	Borehole spacing (m)	Drift spacing (m)
	10	24.50	21.22
	15	21.00	20.00
	30	17.25	14.94

disposal level, i.e., the distance of the disposal areas from certain unwanted rock types like carnallite and anhydrite seams. The maximum allowable temperature is set at 200°C at the interface between the waste canisters and the backfill, consisting of crushed salt rock. Table 4.5 summarises geometrical data for the Gorleben salt dome governed by the aforementioned design criteria.

In the case of borehole disposal of HLW, a drift cross-section of about 5 m by 6 m is required for operation of the disposal machine, the borehole diameter being about 0.6 m. The 8.5 cm annulus between the canister and the rock is backfilled with crushed salt.

For drift disposal of the socalled SNF/Pollux casks, the drift cross-section will be about 4.5 m by 3.5 m, which is sufficient for the horizontal emplacement of the 1.6 m wide Pollux cask.

4.5.5 Engineered Barrier Systems

The host rock represents the most important barrier within the multiple-barrier concept of a salt rock repository employed in both Germany and the US. However, as concisely described below, notwithstanding this very advantageous inherent characteristic of salt rocks, both the German domal salt and the US bedded salt programmes also benefit from EBSs that vary in design and performance requirements as a function of the types of waste involved.

4.5.5.1 Bedded Salt

The WIPP disposal system includes the same engineered barriers as repositories in crystalline rock, i.e. waste canisters and buffer and backfills. The CH-TRUW waste is contained in mild-steel drums and standard waste boxes (SWBs) that are stacked three layers high in the disposal rooms. The RH-TRUW, which may comprise up to 7 080 m^3, is contained in thick-walled steel canisters that will be emplaced in horizontal holes in the ribs/walls of the disposal rooms. The 208 liter steel drums and the SWBs containing the CH-TRUW will be breached early by room convergence because, in only 50 years after closure, the inherent creep characteristics of the repository host rock will have reduced the height of the unsupported disposal rooms from 4 m to 2 m. The steel canisters containing the RH-TRUW may survive the closure pressure somewhat better, but they are also assumed to be breached relatively early in repository history by corrosion [22].

The primary functions of the MgO buffer/backfill are to consume essentially all carbon dioxide (CO_2) produced in the WIPP repository; and control the disposal room CO_2 fugacity/partial pressure (PCO_2) and brine pH within ranges that result in lower actinide solubility than at the ambient pH. In addition, the MgO is able to consume significant quantities of water in any brine that may seep into the disposal room. However, the MgO mainly contributes chemically rather than physically to the long-term containment and isolation of radionuclides at WIPP by driving the repository pH to about nine, giving a basic chemistry in which the actinides are much less soluble than at the ambient pH. Although the MgO is only placed in a portion of the disposal rooms, sufficient amount of MgO is emplaced to react with all the CO_2 possible in the respective waste disposal room, combining with water to form magnesium carbonate/ magnesite ($MgCO_3$), which is an insoluble mineral.

4.5.5.2 Domal Salt

According to the German disposal concept for HLW as illustrated in Fig. 4.41, the main EBSs to be implemented in addition to the geological barrier, consist of back-fills of salt and plugs (seals, dams). In addition, access and transport drifts will be backfilled with crushed salt to minimise remaining voids in the repository and to assure rock stability as soon as possible after the completion of waste disposal operations [22].

The crushed salt in the HLW boreholes will be adequately compacted by the creep of the surrounding host rock within about 10 years. No additional sealing of the waste emplaced in the borehole is required.

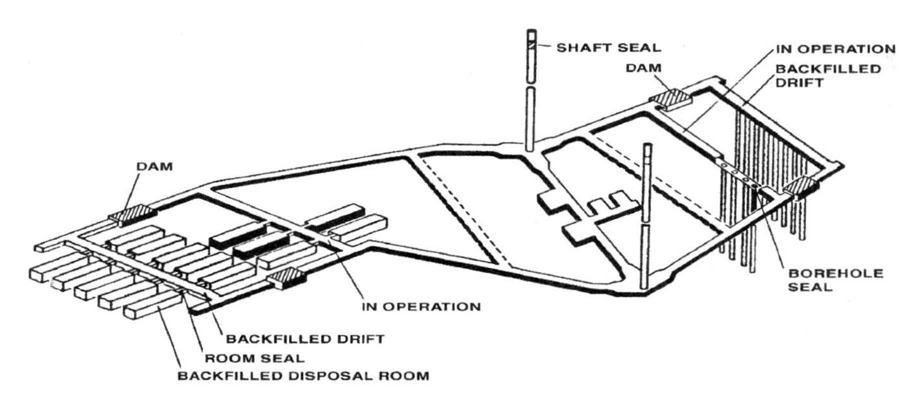

Fig. 4.41 Layout of a domal salt rock repository, including principal seals [22] (By Courtesy of GRS)

A comprehensive performance assessment analysis has shown that radionuclide release into the overburden rock strata and the biosphere is not to be expected if the HLW disposal areas are located closer to the central shaft than the cold disposal areas. The main reason for this is that the heat-producing waste heats the adjacent rock mass, which speeds up convergence and compaction rates of the salt backfill in the borehole seals and drifts in the heated rock mass. The drift connecting the central area and the HLW-area will be closed after 155 years, whereas brine coming from the shaft will reach this drift only after 233 years. The drifts above the HLW-disposal boreholes, into which a limited amount of brine from undetected brine inclusions is supposed to be released, will be closed after 70 years to 80 years, ensuring that the brine coming from the shaft will not reach the disposal areas [22].

4.5.6 Design and Construction

4.5.6.1 Bedded Salt

The design and analyses of the WIPP tests are governed by the large strain and displacements encountered in the salt rock in combination with the elevated temperatures, and by the corrosive environments hosting the tests that were all conducive to instrument failure over the time span of the experiment, which could be as long as ten years or more. Furthermore, salt creep and lack of access to high temperature rooms prevented the replacement of many instruments/gages. Hence, the test designs at WIPP included redundancy of instruments and heaters to assure adequate measurements even if instrument- and/or heater-failures occurred. The WIPP in-situ experiments were frequently preceded by laboratory experiments and always by numerical modelling.

Emplacement evaluations and fluid flow studies provided guidance on the design of the WIPP seal system. Characterisation of the EDZ, which surrounds the excavations and any emplaced plug, was accomplished by gas-flow testing.

The basic construction principle/criterion for the EBSs, as well as the URL and the repository, is to inflict as little damage to the host rock as possible. Hence, virtually all underground openings at the WIPP site were, are, and will be constructed by means of mechanical-mining methods and/or rotary drilling techniques. However, drill-and-blast techniques were used for establishing/developing the initial bailout and equipment-assemblage/storage/staging space adjacent to the shafts at the URL and repository level.

4.5.6.2 Domal Salt

No specific design or construction criterion is needed for the crushed salt backfill used as sealing material in the HLW disposal boreholes and drifts. However, the grain distribution of the crushed salt backfill needs to be controlled to avoid emplacement problems, especially in the narrow (8.5 cm) annulus between the waste

Grain size	Sieve passage
mm	%
31.5	100
16	97.1
8	86.89
4	66.79
2	39.29
1	20.53
0.5	11.26
0.25	6.49
0.125	2.97
0.063	0.49
0.001	0.07

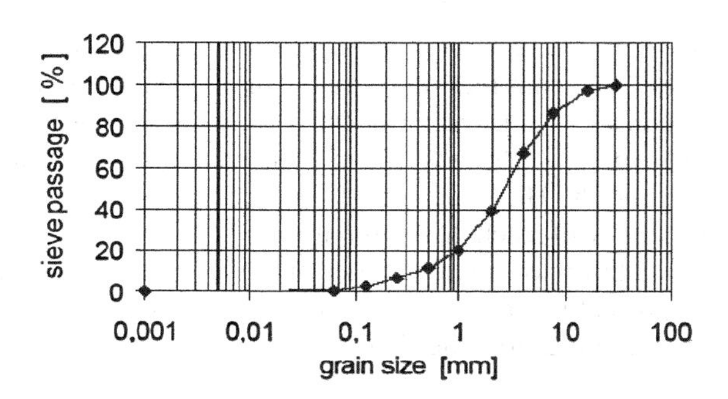

Fig. 4.42 Grain size distribution of crushed salt from mining operations at Asse mine [22]

canister and the host rock in the deep HLW-disposal boreholes. Typical data on the drift backfill produced during the continuous, mechanical mining of the TSDE drifts in the Asse mine/URL are shown in Fig. 4.42, with the maximum grain size being 31.5 mm. However, in the DEBORA borehole annulus experiment, the largest grain size of the crushed salt emplaced without problem in the narrow annulus between the canister and the borehole wall was 10 mm [22].

Plugs separating disposal and other rooms will be installed after the completion of the disposal operations. A preliminary drift-seal design (Fig. 4.43) has been developed. The design includes a short-term seal of sand-asphalt, which is required in the early stage after the installation of the main seal ("dam") to prevent inflow of liquid solutions; and a long-term seal, consisting of pre-compacted salt bricks, and an abutment of salt concrete providing stability against brine pressure in the unlikely case of a complete flooding of the repository. The sealing function of the long-term seal increases with time by its compaction due to drift convergence.

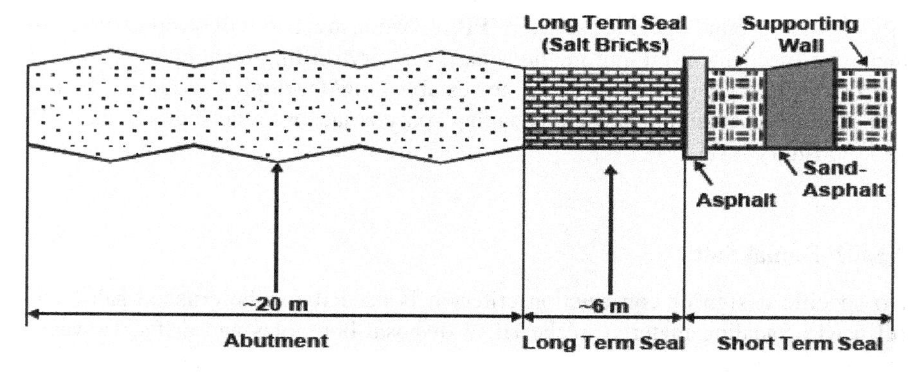

Fig. 4.43 Design principle of a drift seal consisting of salt and asphalt [22] (By Courtesy of GRS)

4.5.7 Conceptual and Mathematical Models

4.5.7.1 Bedded Salt

Recognition that epistemic uncertainty exists within the geologic system, physico-chemical processes within the host rock and disposed waste, as well as the stochastic uncertainty associated with some of these same processes, as well as scenarios for long-term performance, a probabilistic approach is used at WIPP to assess overall disposal/repository system safety. This approach required evaluation of several different scenarios and included the variation of more than 50 near-field parameters, all implemented within the context of the system model for WIPP disposal system. The procedure is summarized in Figs. 4.44 and 4.45.

The WIPP conceptual model comprises a system model, linking both near-field and far-field processes directly through time-dependent processes. The system model consists of a set of models, each of which simulates a principal physical component of the WIPP disposal system. A set of models is used rather than a single, 3-D system model because it would be neither temporally feasible nor financially possible to build such a complete, complex model. In addition, implementing the system model in a set of principal physical models enhances the ability to develop and refine the various models. This flexible approach permits complete separation of near-field and far-field models, while maintaining appropriate interfaces between regions.

The principal physical models used at WIPP are known as "process models" and consist of sets of mathematical equations describing the hydrological, mechanical, and chemical behavior of the disposal system embodied within numerical simulators (codes). Many models derive from widely held theories regarding the relevant physical processes, while others were developed for unique processes occurring at the WIPP site. Although the process models contain significant detail relative to the coarse assumptions frequently made during preliminary site assessments, practical considerations still require a degree of abstractions to move from specific physics to repository-scale processes [22].

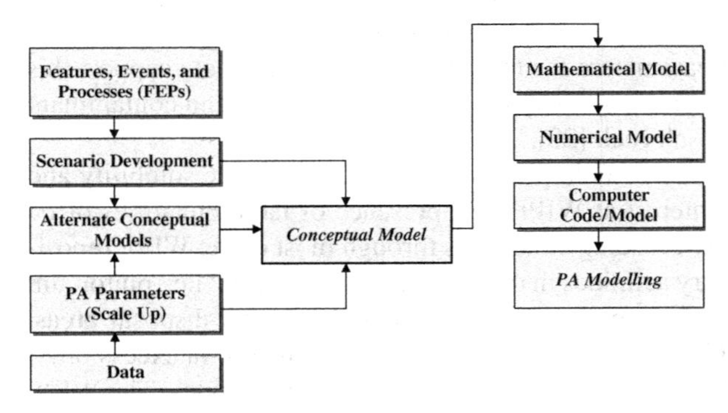

Fig. 4.44 Logic for the progression of code/model development and modelling at WIPP [22] (By Courtesy of US EPA)

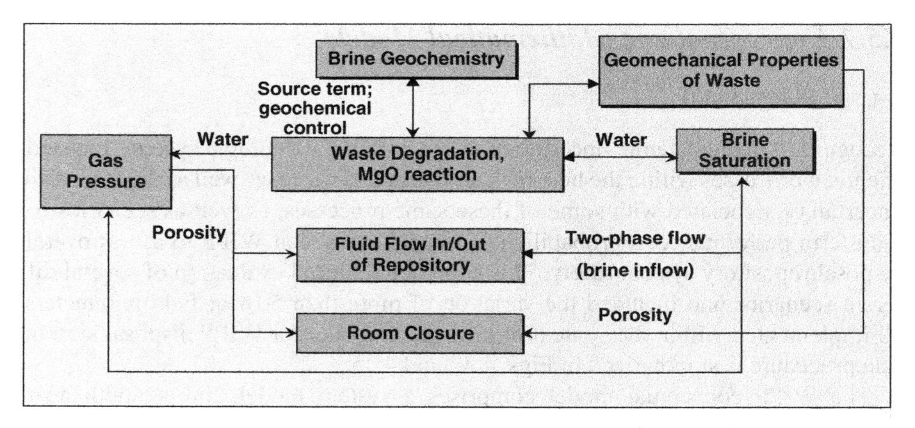

Fig. 4.45 Schematic illustration of major codes, code linkages, and flow of numerical information used in the WIPP [22] (By Courtesy of US EPA)

The WIPP process models were developed during the site characterisation and/or experimental phases. Extensive site characterisation activities, including laboratory and field experiments, supported the iterative model development and refinement process. Laboratory experiments assessing the potential for corrosion of ferrous (Fe) and non-ferrous metals present within the disposal areas and biodegradation of organic materials under WIPP-specific conditions proved of exceptional value in the treatment of the conceptual model for waste degradation. Similarly, large-scale field experiments, coupled with carefully designed laboratory studies, provided the initial data sets as well as validation results for the WIPP creep-closure model, which considers the influence of variable fluid (brine and gas) pressures within the excavated regions, and represents one of the most robust conceptual models within the WIPP PA system. By contrast, multiple field experiments designed to establish the conceptual model for fluid flow in the host salt rock provided results that remain difficult to interpret within the context of existing theories for multiphase flow and transport. An assumption that Darcy's law for flow through porous media applied to the WIPP repository host rock governed the consequent development of applicable two-phase (brine and gas) parameters for this model. WIPP field data support this assumption as a conservative means to simulate brine inflow from and contaminant transport within the bedded salt rock [22].

The conceptual models of near-field processes related to actinide solubility and transport are of particular interest at WIPP. The presence of large quantities of Fe metal will induce chemically reducing conditions through most of the WIPP repository. Speciation of the primary actinides in the projected WIPP waste, i.e., plutonium (Pu), americium (Am), neptunium (Np), and uranium (U) in the disposal areas, which may include up to 17 million tons of isotopes with a half-life in excess of 24 000 years, likely, will be limited to the (+III) and (+IV) oxidation states. The WIPP conceptual model for near-field actinide solubility assumes these conditions, but also considers the possible presence of organic ligands and chelating agents within free brines. Kinetic processes, such as sorption of mobilized actinides on backfill or other waste components, while likely to occur under repository conditions, are omitted from the conceptual model as a simplifying, yet conservative, assumption [22].

The conceptual models for the WIPP EBS rely upon engineering studies of material properties and stress responses for the individual components of each EBS, with particular attention paid to the influence of salt-saturated fluids on the EBS materials. For the closure, this applies to any concrete monolith(s) emplaced in the drifts. The WIPP Shaft Sealing System (SSS) comprises 11 materials within 13 components. Primary SSS barriers include concrete, crushed salt, and compacted clay columns, and an earthen fill to provide stability in the near-surface regions. Effects of the only regulator-certified EBS, the MgO buffer/backfill, are incorporated into the overall conceptual model for chemical conditions in the disposal areas, especially with respect to its effect on actinide solubility and uptake of carbon dioxide (CO_2).

Initial parameter values, uncertainty ranges, and development information are stored in a parameter database, and a single value is used for the parameter in the modelling codes. Parameter uncertainty is handled probabilistically using the Latin Hypercube Sampling (LHS) technique over a range of values for the parameter. To treat parameter uncertainty, the selected parameters were sampled probabilistically, generating sets of input parameters. Each parameter set, or vector, was implemented within a single execution of the near-field models for a simulated period of 10 000 years. Results of this exercise comprise a "probable future state" for the WIPP repository.

Average repository fluid pressure represents one key outcome of the near-field simulations, yielding a range from hydrostatic conditions (~8 MPa) to lithostatic (~15 MPa). The amount of Fe metal is another important dependent variable in the repository near field. Of particular interest is the fact that, for conditions in which no future boreholes intersect the WIPP repository at least 40% of the initial Fe is still present in the disposal areas at the end of the 10 000 year regulatory period due to the low availability of corrosive fluids (brine) within the repository host rock.

Chemical conditions in the WIPP repository were determined to calculate radionuclide solubilities and to estimate gas generation rates by anoxic corrosion of Fe-based metals. The single regulatory-acknowledged EBS at WIPP comprises MgO, which is predicted to consume the CO_2 that could be produced by microbial consumption of the cellulosics, plastics, and rubber (CPR) in the waste. Characteristics of the brine chemistry, such as the fugacity of CO_2 (f_{Co2}), the pH, and the concentrations of organic ligands, continue to be evaluated to assure performance of the MgO.

Site preliminary design validation was evaluated at full scale in the WIPP URL. Room closure constituted one of the primary measurements, which along with many subsequent experiments, helped validate the halite constitutive model for the repository. This multi-mechanism deformation model has been used extensively for compliance calculations, and has empirical parameters for both clean and argillaceous salt. The conceptual model has a time-dependent component and a steady-state (secondary) creep component, using the following material and basic physical parameters [22]:

G = the elastic shear modulus,
Q_i = activation energies
T = the absolute temperature
R = the universal gas constant
n_i s = the stress exponents
q = a so-called stress constant
σ_0 =the stress limit of the dislocation slip mechanism.

Agreement between the calculations and actual performance of the underground openings has been satisfactorily demonstrated for twenty years. However, one of the processes that have not been convincingly reproduced in the modelling, is damage-induced flow in salt, which is dominant in the EDZ. The size and shape of the EDZ around an opening based on a stress invariant criterion are similar to the size and shape derived from sonic velocity studies and from microscopy of core damage. As illustrated in Fig. 4.46, healing of the damaged zone and convergence of the rooms is expected under certain conditions [22].

4.5.7.2 Domal Salt

In the German in-situ research programme on radioactive waste disposal in salt rock formations, the main emphasis was placed on:

- The analysis of the integrity and stability of the host rock, because it was considered the most important barrier, and
- Potential backfill materials envisaged to be used as sealing materials in repository boreholes, drifts, and chambers.

Generic conceptual models of borehole and drift disposal are the main tools used for analyzing and assessing the performance of HLW repository systems in German salt rocks. These conceptual models were used and are further described in the so-called "Systemanalyse Mischkonzept" and in the report on the Backfill and Material Behaviour in Underground Repositories in Salt (BAMBUS), [22].

The conceptual model for deposition holes comprises (1) a 6 m wide and 6 m high disposal drift, (2) a 0.6 m diameter and 300-m deep, vertical, waste-emplacement borehole below the disposal drift, and (3) a stack of 0.43 m diameter HLW canisters of Cogema type 7/86 in the borehole. The disposal drift, the seal area at the top of the emplacement borehole, and the 8.5 cm wide annulus between the canisters and the host rock in the borehole are backfilled with crushed salt. The maximum grain size of the crushed salt is 30 mm in the disposal drift and the seal area, and 10 mm in the borehole annulus. The deposition drifts and holes are initially surrounded by an EDZ with higher porosity and conductivity than the undisturbed salt rock. The conceptual model for drift disposal comprises (1) direct disposal of spent fuel in 4.5 m wide and 3.5 m high disposal drifts, in which 1.54 m diameter and 4.46 m

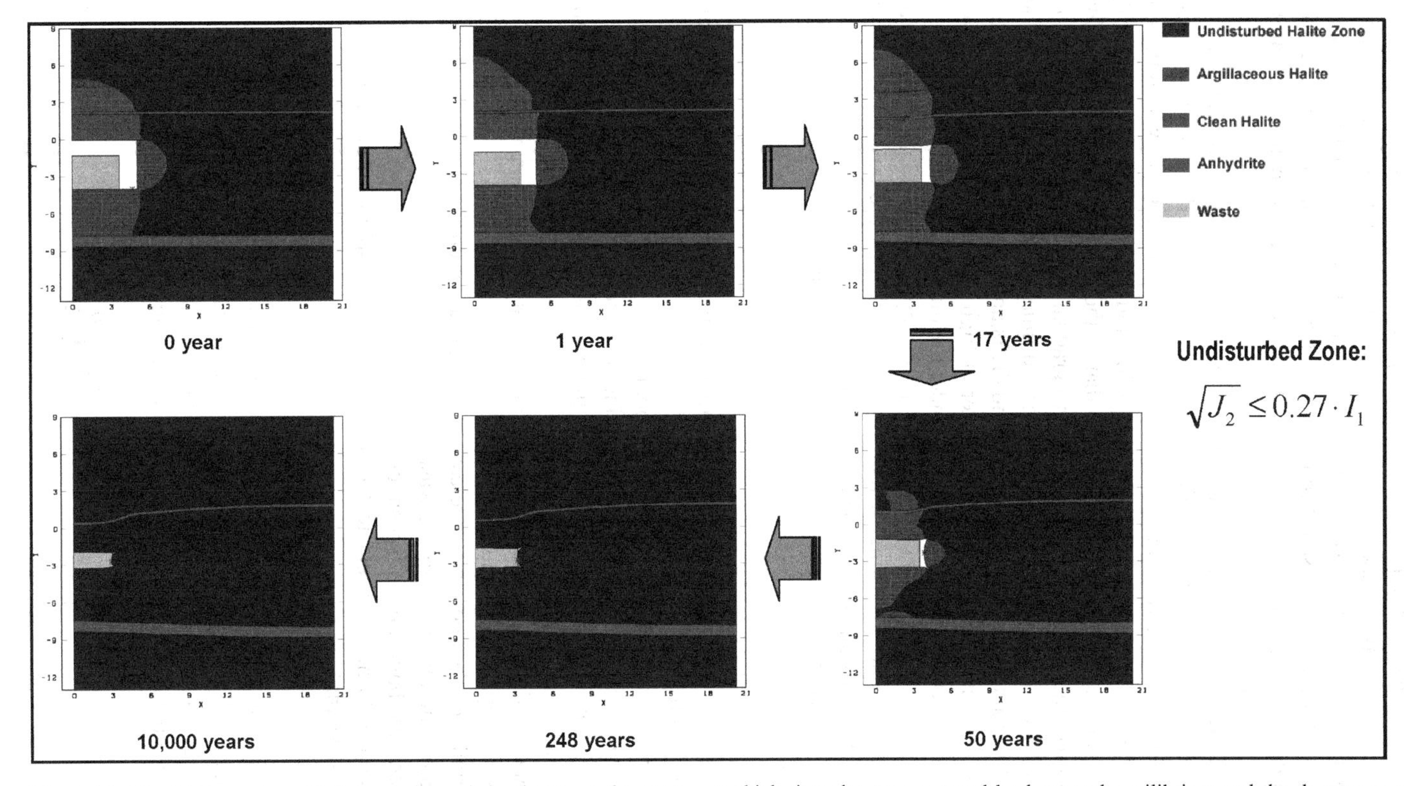

Fig. 4.46 The EDZ grows until the creeping salt impinges on the waste, at which time the stresses trend back toward equilibrium and the damage zone diminishes [22] (By Courtesy of US EPA)

long waste canisters of the POLLUX type are placed at least 1 m apart. The minimum pillar width between the disposal drifts is 10 m. The drifts include a seal area at the entrance and a free space surrounding and separating the canisters, which are backfilled with crushed salt. Similar to the deposition hole concept, the drifts are initially surrounded by an EDZ with higher porosity and permeability than the undisturbed salt rock [22].

Both of the aforementioned conceptual models imply that the inherent creep of the salt rock will gradually close the underground openings. During this process, grain displacement and deformation takes place in the crushed salt, which thereby is consolidated into a gradually denser mass with gradually decreasing permeability that, eventually, is equal to or very close to that of the intact salt rock. The gradual compaction of the crushed salt also results in a gradual stress build-up in the backfill, which stabilizes the underground openings and, ultimately, leads to an almost homogeneous (isotropic) stress distribution around drifts and boreholes. The compaction process is temperature dependent and faster at high temperatures. Potential flow paths for gases and liquids will be sealed under the impact of lithostatic pressure.

Modelling has been made of coupled thermo-hydro-mechanical (THM) processes in the regime consisting of the hot waste packages, the host rock, and the backfill in the drifts and deposition holes [22].

The individual processes considered in numerical simulations of the described conceptual models are:

- Excavation of disposal rooms and resulting change of stress state and resulting host rock deformation.
- Development of EDZ around disposal rooms.
- Waste emplacement and backfilling/sealing of disposal rooms.
- Heat generation of the waste containers and heat dissipation in backfill and host rock in space and time.
- Thermally induced stress and deformation fields in EBS and host rock.
- Backfill compaction and healing of EDZ.
- Thermally induced pore water release into disposal boreholes.
- Gas generation by corrosion of waste packages.
- Gas pressure build-up in disposal boreholes.
- Intrusion of brine from undetected reservoires by two-phase flow in backfilled disposal rooms.

Prediction of the convergence of deposition drifts and holes is made by use of empirical models describing the creep behaviour of the elasto-viscoplastic salt rocks and crushed salt BAMBUS project [22]. Equation 4.1 shows the proposed expression for the creep rate, representing secondary, i.e. steady-state creep, in s^{-1}:

$$\dot{\varepsilon}_s = \mathrm{A} \cdot \exp\left(\frac{-\mathrm{Q}}{\mathrm{RT}}\right) \cdot \left(\frac{\sigma}{\sigma^*}\right)^n \tag{4.1}$$

A = constant (MPa/s)
Q = activation energy (J/mol)
R = universal gas constant: 8.314 (J/(mol, K)
T = absolute temperature (K)
σ = stress (MPa)
σ^* = reference stress 1 (MPa)
n = stress exponent = 5 (-)

For predicting the compaction behaviour of the crushed salt backfill a special constitutive law was applied comprising use of a number of empirically determined parameters.

The codes MAUS/TAUS were used by the DBE in the numerical simulation of the TSDE experiment in the BAMBUS-I project. The code SUPERMAUS was used by the Gesellschaft fuer Anlagen- und Reaktorsicherheit mbH (GRS) for modelling the DEBORA experiments. The code FLAC3D [22] was applied in BAMBUS-II. MAUS and TAUS are two-dimensional (2-D) finite element codes for M and T analyses, respectively. Temperature data calculated with TAUS are used as input data for MAUS and considered in each time increment. SUPERMAUS provides the coupling between the T and M calculations in the sense that not only the thermal influence on the mechanical parameters and quantities, but also the influence of mechanics on the thermal behaviour is taken into account. This is of special importance for the "soft" backfill, which undergoes large changes in its thermal parameters during the compaction process. Therefore, all material parameters for salt rock and crushed salt are given as a function of temperature and porosity, but they are kept constant within a time increment. Generally, SUPERMAUS is a structural code used to analyse/predict the coupled temperature, stress and strain/deformation fields in the elasto-viscoplastic salt rock and crushed salt backfill [22].

For practical application to actual room geometries FLAC3D, a three-dimensional (3-D) finite difference code that includes M and T modules and couples T-M within a time increment, was used. MUFTE (MUltiphase Flow, Transport, and Energy) and TRAVAL are two codes used for the simulation of two-phase flow and water (vapour) transport in the surroundings of heated boreholes in several studies. MUFTE exists in different versions that allow the study of different problems and different stages of development. The one-dimensional (1-D) code TRAVAL is suitable for numerical simulation of water (vapour) flow in low-permeable salt rock around heated HLW boreholes [22].

A field experiment conducted to assess TM processes in crushed salt buffer in disposal drifts and boreholes was the 9 years long TSDE-experiment (Fig. 4.47), for which predictions and evaluation were made by use of the above described conceptual model for drift disposal. It was performed in two parallel drifts at 800 m depth in the Asse mine for measuring drift convergence and buffer compaction. The 70 m long drifts with 3.5 m height and 4.5 m width were 10 m apart. Three electrically heated casks, each with a nominal power of 6.4 kW, were placed in each drift. Crushed salt with an initial porosity of 0.35 was used as buffer material. The temperature of the buffer and surrounding rock, drift convergence, and buffer

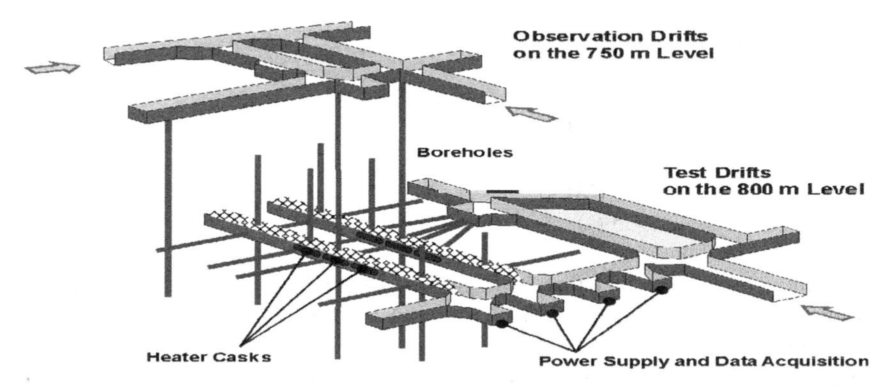

Fig. 4.47 The TSDE test area at the 800 m level of the Asse mine/URL [22] (By Courtesy of GRS)

pressure were measured at several cross sections. Additional parameters of interest were gas release and metal corrosion. The initial host-rock temperature was about 36°C and the initial stress at the test level was estimated to be 12 MPa. The codes MAUS/TAUS were used in the numerical simulation of the experiment [22].

Figure 4.48 shows the theoretically predicted temperature evolution according to 3D finite element technique. These calculations agreed well with recordings, while 2D modelling of the convergence, shown in Fig. 4.49, gave very significant deviation from the field measurements due to partly incorrect conceptual and material models.

The DEBORA-2 experiment (Fig. 4.50) of 14 months duration was performed in a 0.6 m diameter, 15 m deep, unlined borehole at 800 m depth in the Asse mine/URL for measuring borehole convergence and buffer compaction. The heat production of the waste canisters was simulated by four heaters at 1.1 m distance from the hole. The temperature, borehole closure, and buffer pressure were measured at

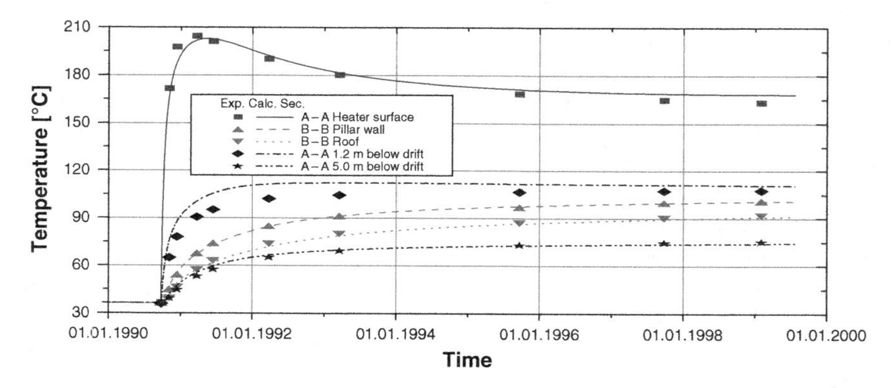

Fig. 4.48 Temperature at selected points of two central cross sections from 3D calculations [22] (By Courtesy of GRS)

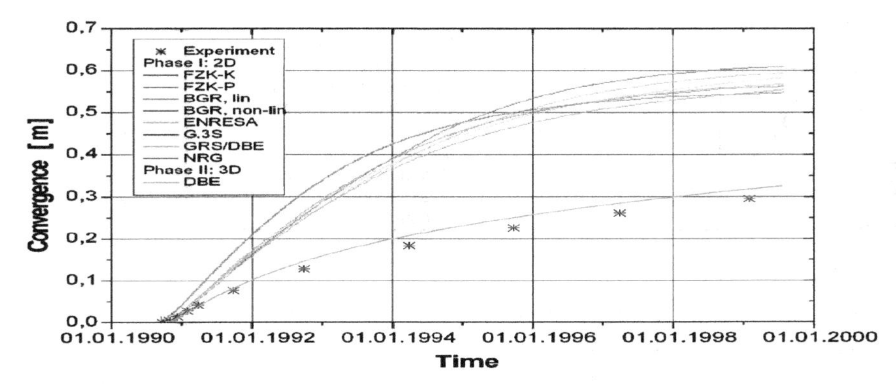

Fig. 4.49 Convergence in the central cross section of TSDE according to 2D calculations. All the predictions exaggerated the movement by nearly 100% in the first 10 years [22] (By Courtesy of GRS)

three levels. The experiment was modelled using the same material model used in the TSDE experiment, and an axisymmetric, finite-element system. The agreement between predictions and measured results was remarkably better than in the TSDE experiment, demonstrating that 3 D modelling is required (Fig. 4.51).

Brine and vapour flow from the host rock into the disposal boreholes was addressed in some of the experiments conducted in the Asse mine. This phenomenon

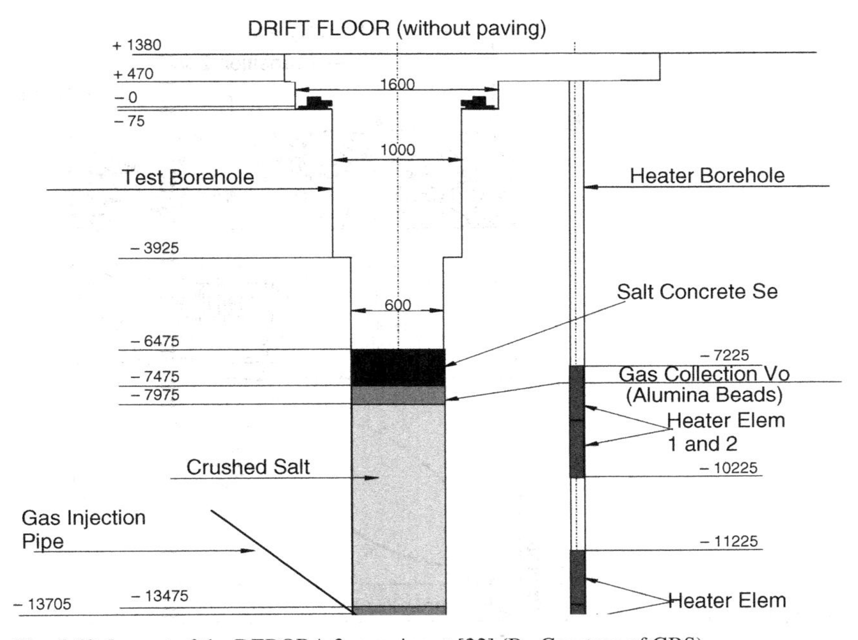

Fig. 4.50 Layout of the DEBORA 2 experiment [22] (By Courtesy of GRS)

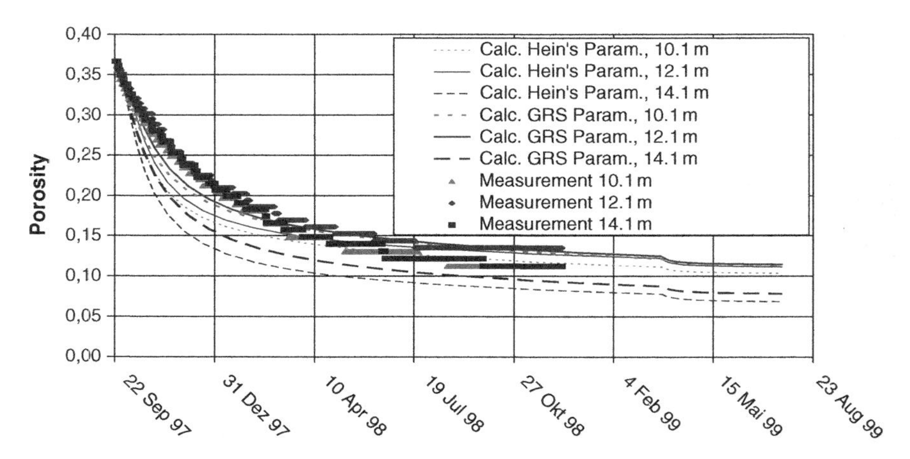

Fig. 4.51 Porosity decrease in the DEBORA borehole backfill. The agreement between prediction and measurements is better than in the TSD experiment [22] (By Courtesy of GRS)

is of importance if the host rock contains significant amounts of brine inclusions or water adsorbed on the crystal boundaries. In the temperature field around HLW disposal boreholes, such water may be driven to the waste canisters and speed up corrosion. Migration of water in the salt is assumed to take place in the form of a moving front (Fig. 4.52).

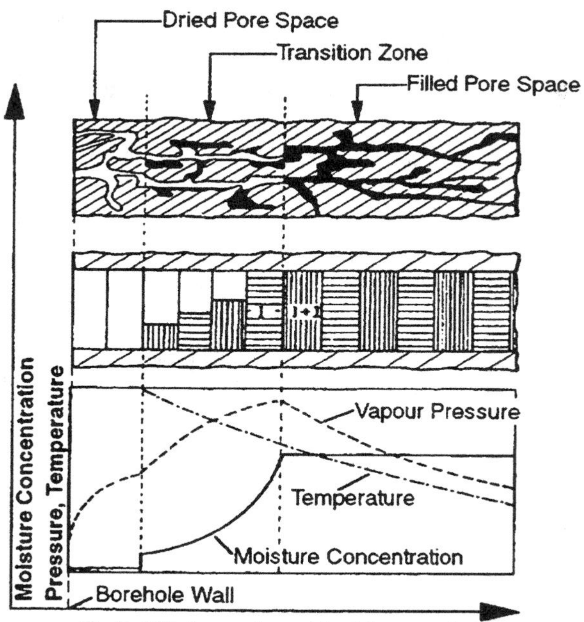

Fig. 4.52 Evaporation front model [22] (By Courtesy of GRS)

The model implies that the brine front moves from the hot borehole into the rock. Behind the front released vapour will move towards the borehole because of the strong gradient of partial pressure. The velocity of the migrating front, u, is given by:

$$u = -\frac{\rho_v}{\rho_w \cdot \phi} \quad v \tag{4.2}$$

u = velocity of the migrating front (m/s)
ρ_v = vapour density in the IP (kg/m^3)
ρ_w = density of water in the IP (kg/m^3)
ϕ = porosity (-)
v = filter velocity of vapour (m/s); defined as volume flow divided by the area perpendicular to the flow.

Comparison of predictions using the code TRAVAL and measurement results were made in the Asse project series, from which it was concluded that vapour transport takes place both in the form of Darcy flow and Knudsen (surface) diffusion in low-permeable rock salt. The following evolution scenarios were predicted:

- Brine inflow into a deposition drift from an instantaneously flooded 500 m high shaft;
- Hydrogen production in a flooded borehole due to corrosion of the HLW steel canisters;
- Spontaneous connection of an unexplored brine pocket with the bottom of the borehole.

The main objective of these modeling exercises was to investigate two-phase flow processes for understanding of the involved mechanisms. The calculations, which were made by use of the finite element code MUFTE, showed that there is still no perfectly valid conceptual model of the complex physico/chemical processes, and that more research is required before reliable prediction can be made of canister corrosion and gas migration and accumulation [22].

4.5.7.3 Conclusive Remarks Concerning Function and Modelling of Repositories in Salt Rock

The most important processes that need to be considered are the rate of convergence of drifts, tunnels and holes, and the migration of vapour and brine in the system of virgin rock, EDZ and buffers and backfills. None of the processes can be accurately modelled today. Thus, convergence can only be roughly predicted, largely based on empirical data, and the same is valid for the true mode of migration of brine and vapour in the backfilled system of virgin and artificially prepared buffer and backfills in drifts and holes. Settlement of hot canisters in salt repository rock can be estimated on the basis of the stress/strain/time relationships worked out for predicting convergence of rooms but the outcome of such calculations can be questioned because of the uncertain parameter data. In a long-term perspective one would expect large, uncontrolled movement of the canisters that can bring them closely together and – depending on the type and age of HLW – yield a critical mass.

One hence lives with uncertainties respecting the risk of formation of very highly pressurized hydrogen gas bubbles in the nearfield and of quick corrosion of the HLW canisters. If these problems can be solved deposition of highly radioactive waste in salt appears to be ideal. However, the problem of non-retrievability remains because it will not be possible to identify the location of the canisters after some time.

4.6 Construction and Performance of Buffers and Backfills in Argillaceous Rock

4.6.1 General

Clay strata represent very good conditions for hosting HLW repositories because of their low hydraulic conductivity and self-sealing ability provided that they do not contain continuous permeable layers of silt and sand. The ongoing R&D work in Belgium, Spain, France and Switzerland presently focuses on solving potential practical problems like rock stability, predicting the long-term waste isolation function of undisturbed and disturbed clay host media (e.g., diffusion experiments, EDZ investigation) and the effect of coupled processes (e.g., heat-induced porewater pressure changes), as well as on gas release of repository-generated gas. Depending on the mechanical properties and the state of stress, the construction of the shafts, drifts and rooms may be more difficult than in other potential host rocks (salt and crystalline), especially for non-cemented normally or slightly overconsolidated clays. Therefore, special emphasis will be given to rock mechanical issues in planning and construction of repositories in this type of geological medium [22].

Sedimentary rock containing clay strata has been proposed as host rock of radioactive waste long ago. Some 15 years ago the British company C.E.T. worked out a deep-hole concept for disposal of large reactor compartments from submarines that was claimed to be accepted by Greenpeace (Fig. 4.53). The large-diameter shafts – 15–18 m – were planned to be constructed off the British North Sea coast with the radioactive units embedded in dense smectite-rich clay and plugged with concrete in the uppermost part. The difficulties with finding suitable densities for balancing the effective pressures caused by the heavy reactor units were much more severe than for the deep hole concept for crystalline rock described earlier in this chapter. For the large shafts the problem was solved by using different bentonite contents and densities at different depths. The construction would have to be completed before starting placement of the waste and was planned to be made by use

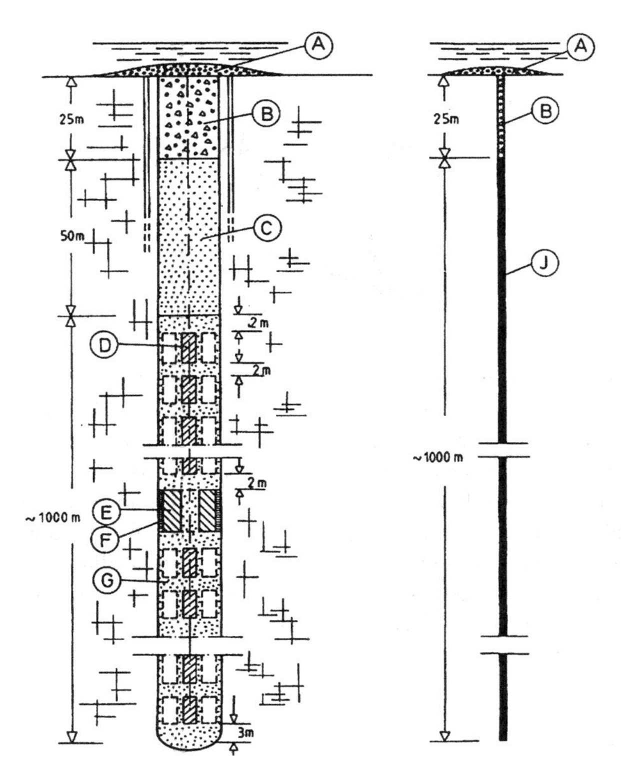

Fig. 4.53 Concept proposed by the British company C.E.T. for disposal at sea of waste in containers and of large reactor compartments. The diameter of the shaft, which extends to 1 km depth, is 15–18 m. (A) Erosion-resistant, coarse fill, (B) Concrete, (C) Bentonite compacted on site, (D) Waste containers, (E) Large containers or large reactor parts, (F) Highy compacted bentonite blocks, (G) On-site compacted bentonite-rich backfill [4]

of freezing for stabilization of the rock. The total volume available for hosting the waste was estimated at 100 000 to 200 000 m^3.

In recent years the same basic principles have been proposed as for disposal of HLW in crystalline and salt rock, i.e. construction of drifts, central areas, shafts and ramps, as well as buffers, backfills, and plugs. We will therefore not go deep into these matters but focus on valuable and less valuable features and properties of disposal in argillaceous rock. One difficulty is the mechanical stability of rooms at depths in normally or weakly overconsolidated clay. Figure 4.54 shows the hoop stress of a rock element located at the periphery of a deeply located tunnel and breakage and tunnel failure is expected if it exceeds the uniaxial compressive strength.

For isotropic primary rock stresses the maximum hoop stress σ is $2D\,\rho$, where D is depth and ρ the density of the rock. For repositories at 500 m depth the hoop stress is up to 20 MPa, if the primary rock stress field is isotropic, while the compressive strenth of normally consolidated clay is much lower than that and failure will occur in conjunction with or very soon after excavation. It will be required to apply rock

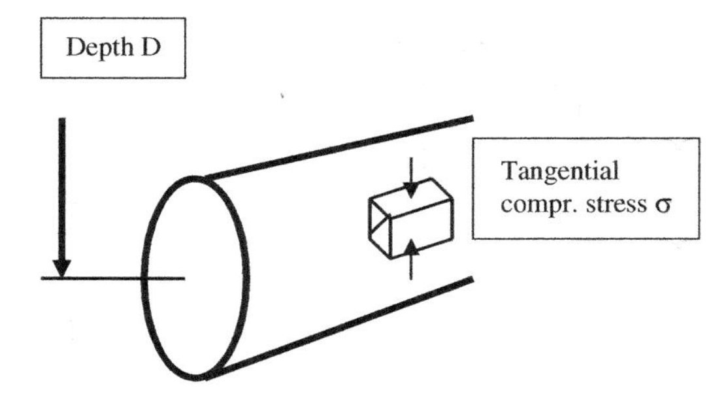

Fig. 4.54 Element at tunnel wall

supports prior to or parallel to the excavation and detailed rock mechanical investigations need to be made to get a basis for the design work. We will see that one of the concepts, the Belgian "Hades", represents such conditions and requires quick placement of supporting concrete liners for avoiding large strain and collapse.

No repository for HLW has yet been built in the world but several underground laboratories have been established for investigating the performance of drifts and shafts and buffers, backfills and plugs. Countries in which disposal in argillaceous rock, including clastic clay, are considered are Belgium, Switzerland, and France. There is presently also a growing interest in Germany in utilizing argillaceous rock as host rock for HLW repositories.

4.6.2 National Concepts

4.6.2.1 General

Sedimentation of dispersed particles or flow of slumping fine-grained soil masses followed by burial, up-lift and exposure to heat, have formed strata with different densities, water contents and mineral contents with more or less anisotropic structure and properties. Two different classes of clay materials are considered as host rock of HLW repositories:

- Normally or slightly overconsolidated[3] soft (or) clays with relatively high water content, behaving as ductile or viscous material with significant creep strain.
- Stiff, overconsolidated or cemented clays (clay shale, claystone or argillite) with quite low water content, deforming and failing like brittle materials (Fig. 4.55).

[3]The expression "poorly indurated" is sometimes used.

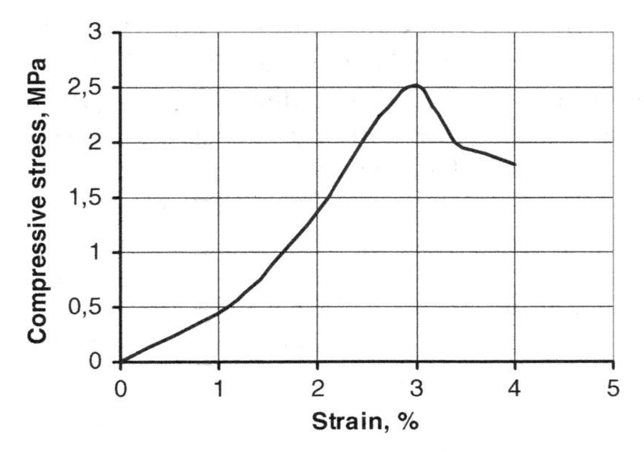

Fig. 4.55 Typical failure mode of slightly cemented, very fine-grained illitic Silurian clay from Gotland, Sweden

Depending on rock parameters (stress field and pore pressure) and the mechanical properties, some of the concepts for the emplacement cells (tunnels or boreholes) require massive liners or rock reinforcements, while others can rely on the strength of the rock and need little ground support during the operational phase. These differences have consequences for the design, construction and operational phases as well as for long-term safety assessments and so have national regulations respecting waste quantities and waste types. Therefore, the national concepts show significant differences, although some basic design elements also show similarities that are clay specific. In most cases clay layers have, in contrast to crystalline rocks or salt domes, a thickness of a hundred to a few hundred meters. This limited vertical extension of the host rock provides some natural limitations on the design of the repository. Most designs place the repository in the centre of the clay layer and keep its vertical extension at minimum for maximising the distance to the boundaries of the clay bed. This principle implies that the repository has the same inclination as the clay layer. Vertical emplacement boreholes, even short ones, like in the KBS-3 V concept for crystalline rocks, are presently considered less suited for clay.

4.6.2.2 Belgium

The Boom Clay Formation, considered as the primary candidate host formation, is of marine origin and Tertiary age, i.e. 30–35 million years old. It has rather uniform chemical and mineralogical compositions but there are variations in grain size, organic matter, and carbonate content, resulting in the typical layering with thin silt layers, representing paths for water flow and migration of radionuclides. The Formation has a downward inclination of 1% in NE-direction and its base is located at some 400–1000 m depth in the part of Belgium where a repository can be hosted. In northeastern Belgium it has been affected by tectonics resulting in series of faults oriented NNW-SSE. At the Hades Underground Research Facility, the overburden pressure is about 4.5 MPa and the horizontal stresses somewhat lower than that. The large-scale hydraulic conductivity of the host formation is about E-11 m/s [22].

The current repository design for disposal of HLW including spent fuel [19] is shown in Fig. 4.56 is based on a horizontal network of disposal galleries. Access to the underground facility is offered by two shafts with about 6 m diameter. The disposal gallery lining is composed of non-reinforced precast concrete vault blocks of the "wedge" type. "Key" blocks, inserted with a certain pressure as a last step, provide the necessary expansion force to keep the ring elements in place (Fig. 4.57). The design of the disposal gallery itself has not been decided yet and alternative concepts are presently being considered. According to the original design the HLW-containers will be placed in a central tube in the gallery axis, with the space between this tube and the gallery lining being backfilled with precompacted clay blocks.

Mock-up and parallel studies have led to identification of a number of unresolved questions regarding the practical implementation of the reference design. Furthermore, selection of materials and dimensions appear to require justification and documentation. A major question concerning disposal in clay host rock, particularly in normally consolidated clay, is how the effective pressure in the host rock just outside the concrete liner will evolve when the heat wave advances through the liner. It is expected that the porewater pressure will rise on the expense of the effective pressure since it would take extremely long time for the porewater overpressure to dissipate because of the very long flow paths up to permeable strata.

The problem of insufficient stability in deep underground construction in argillaceous rock has been fully realized by the designers, which has led to openings with

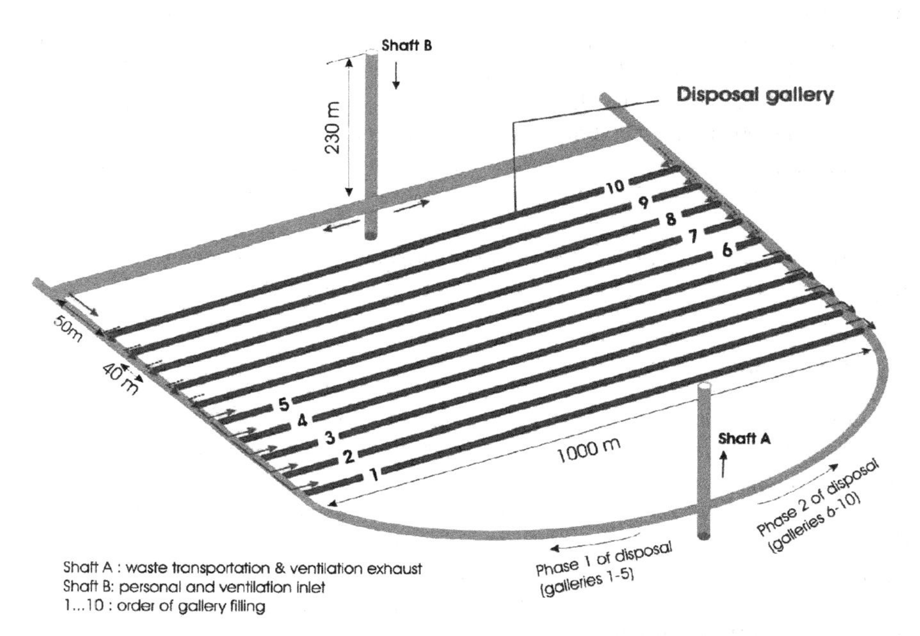

Fig. 4.56 Basic design concept with long emplacement tunnels (example from the Belgium program) for HLW / spent fuel disposal in clay [22] (By Courtesy of NIRAS/ONDRAF)

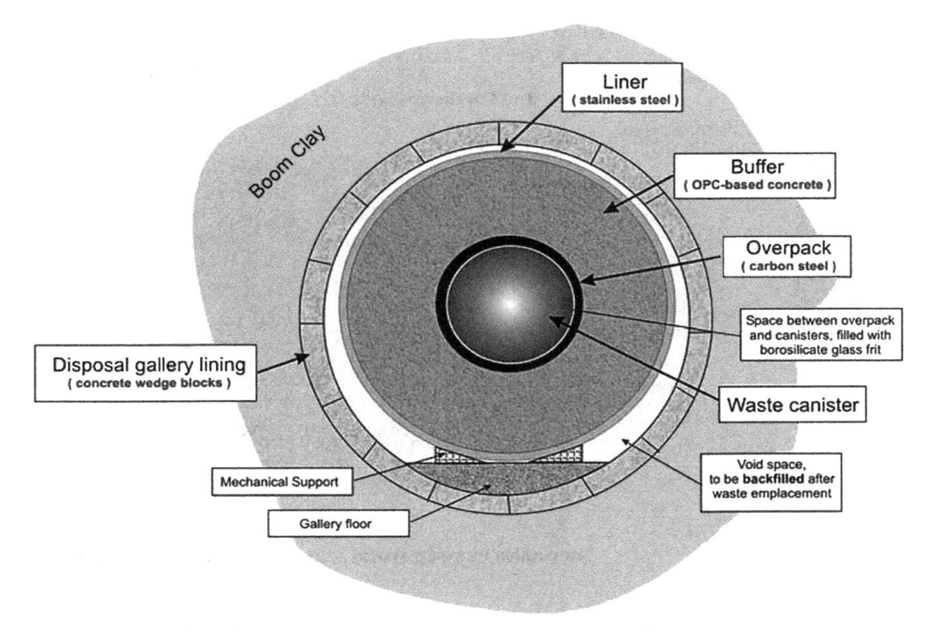

Fig. 4.57 Schematic cross section of the original design (SAFIR-2) of the Belgian concept [22] (By Courtesy of NIRAS/ONDRAF)

small dimensions and planning of the construction work such that the time from excavation to placement of support and backfill is as short as possible. However, there are other problems implying that the original reference design needs further development and modification and three other designs have been proposed, two of them "in galleries" (Supercontainer and sleeve concept, Fig. 4.58), and one "in boreholes". According to the Borehole concept, deposition of the waste packages is made in short vertical holes like those of the KBS-3 V but with steel liners strong enough to resist the pressure exerted by the surrounding clay. The idea is to close each hole with a steel lid and concrete after placing the waste container. At present, the Supercontainer concept with high pH (Portland cement) concrete is the number one candidate with "normal pH" cement as an option. A horizontal version of this concept is also being considered.

A most important, remaining problem concerns the stability of the drifts with canisters some tens to hundreds of years after closure. At this stage the water saturated host rock surrounding the drifts becomes significantly heated, by which the pore pressure increases and the effective pressure drops. This can reduce the shear strength of the nearly normally consolidated rock sufficiently much to cause instability and movement of the lined drifts.

4.6.2.3 Spain

The clay formation that is considered to be potentially capable of hosting a deep geological disposal facility is composed of lutites and high-plasticity clays of lacustrine

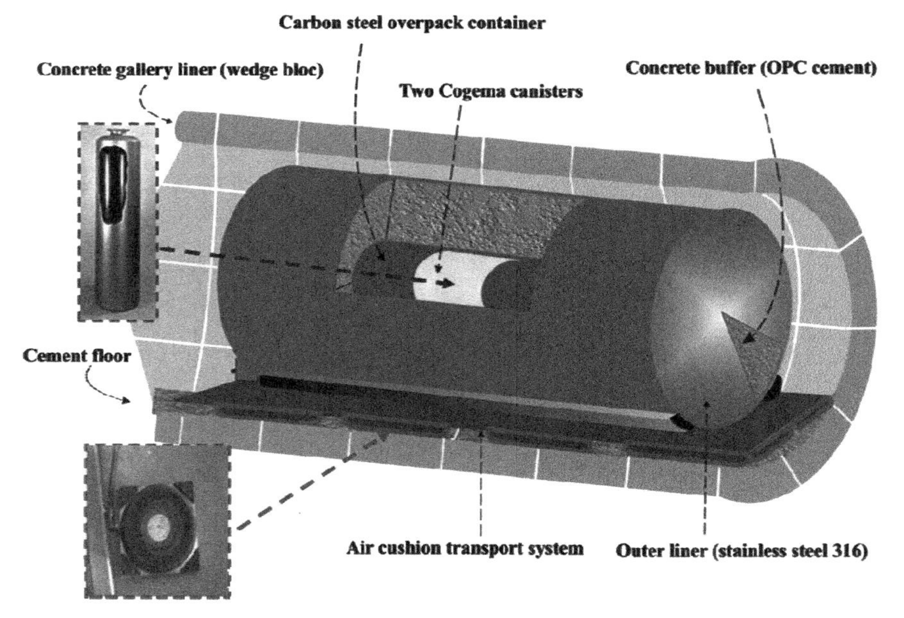

Fig. 4.58 Supercontainer and sleeve concept [22] (By courtesy of NIRAS/ONDRAF)

origin [22]. Its longitudinal extension exceeds 8 km and it measures more than 4 km in the transverse direction. The upper boundary of the layer, which is about 200 m thick, is located some 150 m below the ground surface, the reference depth being set at 250 m. The stress field at this level is characterized by a vertical pressure of about 6.4 MPa, and 6.2 and 4.3 MPa for the maximum and minimum horizontal principal stresses. This pressure constellation would create severe stability problems in normally consolidated clay but will not cause collapse in cemented or metamorphosed clay of shaly type [22].

The rock mass is hydrogeologically very complex with subhorizontal groundwater flow in the more permeable formations above and below the clay formation. Subvertical downward and upward percolation of the clay formation are expected due to hydraulic gradients across it. The average hydraulic conductivity of the part of the formation that can be used for hosting a repository is about 4E-12 m/s.

The repository geometry is similar to the Belgian concept comprising two parts, one where disposal of waste takes place – HLW and ILW are separated – and one where construction is conducted. The parts have independent access routes, systems and services, and different personnel. The design principle of the EBS is illustrated by the section in Fig. 4.59.

The HLW disposal zone occupies a horizontal area of about 2000 000 m^2 and is divided into disposal modules, with their respective disposal drifts that are 500 m long and 2.4 m in diameter.

The canisters are separated in the drifts by 1 m, as a result of which 87 canisters are stored in each, with 42 disposal drifts being required. Within the ILW disposal zone, the ILW may be stored in caverns or silos; however, in view of the low strength

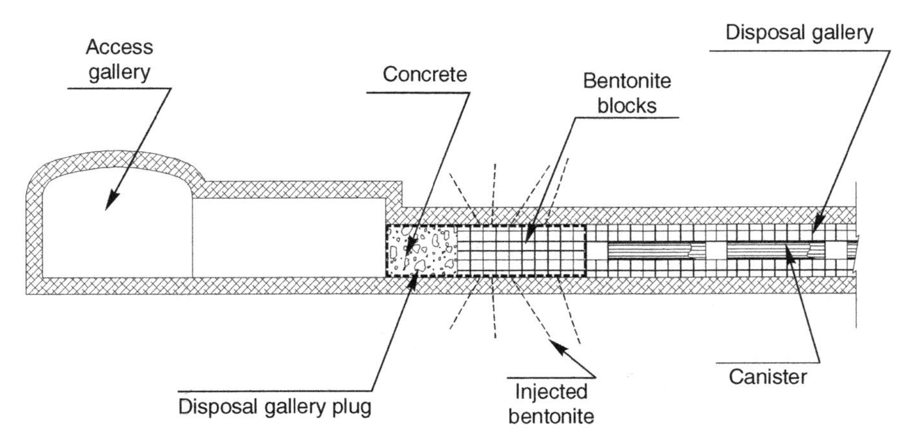

Fig. 4.59 Section of ENRESA's deposition tunnels in argillaceous rock. The clay blocks are in direct contact with concrete bulkheads and the rock tightened around the block masonries by use of clay grout [22]

of clay, in this concept the ILW is also disposed of in small diameter drifts, similar to those in the HLW disposal area. The occupied horizontal area amounts to 60 000 m^2.

4.6.2.4 France

The proposed host formation belongs to the eastern margin of the intracratonic Paris Basin [22]. This marine clay sediment dates from jurassic time (middle Callovian to lower Oxfordian). The layer is flatlying with a slight dip of 1° to 1.5° towards the west and the centre of the Paris Basin. This shale formation has a relatively homogeneous constitution over a large area but the mineralogical composition varies somewhat in vertical direction respecting the contents of carbonates, quartz and type of clay minerals. As in all sediments these variations are related to the sedimentary history and to changes in the seal level.

The physical performance of the host rock is controlled by the microstructure. Thus, recrystallized carbonates occupy voids in the clay mineral matrix, which means that effective porosity is very low and that the clay is cemented and hence brittle. The low porosity explains the very low hydraulic conductivity, 5.6 E-13 to 5.6 E-14 m/s.

A set of preliminary one-level concepts have been defined all of them implying location in the centre of the Callovo-Oxfordian clay formation (Fig. 4.60). The layouts of the potential repository contain disposal zones of different wastes ILW, HLW, and spent fuel. Each zone is divided into modules, for separating the waste types with respect to gas production and organic content, and offering sufficient flexibility with respect to potential variations in inventory or waste management modes. The disposal zones are separated by several hundred meters and the modules by several tens of meters [22].

At the time of the "CROP" project ANDRA proposed that spent fuel and other HLW disposal packages be placed in a carbon steel central tube surrounded by dense

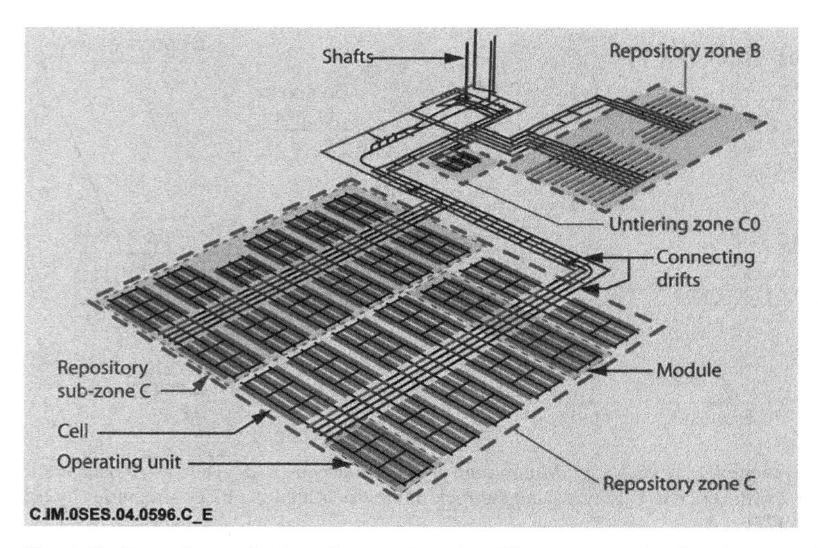

Fig. 4.60 General organization of a repository for different types of radioactive waste in argillaceous rock proposed by ANDRA in the CROP project (Dossier 2005). It is now abandoned [22] (By courtesy of ANDRA)

smectitle clay. For spent fuel, an outer, perforated carbon steel lineer would proivde mechanical stability of the tunnel during the operation period and allow water from the host rock to saturate the buffer (Figure4-61). The carbon steel components were designed to provide total confinement of the waste for about 10000 years, while, for other HLW, the carbon steel overpack would serve for about 4000 years. This waste was planned to be disposed in the form of packages placed in short tunnels of about 40 meters length with 2.5 to 3 m diameter depending on the type of waste. These concepts are now being changed implying that buffer clay may be replaced by concrete except in plug seals.

The repository will also contain long-lived ILW embedded in cement, which provides alkaline conditions (pH>12) using ordinary Portland cement (Fig. 4.62). The cement buffer also provides stability to the underground openings during the operating phases.

All disposal cells are oriented in the direction of the major principal horizontal stress in order to minimize the EDZ around the cavities. Waste will be put in the disposal modules directly after construction, while additional modules are being built. The concepts make it possible to keep the entire disposal infrastructure accessible and ventilated throughout the repository-operating period.

4.6.2.5 Switzerland

The proposed repository host rock is an overconsolidated Jurassic claystone termed Opalinus Clay, including the clay-rich Murchisonae Beds. At the potential site, the formation consists of an approximately 200 m thick sequence of fine sandy claystones and marls with intercalated limestones, calcareous sandstones and iron

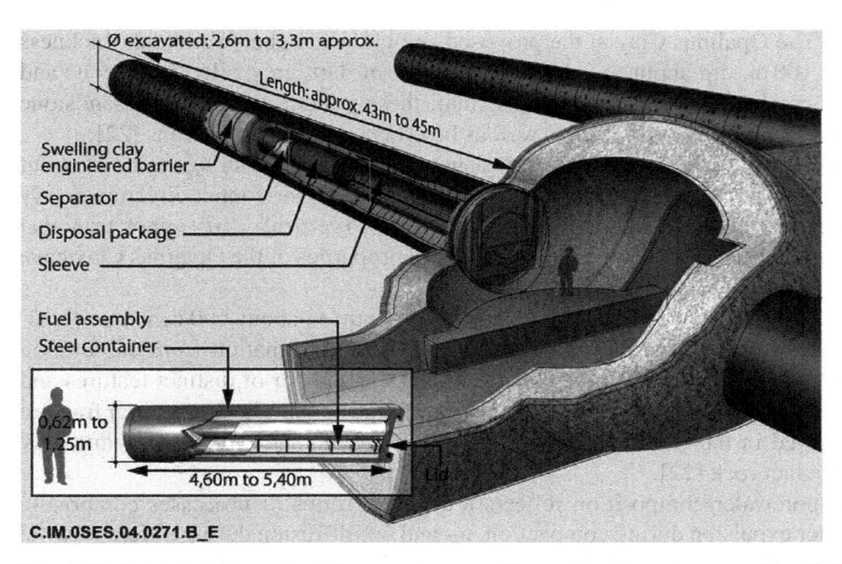

Fig. 4.61 Spent fuel disposal cell in operation according to the concept proposed by ANDRA in the CROP project. The outer, perforated, tube and the inner tube sleeve were planned to be made of carbon steel (Dossier 2005). The concept is now abandoned [22] (By courtesy of ANDRA)

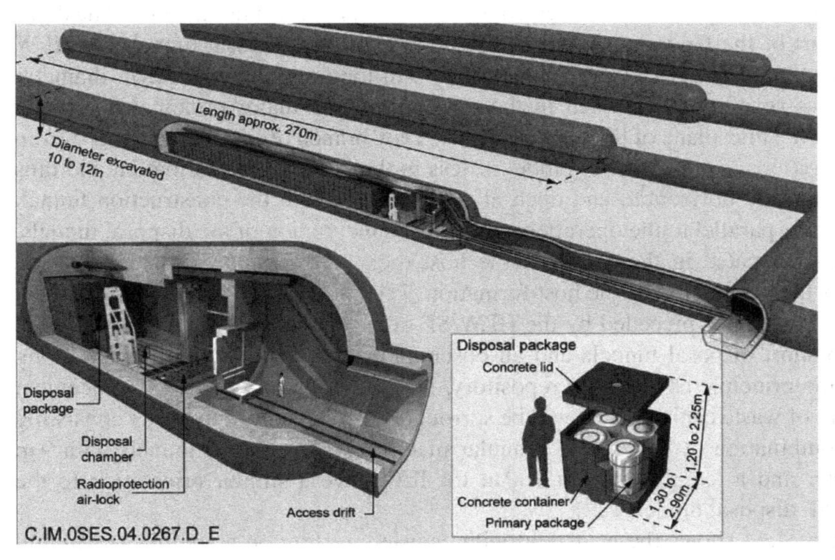

Fig. 4.62 Disposal cells for non-exothermic ILW in clay with concrete liner for stabilizing the tunnels (Dossier 2005), [22] (By courtesy of ANDRA)

oolites. The Opalinus Clay at the proposed site in the Zürcher Weinland (thickness approx. 100 m, dip about 6°) generally consists of dark grey, silty, calcareous and micaceous claystones. To a varying extent, these contain thin silt and sandstone inclusions, lenses and laminae, as well as brownish siderite concretions [22].

The principal stress components measured at the potential siting region at about 650 m depth are: (i) major principal stress 20–23 MPa (horizontal, direction nearly NS), (ii) intermediate principal stress 15–16 MPa (vertical), (iii) minor principal stress 15–16 MPa (horizontal). The mechanical properties of the Opalinus Clay have been found to be strongly anisotropic.

The hydraulic conductivity measured at a depth of about 600 m in the Opalinus Clay varies between 2E-14 m/s and E-13 m/s. Information from exploration boreholes and 3D seismic surveys indicate that the number of distinct features and fracture zones is very limited in the potential siting region. The few natural fracture zones tested in the Opalinus Clay do not show any higher hydraulic conductivity than in intact rock [22].

The porewater composition reflects a complex series of processes comprising porewater expulsion during compaction, in- and out-diffusion during different burial and uplift phases, and rock/water interaction. The salt content is about 10 °/oo. The mineralogy investigations show that the phyllosilicate fraction contains mainly kaolinite, illite and illite/smectite mixed-layer minerals in about equal amounts. The carbonate content is rather high ranging from 10 to 50%, and dominated by calcite. Quartz is also present in relatively large amounts (~20%). The presence of pyrite and siderite, which show no signs of oxidation, indicate the reducing nature and the high redox buffering capacity of the Opalinus Clay. The content of organic matter is about 0.5% [22].

The general layout of the underground structures is shown in Fig. 4.63. The dominant part of the layout is the main repository with the disposal tunnels for HLW including spent fuel. The approximately 800-m-long tunnels with 2.5 m diameter and 40 m spacing are oriented in the direction of the major principal horizontal stress in the mid-plane of the Opalinus Clay. They branch off at an angle of 60° from the operations tunnel, which is more or less in the strike of the formation and thus approximately horizontal, and open at the other end into the construction tunnel, which runs parallel to the operations tunnel. The inclination of the disposal tunnels, which are located in the centre of the host rock layer is approximately 6% and follows the general dip of the host formation. The construction and operation of the main repository is preceded by the HLW/SF pilot repository. This facility consists of two short disposal tunnels and an observation tunnel and is constructed along the same principle as the main repository. It is based on an in-room displacement concept of waste canisters that will be surrounded by smectite-rich buffer consisting of a combination of blocks and granular material. Two disposal tunnels with 9 m diameter and a length of about 100 m for ILW have a similar orientation to the HLW/SF disposal tunnels [22].

Figure 4.64 shows the near-field with canisters resting on assemblies of highly compacted smectite-rich blocks and embedded in clay pellet fill.

The underground facilities are linked to the surface by an access tunnel (ramp) and a construction shaft, which also serves as a ventilation shaft and emergency

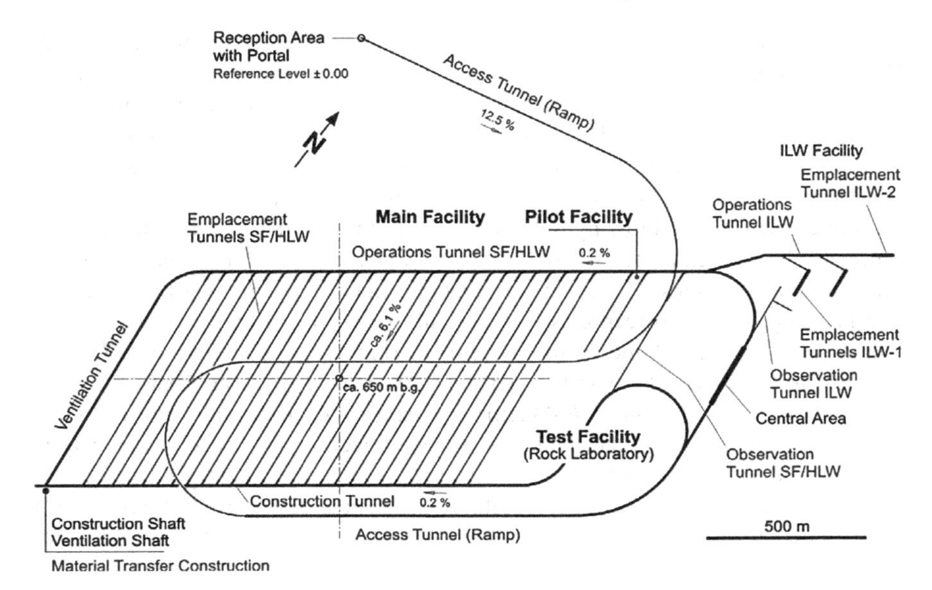

Fig. 4.63 Design principle of the Swiss HLW repository with a spacing of about 40 m of the deposition tunnels and a maximum temperature of the canisters of 150°C [22] (By Courtesy of NAGRA)

exit. The ramp passes over the disposal zone and allows sensors to be introduced (e.g. from boreholes above the disposal tunnels) for monitoring. Due to the operational plan, construction of emplacement tunnels and emplacement of waste are done concurrently. This allows the open period for emplacement tunnels to be kept to a minimum (less than 2 years). The layout of the repository is based on the principle that construction and emplacement can be carried out without interfering with each other by having separate access tunnels and a shaft/ramp permitting, from a radiological point of view, separated protected and non-protected areas.

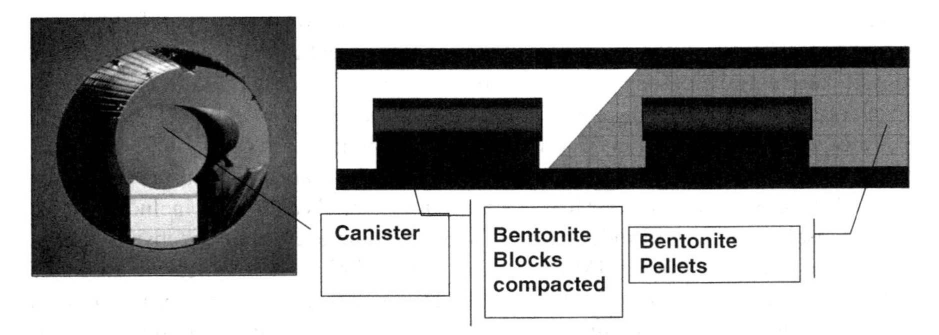

Fig. 4.64 EBS components and their placement in the Swiss concept for argillaceous rock [22]

The Swiss EBS concept is different from most others but similar to the Swedish KBS-3 V concept by making use of dense smectite-rich pellets. One sees from Fig. 4.64 that compacted pellets will occupy a large part of the deposition tunnels and the question has been whether the net density of the block/pellet system will be high enough. The pellets are blown in by use of a conveyor tube that is stepwise moved backwards in conjunction with the filling of the vault. The significant abrasion caused by the blowing helps to increase the dry density, which has been reported to be as high as nearly 1600 kg/m^3 (about 1950 kg/m^3 at water saturation) under certain conditions [12]. Assuming that one can bring pellets into densest possible layering, which corresponds to a porosity (ratio of the space between them to that of the entire volume of the system) of 0.25, they would occupy 75% of the space to be filled. This yields a net bulk dry density of about 1650 kg/m^3 for pellets in water and around 1700 kg/m^3 if a smectitic clay mud with a density of 1200 kg/m^3 can be injected, which is proposed technique. These values can be somewhat increased if the dry density of the pellets is increased beyond 2100 kg/m^3 or if pellets of two sizes are used, a suitable theoretical diameter ratio being 1:3.5 as concluded from preliminary estimates. In practice, however, the possibility of reaching higher dry densities than 1550 kg/m^3 is questionable.

4.6.2.6 Conclusive Remarks Concerning Function and Modelling of Repositories in Argillaceous Rock

Argillaceous rock has the same good isolating properties as salt in the sense that very little water can percolate through these types of rock, in practice groundwater flows only in the small number of rather tight fracture zones representing very low-permeable discontinuities of 2nd and 3rd orders in crystalline rock. This makes the rock itself an excellent barrier to radionuclide migration. In salt rock consolidation and tightening of the EDZ takes place by self-sealing and no water is required for this process. For argillaceous rock, as for fracture-free crystalline rock, the tightness means that very little water can reach the clay buffer, which may hence desiccate and undergo permanent degradation as illustrated in Chap. 3. Naturally, if practically no water will ever reach the buffer there will not be any significant corrosion of the canisters and no way for possibly released radionuclides to leave the nearfield. However, since ultimate wetting and through-flow of the entire repository is expected, significant degradation of the buffer is really a threat to acceptable long-term performance for which one has not yet found a solution.

In practice, the evolution of a repository in argillaceous rock of shale type will involve a number of processes that can change the hydraulic conditions quite dramatically for better or worse. Thus, excavation of tunnels and drifts including those for disposal of HLW will induce an EDZ that may be even more extensive than in crystalline rock also when TBM or road-headder technique is used. The disturbance can be predicted by rock mechanical calculations and leads to the formation of a strongly water-bearing zone in overstressed parts of the rock surrounding excavated rooms. This is illustrated by Fig. 4.65 representing NAGRA's underground rock

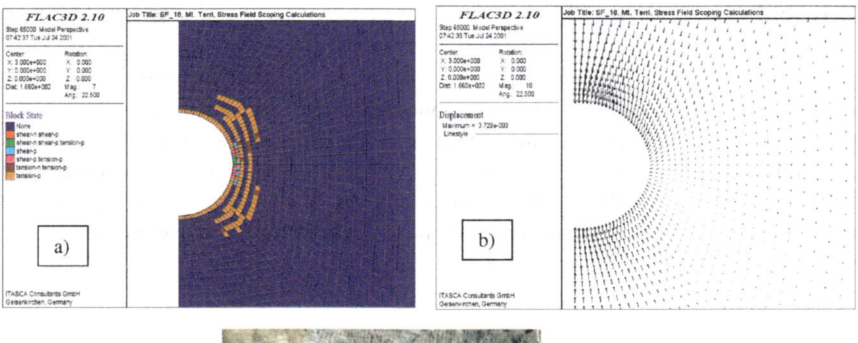

Fig. 4.65 EDZ issues in argillaceous rock. (a) Excavation disturbed zone. (b) Calculated displacements. (c) Observed extensile fractures of the EDZ [22] (By Courtesy of NAGRA)

laboratory at Mont Terri in which tunnels, drifts and niches have been excavated at about 400 m depth.

The disturbance of the rock causes strong fissuring, which promotes water uptake and softening of the clay if the air is humid or water can flow into the construction area. This can cause very significant stability problems and call for quick and comprehensive stabilization by bolting, shotcreting and construction of abutments. Another effect is that the EDZ will serve as a continuous major groundwater conductor in the repository that requires construction of plugs that extend into the rock for cutting it off, i.e. exactly as one has to do in crystalline rock. Hypothetically, the EDZ can self-seal like in salt because clay has a significant creep potential but there is hardly any hope for significant tightening. In fact, a ten- to hundred-fold increase in hydraulic conductivity compared to virgin rock may remain and provide the buffer with sufficiently much water to minimize degradation. However, this requires that the EDZ is not a cut off, which in turn offers paths for radionuclide migration.

Considering finally the other type of clay for potential use, i.e. normally or near normally consolidated clay like the Boom formation we found in examining their behaviour that the matter of geotechnical stability is critical and remains to be investigated in more detail.

4.7 Borehole Sealing

Recently, the matter of possible hydraulic short-circuiting of repositories by deep boreholes drilled before, during or after construction of repositories for site investigation, rock mechanical investigations and geochemical analyses, has come in focus and much is presently being done to investigate the role of deep boreholes and techniques of plugging them. We will examine some of the work being done in crystalline rock for which the issue is much more important than for the other rock types.

4.7.1 The SKB/POSIVA Study

The principle of borehole sealing that was defined early in the project is proposed to comprise construction of tight seats in the parts of boreholes where the rock has few fractures and a low hydraulic conductivity, and filling of those parts that intersect permeated fracture zones with physically stable material that does not need to be very low-permeable but that must be chemically compatible with the tight seals [12]. These consist of smectite-rich clay in the form of highly compacted blocks or pellets, while the fillings separating them consist of silica concrete with little cement of low-pH type and ballast (aggregates) of quartz sand and quartzite. This type of concrete is largely inert with respect to chemical interaction with smectite clay and can be allowed to loose the cement component without jeopardizing its function as fill by using a gradation of the ballast grains that prevents the large majority of them to migrate into the rock fractures. This risk is largely eliminated by stabilizing the intersecting fracture zones by casting cement-poor concrete in reamed parts followed by re-boring. Hence, stabilization and subsequent rinsing and cleaning of borehole walls are prerequisites for plugging, which comprises segmentwise placement of clay plugs and casting of silica concrete.

4.7.2 Tight Seals

Several clay plug types are being investigated in the project because the different conditions that are met with in practice call for plugs with different properties and placeability. Thus, while very deep holes require a safe technique that is presently represented by plugs of what is called the "Basic" type with expandable clay confined in a porous tube, other versions are being developed for both deep and short holes using ring-shaped clay blocks and pellets. We will confine ourselves here to deal with the firstmentioned, which has been successfully used for plugging deep parts of a more than 500 m long vertical hole in Finnish crystalline rock, cf. Sect. 3.5.6.

The tight seals of expandable clay must be confined at the upper end of the holes, which can be produced by use of long-lasting silica concrete or metal plugs anchored in recesses made by reaming. Foreseeable climatic changes and tectonic impact are expected to erode the rock to a considerable depth, up to 100 m by two glaciation/deglaciation events, which means that the tight, lower parts must still be intact in the considered 100 000 year perspective. Hence, the plan should be to seal holes extending from the ground surface by materials with at least the same mechanical strength and erosion-resisting potential as the rock down to this depth. This can be made by backfilling the holes with densely compacted moraine and stone fills and suitably placed "mechanical locks" of silica concrete and metal plugs. Seals of the lastmentioned two types were designed and tested in the project (Fig. 4.66).

One realizes that this type of seals can, and maybe have to be constructed in the upper part of KBS-3 V and KBS-3i deposition holes for preventing the EBS inventories to be pushed up by rapidly increasing water pressure at the base of the deposition holes.

Figure 4.67 shows the investigated tight, clay-based plug types intended for placement between permeable rock zones intersected by boreholes. They are primarily planned to be used in boreholes with diameters ranging from 56 to 100 mm but at least the "Basic" and "Pellet" plugs can be used in much wider holes. The ultimate density of the saturated and fully matured clay should be on the order of 2000 kg/m^3, corresponding to a dry density of 1590 kg/m^3, for fulfilling the criterion that the hydraulic conductivity must be lower than that of the surrounding rock.

The "*Basic*" plug concept was developed in the early eighties and has been used successfully for plugging short and long holes as demonstrated by field experiments performed in conjunction with the international Stripa Project [23]. It has been used for sealing boreholes from a drill rig at sea for the SFR repository at Forsmark, which demonstrated the feasibility of the method. The plugs consist of perforated copper tubes that contain well fitting blocks of highly compacted smectite-rich clay that migrates through the perforation and ultimately embeds the tube that is left in the hole. The density of the clay formed between the tube and the rock is very low in the earliest maturation phase but solidifies and encloses the tube with a density that is determined by the geometry and time of maturation. The plug segments can be a few meters long and connected by a screw mechanism, bayonet clutch or by clinching.

In principle, the method is very safe since one has complete control of the position of each individual segment and good opportunities to pull up those that may not be placeable because of unforeseen roughness of the borehole walls or of a too small radius of curvature of the hole.

The perforated tubes with the dense clay blocks placed in them must have sufficient strength to be handled in the placement phase without being deformed or break. This puts a limit to the degree of perforation and to the thickness of the tube walls and a comprehensive study was made in the early phase of the project for finding an optimum geometry with respect also to the rate of maturation of the clay component. Four metals were considered: copper, navy bronze, titanium, and steel, and they were all found to be acceptable but copper was selected because its chemical stability is documented and its impact on smectite clay negligible. Also, the deformability of copper tubes can adaptate well to the curvature of the holes in the placement phase.

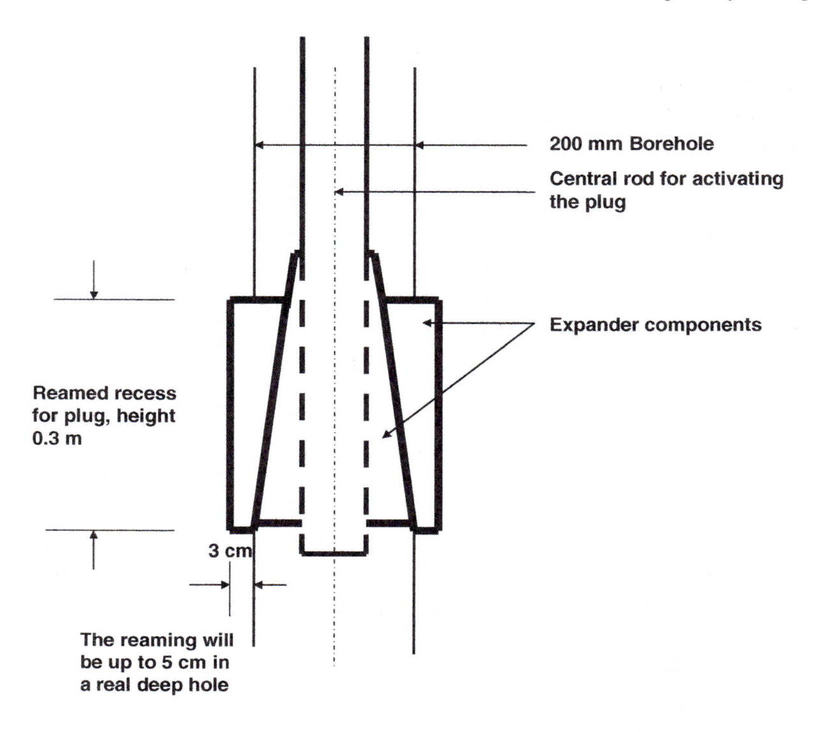

Fig. 4.66 Mechanical seals at the upper ends of deep boreholes. *Upper*: Copper plug of expander type for placement and prestressing in reamed boreholes. *Lower*: Sectioned concrete plug (porphyry-like) cast in reamed boreholes with 200 mm diameter [12]

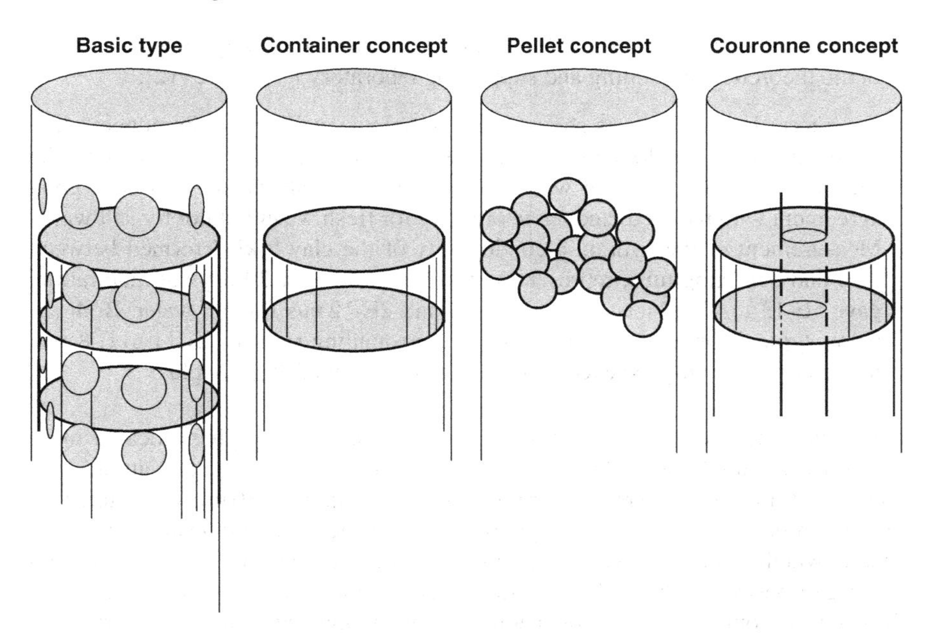

Fig. 4.67 The four clay plug concepts. *Left*: The "Basic" plug with highly compacted MX-80 clay columns confined in perforated copper tubes. *Second left*: The "Container" plug with highly compacted MX-80 clay blocks contained in a cylinder attached to drilling rods and released when the tip of the cylinder is in the desired position. *Third left*: MX-80 pellets poured into the hole without compaction. *Right*: The "Couronne" plug with annuli of highly compacted MX-80 clay stacked around jointed copper rods that are pushed into the hole [12]

The degree of perforation is a determinant of the rate of maturation of the clay component and of the compressive and tensile strength of the tubes. For plugs in deep holes the high water pressure implies that access to water for maturation of the clay is not a limiting factor while geometrical constraints in the form of the size and spacing of the perforation holes control the rate of migration of clay to form a clay gel in the gap between tube and rock. For plugs close to the ground surface or located in those part of a repository where the water pressure is low in the construction and waste placement phases the ability of the rock to give off water to the clay determines the maturation rate.

Earlier field experiments with 50% perforation ratio and 10 mm hole diameter and 2 mm tube wall thickness gave acceptable performance with respect to the maturation rate of the clay and to the mechanical performance of the tubes and these measures were found to be at optimum in the project [12,23].

The rate of maturation of the clay component is very important: if it is very low the state of sealing can be strongly delayed and make the construction of intermediate concrete plugs difficult, while it can cause difficulties by erosion and friction resistance at the placement if it is very high. Considerable efforts have been made to develop a theoretical model of the maturation process but the complexity provided i.a. by the geometry of the perforation holes has caused difficulties and the model is therefore only approximately valid. The most important practically important results from the theoretical modelling and supporting laboratory tests were [12]:

- Long term laboratory tests showed that complete swelling and homogenisation is obtained after 10–20 days. The measured *mean* swelling pressure against the rock was 2.8 MPa for fresh water and 0.6 MPa for salt water. This is in good agreement with the predicted *mean* pressure for fresh water but not for salt water.
- Measurement of the hydraulic conductivity of the clay "skin" formed between tube and confining tube, representing the condition after 20 days of maturation, gave 5E-13 to 9E-13 m/s for fresh water and 2E-12 m/s for salt water. Both are below typical conductivity values for the surrounding rock, i.e. E-11 to E-8 m/s, hence demonstrating the excellent sealing efficiency of the clay plugs.

For unlimited access to water, which is the case for clay plugs located more than a few hundred meters from the ground surface, complete water saturation is reached within less than a month for plugs in holes with up to 80 mm diameter. This does not imply that complete homogenization is reached in this time and a practical issue is whether the clay, which has somewhat different density in the center and in the gap between tube and rock after reaching complete water saturation, will ultimately become totally homogeneous or have permanent differences in density. The laboratory studies suggest that they will be permanent.

If there are differences in water pressure along a plugged borehole, which will be the case in holes extending from tunnels or shafts where the water pressure will be low until the repository has been backfilled, the plugs will be exposed to a pressure gradient that can cause piping before the clay has matured sufficiently. Tests have been made for investigating this, the main results being the following [12]:

- After 8 hours of maturation under 500 kPa uniform water pressure the critical pressure for piping was 700 kPa, after 19.5 hours 900 kPa and after 41.5 hours 1700 kPa. The successive increase illustrates the maturation rate under prevailing test conditions, i.e. with unlimited access to (tap) water.
- Immediately after the last piping test extrusion of the plug was started. The plug began to be displaced but could not be extruded at the maximum capacity of the hydraulic jack, 7 t (70 kN), possibly because of deformation and locking of the perforated tube in the confining bigger tube.
- The extruded plug had a homogeneous appearance except for some small, local heterogeneities (Fig. 4.68).
- Samples of the clay "skin" had a density and water content of 105% at 0.5 m distance from the pressurized end, 94% at 1.25 m distance and 93% at 2 m distance from this end. The clay core inside the tube had a density that was unchanged from the initial state.

In the placement of clay plugs in long boreholes water will flow along them, which causes erosion. A way of limiting the loss is to use very dense clay blocks like the ones used in erosion tests with blocks obtained by compacting MX-80 powder

Fig. 4.68 Appearance of "Basic" plug extruded after 3 weeks of maturation. The perforation can be seen through the thin "skin", which would have become denser and less transparent had the plug have access to water not only at the ends but also over the periphery [12]

that had low water content (6%) under a pressure of 250 MPa. The initial dry density of the blocks was around 2100 kg/m^3 and the loss of clay recorded in the lab tests, simulating placement of clay plugs to 1000 m depth in 3 hours, was 4–5%. The lowest net density of 1000 m deep holes will theoretically be about 1900 kg/m^3, corresponding to a hydraulic conductivity and swelling pressure in salt groundwater of E-12 m/s, and 1.5 MPa, respectively. Plugs placed at 500 m depth, corresponding to the repository level, will undergo less erosion and are estimated to have an average density at water saturation of 2000 kg/m^3.

The rate of maturation of the clay plugs is mirrored by the time-dependent growth of the shear resistance when displacing them by applying an axial force. The resistance to displacement is determined by the shear strength of the clay, which was predicted to be 50 kPa after 6 hours, 200 kPa after 12 hours, 325 kPa after 24 hours, 600 kPa after 48 hours and up to 800 kPa after 96 hours depending on the salt content. The axial force that displaces the plug is equivalent to the product of the area of the clay/rock contact and the shear ("bond") strength, which means that the force yielding displacement can be defined if the borehole diameter and the length of the plug, and the shear strength of the "clay skin" are known.

The actual rate of maturation was determined by extruding plugs of about 3 m length from boreholes with 80 mm diameter and 5 m long boreholes at 450 m depth [12]. The holes were filled with tap water before the plugs were placed in the hole and a hydraulic jack placed and secured at the lower ends of the holes which were connected to equally long, inclined holes through which the tubing from the jack was led up to the pump on the floor of the room (Fig. 4.69). Mechanical strain gauges were installed for measuring the displacement of the released packer resting on the upper end of the clay plug at the loadings.

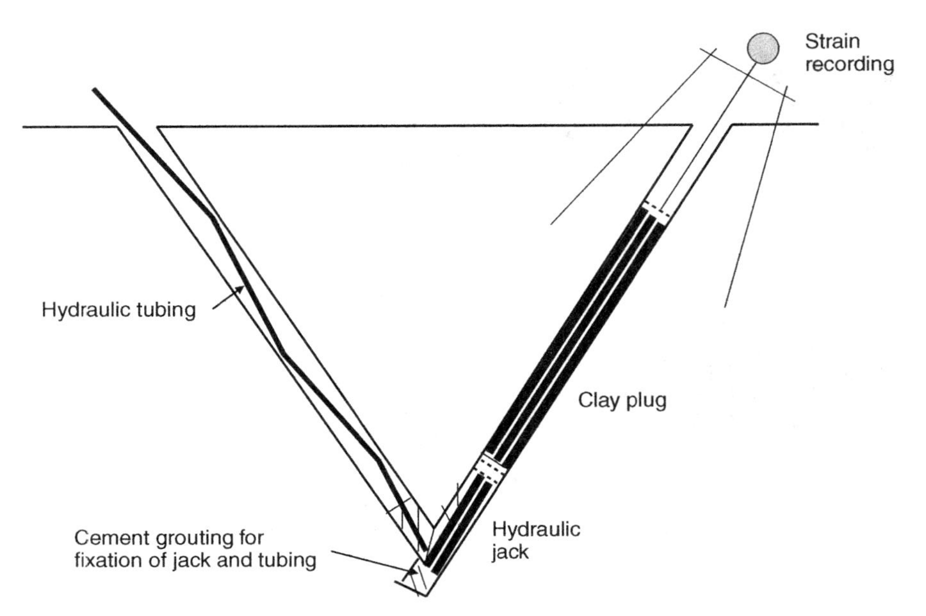

Fig. 4.69 Arrangement for measuring clay/rock adhesion by extruding the plugs in the 5 m long holes and measuring the displacement [12]

The plug was loaded from below and the force determined as given in Table 4.6, which also shows the evaluated average shear stress when the plug started to be displaced for the respective load.

The low shear strength in the first few days confirms that the perforated tube significantly delayed the skin formation and caused microstructural heterogeneity in the first few days. However, it is obvious that the force required to displace a plug of this length is appreciable already after one day and that delay in placing long plugs can cause problems. After four days a 10 m long plug would require a force of about 30 t (300 kN) to move it.

The following major conclusions from the various test series were drawn:

Table 4.6 Actual force required for displacing the "Basic" plug and the mobilized shear resistance (strength) of the "clay skin" [12]

Time after placement, h	Load required for displacement, t*	Maximum displacement, mm Upper/lower	Average shear stress exerted on the "clay skin", kPa
24	1.16	0.45/1.2	15
48	2.66	0.52/5	35
96	9.96	0.05/-	132
720	3.50	0/>10	>133

* Multiplication by 10 gives the force in kN.

- After 4 hours in low-electrolyte water, clay from plugs of MX-80 penetrates the perforation but will not form a homogeneous paste around the tube, while after 8 hours such a paste is formed.
- The clay paste formed around the perforated tube will cause resistance to insertion of clay plugs in boreholes, which must therefore be made within a limited period of time. In salt water the clay plug must be placed within one hour, while in electrolyte-poor water the corresponding time is 5–10 hours.
- For avoiding very rapid expansion and erosion of the clay gel formed early in the space between tube and rock the water in the borehole must be poor in electrolytes. In practice, this means that the natural water should be replaced by tap water.
- After a few weeks the clay between the tube and the rock becomes dense and after several months the entire clay mass tends to become homogeneous and sufficiently dense for providing the required tightness. Complete homogeneity may require years or decades and may in fact never be reached.
- The long term tests show that very significant swelling and homogenisation are obtained after 10–20 days. The measured mean swelling pressure against the rock for the initial dry density 1905 kg/m^3 of the clay plug core was 2800 kPa using fresh water and 600 kPa for strongly brackish water. Measurement of the hydraulic conductivity of the clay paste between tube and rock showed that it was lower than 9E-13 m/s for saturation and percolation with fresh water and 2E-12 m/s for saline water.
- Only clays with Na as major adsorbed cation should be used since Ca-saturated expansive clays behave like Na clay in very salt water, i.e. the freshly formed soft skin coagulates and settles in the space between tube and rock.
- For blocks prepared by compacting MX-80 powder under 250 MPa the clay migrates slower through perforation than ordinarily compacted clay.
- The slow maturation compared to unshielded plug types (Container, Coronne, and Pellet) shows the retarding impact of the perforated tube and hence its erosion-protecting ability in the placement phase.

Some practical rules for the placement were proposed:

- A practical solution for use of copper tubes implies 24 m long segments consisting of jointed 3 m long parts. The tensile strength is sufficient, assuming 4-fold safety, and safe attachment to the drill string can be achieved. The thickness of the tube wall should be 2–3 mm and the outer diameter about 6 mm smaller than the diameter of the hole.
- Each 24 m segment is lowered into the desired position, i.e. in the space between two stabilized fracture zones, and left there. Several segments can be placed in series without coupling them together. Their weight guarantees that they will rest on the underlying ones without moving in the axial direction.
- Before a clay plug segment is emplaced a previously cast silica concrete plug must have hardened sufficiently much to be able to carry it. With a suitable concrete recipe this would take about 24 hours.

4.7.3 The "Container" Concept

The bearing idea of the *Container* concept is to use a container tube of stainless steel for bringing the clay plug segment-wise to the planned depth of placement and let it slip out by opening a valve at the lower end of the tube. The container will be exposed to a uniform water pressure of 10 MPa in a 1 km deep hole and has to resist this pressure and still be perfectly tight, which determines the thickness of the container wall.

4.7.4 The "Couronne" Concept

The "*Couronne*" concept was proposed a few decades ago and has been used in at least one project [12]. The bearing idea of this concept is to use a plug that consists of a central rod around which tightly fitting annular clay blocks are stacked. Submerging the plug in a water-filled hole will instantly cause dispersion as in the "Container" case and erosion will be substantial if the plug is brought down in deep holes. Delay in placement caused by jointing plug units to form a continuous train of units will cause intermittently increased dispersion and contribute to make the finally matured plug heterogeneous. One therefore may have to restrict the use of this plugging technique to holes with a depth of less than 100 m. Shorter holes, i.e. with a length of a few tens of meters can be drained so that the plug can be placed under dry conditions. Suitably arranged flanges can make it possible to use this technique for plugging of holes of any direction.

The central rod is left in position after placing the plug and should hence be chemically compatible with the clay, for which copper is proposed to be most suitable. Detailed design with respect to geometrical features and strength of the metal rod must be made for the individual cases, the most important matter being to reach the required ultimate density of the clay.

4.7.5 The "Pellet" Concept

Pellets have been used for borehole sealing in different contexts. Thus, NAGRA has conducted several experiments in downwards and upwards oriented boreholes and oil companies in the US have made experiments with rather large pellets using unisized pellets or mixtures of pellets of two sizes for sealing of abandoned oil and gas production holes. A matter of significance is whether the ultimate dry density of the pellet plug needs to be as high as specified for fulfilling the criterion that the hydraulic conductivity must be lower than that of the surrounding rock, i.e. on the order of 2000 kg/m^3, corresponding to a dry density of 1590 kg/m^3. None of the experiments referred to in the literature gave densities of this magnitude for which it is believed that a combination of load and vibration is required.

In practice, it seems that the pellet plugging technique is suitable for sealing holes with a length of up to about 20 m, with a further requirement that it should be

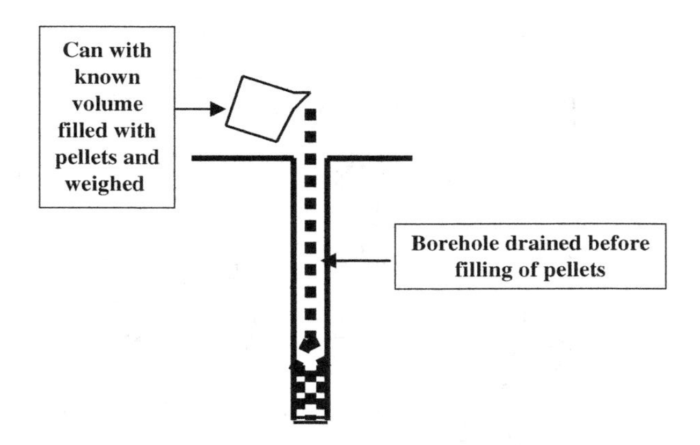

Fig. 4.70 The "Pellet" concept

used where the sealing function is not very critical, i.e. not in holes extending from deposition tunnels and definitely not in the near-field of canisters. Homogeneous pellet fills require that the boreholes are dry (Fig. 4.70).

In the present study pellets with 10% water content were poured in a 5 m long tube in the laboratory for piping tests and in a borehole with 80 mm diameter giving the air-dry fill a density of 1150 kg/m^3 corresponding to a dry density of 1035 kg/m^3 and a density at complete water saturation of 1650 kg/m^3.

Figure 4.71 illustrates schematically the maturation process of a pellet fill when water enters it. The pellets contain large numbers of smectite particles and they expand on hydration and release aggregates that rearrange to form soft gels that become consolidated under the pressure exerted by the expanding grains although variations in density will remain also after complete water saturation. Pellet fills have a large number of continuous voids that are instantly filled with inflowing water and they contain numerous continuous, rather permeable flow paths also after maturation. Their sealing potential will therefore be less good than of the other plug types and the variation in density means that their hydraulic conductivity will be more sensitive to saline water.

The very large voids in a pellet fill will be occupied by water very soon after filling the hole with air-dry pellets. The main difference between the "Pellet" plug and the other plug types is that the latter become wetted from their outer boundary while the "Pellet" plug has a high degree of water saturation in all parts from start, i.e. on the order of 70–80%. This means that the suction, which is the main cause of saturation in the absence of high water pressure, is rather low. Another major difference is that "skin" formation is much less developed in "Pellet" plug. Both phenomena combine to cause a very slow increase in the degree of water saturation beyond the initial value even if the rock can offer unlimited amounts of water for the plug's maturation (Fig. 4.72).

The following major conclusions from the various test series of the "Pellet" concept were drawn:

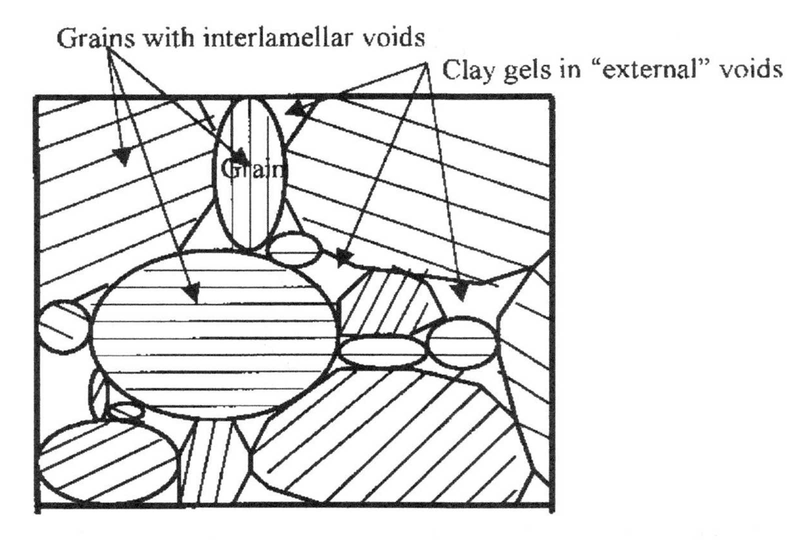

Fig. 4.71 Pellet fill. *Upper*: Schematic picture of the microstructural constitution; the grains take up water from the boundaries by which they expand and give off clay particles that form gels in the voids between the expanded grains. *Lower*: Pellets with 10% water content in 30 mm cell

- The expected quicker initial water saturation and homogenization of the "Pellet" plug than of the other types were verified.
- The low density of the "Pellet" plug means that its bulk hydraulic conductivity is sensitive to high salt contents, and particularly to Ca, in the groundwater. Thus, for an average bulk density of 1650 kg/m^3 saturation and percolation with

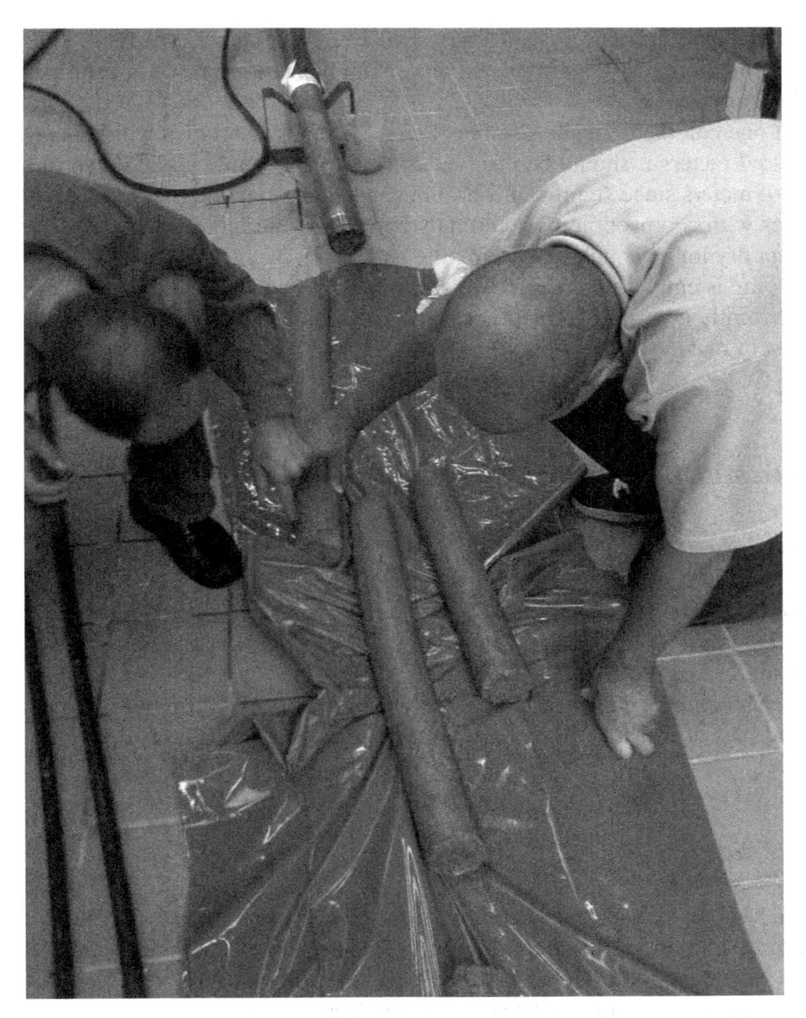

Fig. 4.72 Appearance of the "Pellet" plug extruded from the tube used in the laboratory for simulating a borehole in the "piping" test. Notice the coherence of the plug which broke by mechanical strain at the expulsion

salt-free water the conductivity is estimated at 2E-12 m/s and around E-10 m/s for ocean water.

- Placing of pellet fills requires that the borehole is drained until placement is started since this gives good initial homogeneity and offers a possibility to make some slight compaction. Groundwater will enter from water-bearing fractures and may produce irregular wetting of the fill, which can lead to local expansion and some variation in density along the hole.

- Only clay pellets with Na as major adsorbed cation should be used since the initially formed microstructure will be more homogeneous than if Ca-saturated clay pellets are used.
- Although one can theoretically use the "Pellet" technique in holes with a depth of a hundred meters it should be reserved for holes with a maximum depth of a few tens of meters since significant variations in density can otherwise occur and difficulties with compaction most certainly appear. Furthermore, deep holes can not be kept dry long enough for placing the pellet fill.
- Before a plug is emplaced a previously cast silica concrete plug must have hardened sufficiently much to be able to carry it. With a suitable concrete recipe this is a matter of one day.

4.8 Stabilization of Fracture Zones in Boreholes

4.8.1 Principle

Fracture zones intersecting boreholes need to be stabilized for avoiding rock fall in the plugging phase. The technique is to ream the holes where fracture zones are intersected and to cast concrete that is left to harden, after which re-boring is made to the original borehole diameter. The hole is then ready for casting concrete in the stabilized part of the hole. The recipe of the cement-based material is given in Table 4.7 [22].

4.8.2 Material

The material has paste consistency and hardens to give a compressive strength of at least 10 MPa in 24 hours. This strength is sufficient to support the rock so that the stabilization work can proceed in other parts of the hole after one day. pH is less than 11. It was used in the plugging of the about 500 m deep borehole OL-KR-24 hole at Olkiluto.

Table 4.7 Concrete for stabilizing boreholes ("lining of reamed hole"), [12]

Components	Amount, (kg/m^3 concrete)	Manufacturer
White cement	514.26	Aalborg Portland
Silica Fume	342.84	Elkem
Fine ground, α-quartz M300	133.2	Sibelco
Fine ground, α-quartz M500	107.5	Sibelco
Superplasticizer Glenium 51*	8 (dry content)	Degussa
Fine quartz sand, < 250 μm	324.4	Askania
Coarse quartz sand < 500 μm	488.1	Askania
Glass fibers, 6 mm	53.6	Saint Gobain
Water	244.27	local

* Other superplasticizers can be considered as well: Set Control II, SP-40, Mighty 150.

The practical outcome of the tests was interpreted as follows:

- A reaming tool of the type worked out for the project is required. Its performance at large depths needs to be tested and reconfirmed.
- The long-term performance of the concrete with respect to filtering for eliminating loss of fine quartz particles into the stabilized fracture zones has to be demonstrated.
- The long-term performance of the concrete with respect to its chemical interaction with nearby clay plugs has to be investigated.

4.8.3 Construction of Concrete Plugs in Stabilized Parts of Boreholes

The technique is to cast concrete in the stabilized borehole and leave it to harden after which clay plugs are placed on it, reaching up to the next stabilized part etc.

The general principle of the Borehole Plugging Concept is to make the boreholes at least as tight as the surrounding rock and to seal them so that they do not serve as conductors of radionuclide-bearing water that may ultimately emanate from the repository. Where the holes intersect water-bearing fractures zones, plugs do not need to have a low conductivity but must be physically stable for supporting the surrounding rock and the clay plugs that rest on them or are located below them. In the construction phase they must be coherent and soon become strong enough to carry clay plugs without settling.

The property of reaching a relatively high mechanical strength rather quickly requires use of a cement binder and for minimizing negative impact on contacting clay plugs the cement content has to be very low and low-pH cement utilized. Likewise, the amount and type of the superplasticizer, that is required for making the concrete sufficiently fluid, will be at minimum. The cement is not relied on for long periods and it is assumed that it will be dissolved and lost, which requires that the physical stability of the remaining quartz fill is sufficiently stable to provide the rock and neighbouring clay plugs with adequate support. For this purpose the grain size distribution of the quartz grains is of Fuller-type, implying that smaller grains fill up the space between larger grains, a principle that gives a high density and prevents small particles from moving with percolating water.

A preliminary recipe, worked out by the Swedish Cement and Concrete Institute, CBI, and yielding pH in the range of 10–11 has been selected for use in the project (Table 4.8), [12].

The concrete is self-compacting and, compared to normal concrete, it is fairly viscous, i.e. like syrup. Application of the "Basic" type plugging in a 500 m deep hole at Olkiluoto in Finland has shown that it is possible to construct a plug with low pH concrete in a borehole even at a fairly great depth.

The evolution of strength has been tested and led to the conclusion that the strength increases slowly, especially at low temperatures. However, the ultimate

Table 4.8 Composition of bore hole concrete [12]

Components	Kilograms per cubic meter of concrete
White cement (Aalborg Portland)	60
Water	150
Silica Fume (Elkem)	60
Fine ground α-quartz (Sibelco)	200
Fine ground cristobalite quartz (Sibelco)	150
Superplasticizer (Glenium 51 Modern Betong)	4,38 (dry weight)
Aggregate 0–4 mm (Underås, Jehanders Grus)	1679

strength is higher than that of normal construction concrete that normally contains more than 300 kg of binder (cement).

4.8.3.1 Performance

The material has self-compacting consistency and hardens to give a compressive strength of at least 10 MPa in 24 hours. This strength is sufficient to support the load of a 10 m clay plug segment. pH is not less than 11.

References

1. SKB, 2007. Fud-program 2007. Program för forskning, utveckling och demonstration av metoder för hantering och slutförvaring av kärnavfall. SKB, Stockholm.
2. Pusch, R, Borgesson L, Ramqvist G, 1987. Final Report of the borehole, shaft, and tunnel sealing test – Volume III: Tunnel plugging. Stripa Project Technical Report 87-03. SKB, Stockholm.
3. Svemar C, 2005. Prototype Repository Project. Final Report of European Commission Contract FIKW-2000-00055, Brussels, Belgium.
4. Pusch R, 1994. Waste disposal in rock. Development in Geotechnical Engineering, 76. Elsevier Publ. Co. ISBN: 0-444-89449-7.
5. Coons W, Bergström A, Gnirk P, Gray M, Knecht B, Pusch R, Steadman J, Stillborg B, Tokonami M, Vaajasaari M, 1987. State-of-the-art report on potentially useful materials for sealing nuclear waste repositories. Stripa Project, Technical Report 87-12. SKB, Stockholm.
6. Onofrei M, Gray M, Pusch R, Börgesson L, Karnland O, Shenton B, Walker B, 1992. Sealing properties of cement-based grout materials. Stripa Project, Technical Report 92-28. SKB, Stockholm.
7. Alcorn S, Coons W, Christian-Frear T, Wallace M, 1992. Theoretical investigations of grout seal longevity – Final Report. Stripa Project, Technical Report 92-23. SKB, Stockholm.
8. Börgesson L, Pusch R, Fredriksson A, Hökmark H, Karnland O, Sandén T, 1991. Final report of the rock sealing project – Sealing of the near-field rock around deposition holes by use of bentonite grouts. Stripa Project, Technical Report 91-34. SKB, Stockholm.
9. Pusch R, Börgesson L, Fredriksson A, Markström I, Erlström M, Ramqvist G, Gray M, Coons W, 1988. Rock sealing – Interim report on the rock sealing project (Stage I). Stripa Project, Technical Report 88-11. SKB, Stockholm.
10. Chandler NA, Martino J, Dixon DA, 2002. The tunnel sealing experiment. In Trans. 6th Int. Workshop pn Design and Construction of Final Reprositories: Backfilling in Radioactive Waste Disposal. ONDRAF/NIRAS, Brussels, Belgium 2002, March 11–13.
11. Pusch R, Karnland O, Hökmark H, Sandén T, Börgesson L, 1991. Final report of the rock sealing project – Sealing properties and longevity of smectitic clay grouts. Stripa Project Technical Report 91-30. SKB, Stockholm.

12. Pusch R, Ramqvist G, 2007. Borehole Project – Final report of Phase 3. SKB Report R-07-58. SKB, Stockholm.
13. Gunnarsson, D, Börgesson L, Hökmark H, Johannesson LE, Sandén T, 2002. Installation of the Backfill and Plug Test. in, Clays in natural and engineered barriers for radioactive waste confinement: Experiments in Underground Laboratories, ANDRA Science and Technology Series Report, ISBN: 2-9510108-5-0, Chatenay-Malabry Cedex.
14. Pusch R, Börgesson L, 1992. PASS – Project on Alternative Systems Study. Performance Assessment of bentonite clay barrier in three repository concepts: VDH, KBS-3 and VLH. SKB Technical Report TR 92-40. SKB, Stockholm.
15. Pusch R, 1994. Rock mechanics on a geological base. Developments in Geotechnical Engineering, 77. Elsevier Publ. Co. ISBN: 0-444-89613-9.
16. Harrison T, 2000. Very deep borehole. SKB Report R-00-35. SKB, Stockholm.
17. Kozlovsky Ye A, 1984. The world's deepest well. Scientific American. Vol. 251, No. 6 (pp. 106–112).
18. Popov V, Pusch R, (Eds), 2006. Disposal of hazardous waste in underground mines. WIT Press, Southampton, Boston. ISBN: 1-85312-750-7.
19. Autio J, Börgesson L, Sandén T, Rönnqvist P-E, Berghäll J, Kotola R, Parkkinen I, Johansson E, Hagros A, Eriksson M, 2007. KBS-3H design description 2006. Report SKB R-08-32.
20. Autio J, Saanio T, Tolppanen P, Raiko H, Viena T, Salo J-P, 1996. Assessment of alternative disposal concepts. POSIVA Technical Report POSIVA-96-12.
21. Seidler WK, Bosgiraud JM, 2005. Common features of design studies. Interim Report of EC Project Contract Number F16 W-CT-2004-508851.
22. Svemar C, 2005. Cluster Repository Project (CROP). Final Report of European Commission Contract FIR1-CT-2000-20023, Brussels, Belgium.
23. Gray MN, 1993. OECD/NEA International Stripa Project Overview Volume III: Engineered Barriers, Published by Swedish Nuclear Fuel and Waste Management Co. ISBN 91-971906-4-0. SKB, Stockholm.

Chapter 5
Underground Laboratories (URLs)

5.1 Needs and Objectives

Naturally, experience from construction of repository-like rooms at relevant depth is a necessary prerequisite for working out and proposing doable and functioning repository concepts, and the corresponding insight in use of materials for constructing engineered barriers (EBS) is required for designing canisters, buffers and backfills. Likewise, experience from modelling the performance of rock and engineered barriers is needed for predicting the integrated performance of the two in order to find optimal dimensions etc of the various barrier components through parametric analyses.

The problem is that repository technology is not an academic issue, it involves a lot of practical estimates and measures and admittance of the fact that even the fundamental physical behaviour of rock and clay soils is not too well understood or known. Hence, a lot of empirical relationships and simplifications have to be found and applied in the scoping calculations and in later analyses with presumed higher accuracy. Since blind predictions can lead to totally wrong conclusions some coupling of the actual performance of the underground space with EBS to the theoretical models has to be made, which requires field experiments in underground laboratories (URLs). They are needed already in the first phase of planning the design and construction of a repository, which shall start with characterization of the rock mass in question and of its structural constitution, followed by estimating the rock mechanical performance and the hydraulic ("geohydrological") and chemical behaviour. The problem is that structural features and various physical and chemical rock properties are site specific and can be quite different from those of the finally selected part of a rock mass for constructing rooms, tunnels and shafts. This is a less important issue for engineering barriers, except with respect to the potential of the rock to provide clay buffers and backfills with water for saturation.

A major question that was much discussed in the CROP project [1] was if instrumentation can affect the buffer and backfills, and how the heat production for simulating the decay of the radioactive waste shall be made. A related question is if

R. Pusch, *Geological Storage of Highly Radioactive Waste*,
DOI: 10.1007/978-3-540-77333-7_6,

HLW should be used for running realistic experiments or if one has to be content with using electrical heaters for simulating the heat production. We will touch on both issues here.

5.2 National Underground Laboratories in Crystalline Rock

5.2.1 General

Five large-scale URLs in crystalline rock have been or are being used and they have all provided very valuable information for calibrating theoretical models and for ruling out less practical construction techniques. Also, they have been of particular value for development of site investigation techniques and for conducting experiments with buffers and backfills. They are the Swedish Stripa mine and Äspö test site, the Swiss Grimsel test site, the Canadian Pinawa test site and the Finnish Onkalo test site. We will consider them here for getting an overview of the testing principles and selection of experiments.

5.2.2 Stripa (SKB)

The *Stripa URL* was excavated at 300–400 m depth in granitic rock about 200 km NW of Stockholm when mining of haematite ore ended in 1987. It was used for a joint Swedish/American research project (SKB/LBL[1]) started in the late seventies and was followed in 1981 by the international Stripa Project. The area was characterized with respect to the large-scale structure and hydrology and to rock mechanical conditions in the first phase and a number of experiments were performed on various scales in the URL in the second phase that ended in 1992. The main ones related to EBS were [1,2]:

1. *Characterization of near-field rock.* This work comprised structural identification and hydrological and geochemical characterization by logging in shallow and deep boreholes and by cross-hole investigations. Rock stress measurements were included as well.
2. *Buffer mass test.* This experiment, which represents the KBS-3 V concept with canisters embedded in highly compacted bentonite (buffer) clay, took place in a blasted drift. Focus was on the thermal, hydraulic and mechanical (THM) evolution of the buffer and backfill components of the EBS.
3. *Borehole, shaft and tunnel plugging.* The experiments comprised sealing of boreholes with up to 100 m length and recording of the sealing effect, plugging of a 1 m diameter shaft for investigating the effect of excavation damage, and backfilling of a blasted tunnel with 20 m^2 cross section for the same purpose.

[1]Lawrence Berkeley Laboratories, California, USA.

4. *Sealing of fractured rock.* This project comprised experiments on different scales for investigating if the zone of excavation damage can be sealed by grouting of short holes and if natural fracture zones can be effectively sealed by grouting of medium-long holes drilled in them. The study also comprised grouting of large-diameter boreholes using "megapacker" technique and "dynamic" injection of grouts.

The major results from the experiments are briefly described under the respective headings in this chapter.

5.2.3 Äspö (SKB)

The Äspö URL was excavated in complex but granite-dominated rock about 200 km S of Stockholm for the purpose of demonstrating all the major steps to be taken in constructing and assessing the performance of a repository in virgin rock. It comprised comprehensive structural and hydrological characterization of the repository rock in conjunction with construction of a blasted helical ramp to about 400 m depth and several blasted and TBM-bored drifts in which a number of EBS-related experiments are being conducted. The construction started in 1993 [1].

The main experiments related to EBS were [1]:

1. *Characterization of near-field rock.* This work comprised structural identification and hydrological characterization by logging in short and long boreholes and by cross-hole investigations, as well as rock stress measurements.
2. *Prototype repository project.* This full-scale experiment, which simulates a more than 50 m long part of a KBS-3 V repository with actual copper/iron canisters embedded in highly compacted bentonite clay, is conducted in a TBM-bored drift. Focus is on demonstrating construction of buffer and backfill components and to record and evaluate the THMC evolution of the near-field rock, buffer and backfill.
3. *Backfill and plug test.* The experiment comprised backfilling and recording of the tightness of the backfill in a blasted drift and of a large concrete plug at its outer end, extending into the surrounding rock for sealing off the excavation-damaged zone.
4. *Long-term experiments* of KBS-3 V type for investigating the chemical stability of canister-embedding smectite-rich buffer.
5. *Manufacturing and placement of canisters.* Uniaxial compression of clay powder for manufacturing large buffer blocks under loads of up to 30 000 t (3E2 MN). Pilot tests of manufacturing "tea-cup type" blocks of half required size by isostatic compression technique.
6. *Predictive modelling* of the performance of clay buffers and backfills.

The major results from the experiments are briefly described under the respective headings in this chapter.

5.2.4 Grimsel (NAGRA)

The Grimsel URL was excavated at about 400 m depth in rather homogeneous and tight granite by use of TBM technique in the early 90s. Xenolithic seams of permeable rock and widely spaced water-bearing fractures and fracture zones of long persistence, representing 3rd and 4th order discontinuities, have been identified. The laboratory has hosted various national experiments as well as ENRESA's comprehensive FEBEX project. Test activities in the form of experiments performed by NAGRA have been preceded by comprehensive structural and hydrological characterization of the rock mass on various scales.

The main NAGRA experiments related to EBS were [1]:

1. *Rock mechanical tests* for determining the rock stress and stability conditions with special respect to temperature by performing heater experiments.
2. *Plug test.* The experiment comprised construction of a large concrete plug and recording of its tightness with respect to water and gas pressures.
3. *Predictive modelling* of the performance of clay buffers and backfills.

The experiments made as part of the FEBEX project have involved:

1. *Inflow tests for hydraulic characterization* of the TBM tunnel that was extended in order to host a test with horizontally oriented canisters embedded in highly compacted clay.
2. *Design, construction and performance of a full-scale buffer clay experiment* representing the ENRESA concept with horizontally oriented canisters embedded in highly compacted smectite-rich clay.
3. *Plug test.* The experiment comprised construction of large concrete plugs and recording of their tightness with respect to water and gas pressure.
4. *Predictive modelling* of the performance of clay buffers and backfills.

FEBEX was an international project operated at Grimsel for testing the emplacement geometries considered by NAGRA, ENRESA and, to a degree, the KBS-3H concept [3]. The concrete plug portion of FEBEX measured approximately 2.7 m in length in a circular borehole of 1.9 m diameter. It was keyed in the rock to a depth of 0.4 m along about 1.6 m of its length. Excavation was accomplished using saws followed by mechanical disintegration for minimizing damage. The concrete used was a low-heat, low-shrinkage mixture that was cast in 3 layers without reinforcement. As a result it did not provide a tight contact with the rock at the top. It was designed to resist a combined swelling and hydraulic pressure of 5 MPa on its upstream face but there were no requirements for it to provide evidence of either gas or water tightness.

The buffer material used for the FEBEX project consisted of sector-shaped blocks of highly compacted montmorillonite-rich Spanish bentonite. It was placed, as in all other field- and Mock-up tests of this type, around heaters, simulating canisters with reprocessed HLW. The clay was prepared by compressing clay powder to a dry density of maximum 1600 kg/m^3 in order to limit the swelling pressure to

5 MPa. Considerable leakage past the concrete bulkhead was reported in this test, predominantly via cable ports, but it was not quantified.

5.2.5 Pinawa (AECL)

The AECL underground facility at Pinawa was constructed in the mid 80s by drill- and blast-reaching down to about 400 m depth in rather homogeneous, very tight granitic rock. Comprehensive structural characterization revealed the presence of three very big subhorizontal fracture zones (2nd order type), a few fracture zones of 3rd order, and a low frequency of water-bearing fractures of 4th order. Stress measurements gave a good picture of the impact of stresses on the hydrology and stability, the latter suggesting critical stability conditions at depth, which was confirmed when the excavation had reached deeper than about 200 m.

The main experiments related to EBS were [1]:

1. *Characterization of near-field rock.* This work comprised determination of the character of the walls of full-scale vertical deposition holes made by water jet excavation, and hydrological characterization.
2. *Buffer tests.* Full-scale experiments were performed in deposition holes comprising preparation and placement of canisters and buffer clay, which was investigated with respect to the evolution of temperature and hydration rate. The buffer consisted of mixtures of bentonite powder and quartz sand. It was compacted on site.
3. *Backfill test.* A shaft was backfilled for investigating the achievable density by use of a special compaction technique.
4. *Plug test.* The experiment comprised construction of two plugs, one of concrete and one of steel, for recording their tightness at pressurizing with water, and for characterizing the EDZ.
5. *Predictive modelling* of the performance of clay buffers and backfills.

5.2.6 Onkalo (POSIVA)

The URL rooms were excavated in the Olkiluoto area at about 200 m depth in rather homogeneous and tight gneiss (tonalite) by use of drill-and-blast technique in the mid ninetees. Deposition holes with dimensions corresponding to those of the KBS-3 V concept were drilled by use of a new TBM-type technique with removal of the muck by vacuum. Comprehensive structural and hydrological analyses were performed resulting in detailed maps of the walls of the holes, and results from inflow measurements.

The main experiments related to EBS were [1]:

1. *Rock stress measurements* for determining the stability conditions with special respect to possible future heater experiments.

2. *Correlation of rock structure and water inflow* into the deposition holes as a basis of modelling of the hydration rate of clay buffer.
3. *Sampling of rock in the deposition holes for determination of gas and hydraulic conductivity* of the most shallow part, i.e. the boring-disturbed rock.

5.3 National Underground Laboratories in Salt Rock

5.3.1 General

In Chaps. 3 and 4 we distinguished between bedded and domal salt and we will do the same in the present examination of URLs in salt rock. Three large-scale URLs have been used for developing investigation techniques and for conducting experiments with construction of repositories in salt and buffers and backfills: the US WIPP repository serving as a large-scale URL, and the two German URLs in salt domes at Gorleben and Asse. We will describe them briefly here for providing overviews of the testing principles and selection of experiments.

5.3.2 Bedded Salt

The US programme for siting and development of deep geological repositories in bedded salt rock has used both on-site and off-site URLs since the early seventies to obtain the large-scale rock mass information required to support the design, construction, and operation of deep geological repositories for safe disposal of long-lived radioactive waste. However, since 1987, only the WIPP URL has been used to support the development of repositories in salt rocks [1].

The primary initial objective of this URL was to facilitate the acquisition of large-scale, in-situ data required for:

- Corroborating and upscaling the data obtained by laboratory- and surface-based tests.
- Refining underground experiments, testing and excavation techniques, conducting repository design, and developing process and system models.
- Credible prediction of the long-term safety of the proposed TRUW repository.
- Acquiring information on HLW disposal in salt rock. The suite of large-scale in-situ experiments conducted in the WIPP URL gives the basis of selecting testing and excavation techniques, repository design, and principles of working out process and system models.

The location of the WIPP URL was selected in 1974 based upon the information obtained through investigations from the surface at and adjacent to the site. A basic siting criterion was that the URL and the potential repository would be located reasonably close together. An important design criterion was that ground support should be minimized to optimise the inherent "self-healing" characteristic of rock salt. The construction of the first of four shafts and the WIPP URL commenced in

1982, and was essentially completed in 1986. A suite of large-scale, in-situ tests were conducted in the URL between 1983 and 1995 after which portions of the WIPP URL have been closed in stages.

A large number of different types of instruments were used in the various experiments conducted in the WIPP URL to acquire the data required for the design, construction, certification, operation, closure, and for recording the post-closure performance of the WIPP repository (Fig. 5.1). Many of the tests conducted in the URL involved high temperatures that simulated HLW disposal.

The large number of experiments conducted at the WIPP site during the past 30 years comprised thermal/structural interaction tests, plugging and sealing tests, waste package performance tests, and hydraulic testing. More than 4200 instruments and gauges were used in the experiments, which had the primary purpose to develop an adequate understanding of the physical processes and derivation of parameters needed for designing, constructing, operating, and decommissioning a deep geological repository for safe disposal of TRUW and DHLW in salt rock. With time, the use of the codes and models that supported the prediction of the behaviour and safety of the EBS and the WIPP repository for at least the 10 000 year regulatory period became increasingly important.

The major test series included the following large-scale in-situ experiments, which essentially addressed the mechanical behaviour of rock salt as influenced by excavation effects, stress, and thermal loading, and interactions induced by waste emplacement:

- An 18 W/m^2 Mock-up Test.
- In-situ stress determination by hydraulic fracturing.
- Heated axisymmetric pillar test.
- Scale-effect tests.

The tests were designed to produce data for development and validation of numerical codes, predictive models, and calculation techniques used for the design work. A particularly important matter was creep for which rod-extensometers proved to be the most useful instrument. Convergence measurements provided direct information on the changing dimensions of the excavated openings. Because

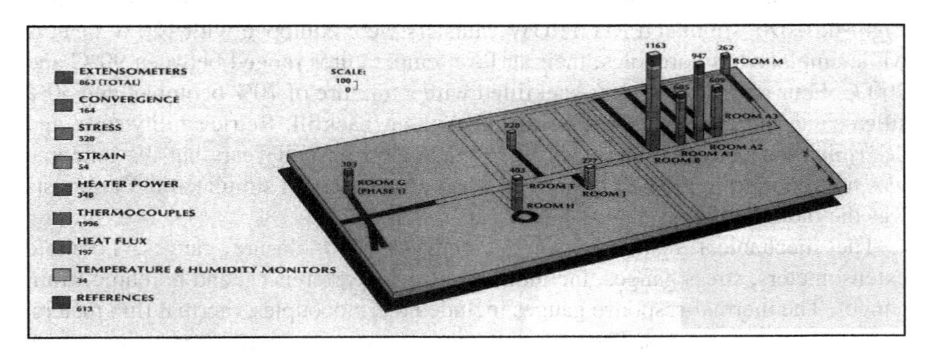

Fig. 5.1 Instrument distribution in the WIPP URL [1]. (By courtesy of DOE)

the dimensions of the excavations were critical parameters in the modelling, the experimental rooms were excavated with precision and laser techniques employed to put all these relative movements on an absolute basis. Extensometers and convergence gauges proved to be more useful than the large number of stress gauges used due to the creep of the salt around the stress gauges, which compromised the recorded stress value. This experience agrees with the conclusions from comprehensive measurements in clastic clay.

The hydraulic tests included the following groups of experiments:

- Structural characterisation, including hydraulic and gas conductivities comprising moisture transport and release.
- Construction and performance assessment of seals, including plug tests in tunnels, shafts and boreholes.
- Backfilling of salt rock; design and emplacement.
- Fluid and gas pressure testing to determine the leak rates around and through the various seals.
- Periodic measurements for evaluation of the progressive creep closure around seals and of the EDZ.
- Coring of salt-seal interfaces for examination in the laboratory of chemical interactions and bond strength. Samples taken at termination of certain experiments.

Hydraulic testing was made by use of conventional electronic instrumentation to record water levels and pressure fluctuations in 10 to 30 deep boreholes. Testing in some salt layers presented unique problems due to the very low conductivities, and special packer systems were designed and used to assure a tight borehole seal. Both pressure build-up and pressure decay measurements were used to calculate the conductivity of virgin rock and EDZ. Pure virgin halite had conductivities of less than E-14 ms, which was about the limit of the testing capability, but it was appreciably higher, i.e. up to 7E-11 m/s in clay-rich salt. In contrast to what causes problems in crystalline rock, water inflow during tunnel and shaft construction did not cause problems in the WITT.

A total of 174 208 liter standard oil drums filled with simulated CH-TRUW were emplaced in a big room. Some were immersed in a brine pool and others covered by backfill of salt and salt/bentonite while some were subjected to humid air only. Eight full-size, simulated RH-TRUW canisters were equipped with 120 W heaters while emplaced in boreholes; their surface temperatures ranged between 90°C and 200°C. Four of the holes were backfilled with a mixture of 70% bentonite and 30% silica sand and the other four holes left without backfill. Retrievability tests and determination of corrosion attack were made after several years but the outcome was not reported in the CROP project. The only aspect not simulated in these tests was the radiation field.

The mechanical response gauges included room-closure gauges, borehole extensometers, stress gauges, inclinometers, survey references, and borehole strain gauges. The thermal-response gauges included thermocouples, thermal flux meters, and heater-power gauges. The environmental gauges comprised air velocity gauges and air temperature thermocouples that were installed to evaluate heat loss from the test rooms. The measurement accuracy of the various types of closure gauges, extensometers, and thermocouples was evaluated and it was found that these gauges all had an uncertainty less than +1% of the gauges' full-scale range.

Remote, electronic measurement was the principal recording technique but the data were checked by strategically located manual measurements. Redundancy was also applied to the main cable plant, carrying the multiplexed data to the surface, by a totally independent cable plant in a different shaft. Most of the gages used in the WIPP URL were electronic that could be remotely monitored by an automated DAS.

5.3.3 Domal Salt

No special URL has been designed or constructed in domal salt to support the rock characterisation and the understanding of the safety and performance of a specific repository site. Instead, the German government has owned and used the Asse mine since 1965 as a URL to perform in-situ testing of repository concepts in salt domes. In the framework of these research activities, about 130 000 200 l drums containing LLW from research and industry activities and about 1 300 200 l drums containing non-heat-producing MLW were disposed in the Asse mine/URL. The main objective of these tests was the development of optimised transport, unloading, emplacement, and backfilling techniques [1].

After Asse became a research mine in 1965, the first experiments were conducted in boreholes in the existing non-backfilled rooms on the 490 m and the 750 m levels. Later on, test rooms were excavated in areas with specific geological features, e.g., mineral and moisture contents, to enable the investigation of special effects and processes. One such example is a temperature test conducted in the salt with high polyhalite ($K_2Ca_2Mg(SO_4)_4 \times 2H_2O$) concentration and, thus, higher water content. A heater test was conducted here to investigate the heat-induced decomposition of polyhalite and the accompanying release of the contained crystalline water into HLW disposal boreholes.

Based on more than 30 years of German experience, the most important processes that have been investigated in URLs and repositories dedicated to the safe disposal of HLW in salt rock formations are:

- Temperature (cyclic).
- Stress, strain, and strength.
- Chemical conditions.
- Hydraulic and gas conductivities.

Temperatures were recorded by thermocouples and resistance temperature detectors (RTDs). The platinum-type RTD was more accurate than the thermocouple, especially at moderate temperature.

The closure of underground openings was determined by convergence measurements. The extensometers were installed immediately after excavation to ensure that data on the very significant deformations caused by the initial primary creep were obtained. The maximum measuring range was up to 400 mm at temperatures up to 180°C.

Deformation of deposition holes was recorded by multiple point glass-fibre rod extensometers sliding inside protecting PVC tubes for measuring axial and radial strain.

Glötzl-type hydraulic pressure cells oriented in different directions were used for stress measurement, which meets difficulties because of the inherent stress release due to creep. After installation, pressure cells take on stress only gradually and it is concluded that the actual stress is not approached until after E8 days or a couple of million years. Pressure development in the EBS experiments was monitored by electro-pneumatically operated, hydraulic Glötzl pressure cells and "Absolut Widerstandssprung Druckmesskissen" (AWID) cells. The latter can be operated at high temperatures with high accuracy. Large flat jacks and overcoring were also used to determine stresses.

Geoelectric surveying, being a non-destructive method, was used for measuring the water content in the rock mass and backfilled underground openings. The resistivity changes in surveyed areas were related to hydration or desiccation effects and could be measured by use of special electrode arrays. This method was also used for measuring the successive consolidation of the EDZ, and it was also successfully used for recording water saturation of clayey backfills in SKB's Prototype Project in the Äspö URL.

Newly prepared test rooms were located far away from old mine excavations to avoid influences of the disturbed stress field. For example, Fig. 5.2 shows

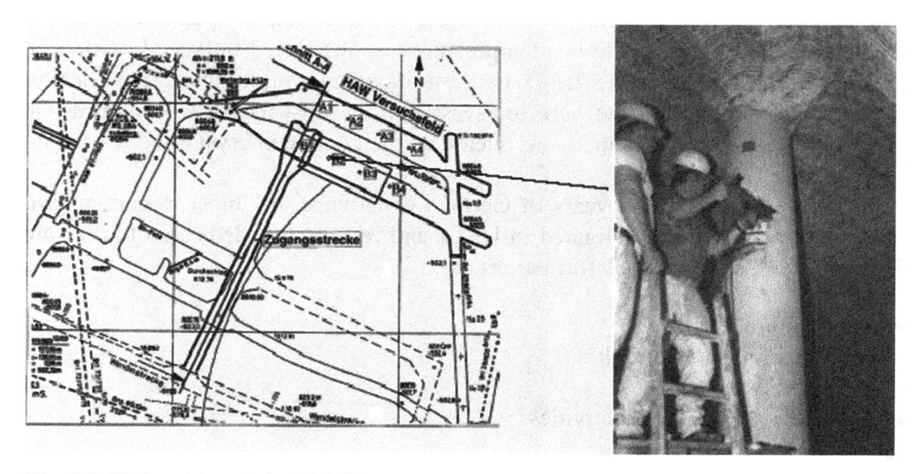

Fig. 5.2 Undermining of the DEBORA experiment in the Asse mine. *Left*: Test field with access drift to the experiment area 800 m deeper down. *Right*: Excavation for retrieval of corrosion experiment details [1], (By courtesy of GRS)

the uncovered heater liner of the so-called DEBORA-2 experiment covered by compacted crushed salt backfill and technicians removing the remaining backfill from the liner for examining a corrosion experiment [1].

5.4 National Underground Laboratories in Argillaceous Rock

5.4.1 General

Currently, three underground rock laboratories are in operation or under construction in argillaceous rock and dense, clastic clay. They are [1]:

- *Hades* URL at Mol, Belgium in Boom Clay (normally consolidated, clastic clay).
- *Mont Terri* International URL at St. Ursanne, Switzerland in Opalinus Clay. (claystone).
- *Meuse / Haute Marne* URL at Bure, France, in Callovo-Oxfordian argillite (under construction in 2003).

The "*HADES*" URL is located at about 220 m depth in the Boom Clay Formation, which extends from 190 to 290 m depth. The most abundant minerals in Boom Clay are clay minerals, quartz and feldspars. The major clay mineral is illite, but also kaolinite, mixed-layer swelling minerals (illite/smectite), and chlorite are present. Traces of glauconite and pyrite are found in carbonate concretions and in nodules and tubes around organic remains. No significant variation or trend has been observed with respect to depth. The Boom Clay is a nearly homogeneous stratum with a hydraulic conductivity of 2E-12 m/s (vertical) to 4E-12 m/s (horizontal).

The *Mont Terri* underground research facility is being used for examining a number of geological and technical issues but at the termination of the CROP project these studies were still in the conceptual development and planning stages. The URL is situated in the Opalinus Clay, a formation consisting mainly of silty and sandy claystones. The formation can be characterised as strongly overconsolidated (cemented) clay with 40–80% of clay minerals and micas, 10–40% quartz, 5–40% calcite, 1–5% siderite, 0–2% pyrite and 0.1–0.5% organic carbon. The URL is located in the Folded Jura, a mountain chain caused by tectonic deformation in connection with the Alpine orogeny. The main tectonic structure of the area is the Mont Terri anticline. The hydraulic conductivity of the Opalinus Clay at the Mont Terri URL is on the order of E-12 m/s to E-13 m/s. No significant differences in hydraulic properties are observed in the different facies of the formation or in the tectonically induced fractures and fracture zones of 4th and 3rd orders.

The *Meuse/Haute Marne* site is located in the Parisian Basin, in the northern part of the Haute-Marne and in the south part of the Meuse Départements. The potential host rock is the Callovo-Oxfordian formation formed about 150 million years ago. It has its center at of 450 ± 100 m depth and a thickness of 110 to 150 m. The Callovo-Oxfordian formation is largely homogeneous and dips towards NW by about 1°. It overlies Dogger limestone and is covered by the Oxfordian and Kimmeridgian. No steep low-order discontinuities have been identified at the URL site, named Bure, but subvertical structural features exist elsewhere in the Callovo-Oxfordian. The major horizontal stress in the Callovo-Oxfordian strikes N155°E and is equal to 1–1.4 times the vertical stress. The minor horizontal stress is on the same order of magnitude as the vertical stress, implying that the stress conditions in the entire mass are largely isotropic.

The cemented Callovo-Oxfordian argillites are brittle but are reported to show significant creep. Their hydraulic conductivity ranges between E-14 and E-13 m/s. Chemically, the Callovo-Oxfordian formation is reducing and it is reported to have a strong buffering capacity with respect to alkaline disturbance (by concrete). Except for the large 1st order-type regional faults of the Marne the investigated sector has an appreciable geodynamic stability. Thus, the area has a very low deformation rate and a very low seismic potential [1].

5.4.2 Research and Development

5.4.2.1 Rock Structure

In contrast to the huge structural variations in crystalline rock, which make it difficult to model hydraulically and geochemically and to find proper locations for deposition tunnels and holes, the macroscopic heterogeneity of argillaeous rock can be insignificant. However, on the detailed scale there can still be very large variations and this issue has been in focus of the respective Organizations and led to detailed geological surveys for investigating:

- Gaps in the sedimentation using chronological reconstitution methods.
- Thin layers using new 3D seismic processing methods.
- Spatial distribution of physico-chemical properties.
- Discontinuities of all orders.

5.4.3 Activities in URLs in Argillaceous Rock

5.4.3.1 HADES

The current facility "HADES" (High Activity Disposal Experimental Site) consists of underground galleries with a total length of some 200 m at a depth of 223 m (gallery axis), and an inner diameter ranging from 3.5 to 4 m [1]. An overview of the current and planned underground infrastructure is shown in Fig. 5.3. After construction of the first access shaft in 1980, in which both aquifer and clay had been conditioned by freezing, the first gallery ("URL"), 35 m long, was constructed in 1982 in frozen clay and lined with cast iron segments. A 65 m extension was excavated in 1987, without prior freezing of the host rock. It was lined with concrete lining segments and later backfilled.

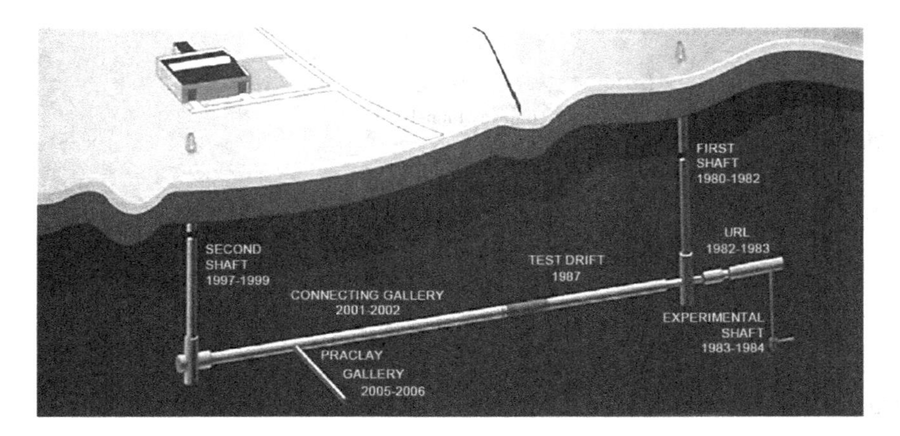

Fig. 5.3 Construction history of the HADES underground research facility [1]. (By courtesy of CEN/SKN)

From 1995 to 1997, a second access shaft and galleries extending from it were constructed, applying freezing where sand layers and a 190 m deep transition zone were intersected. Excavation was made manually and by using road headers. Blasting was also tried but was abandoned because of the disturbance caused. A watertight lining was needed in both shafts and a polymer (PE) membrane was placed between two on-site cast concrete linings in one of the shafts, while prefabricated concrete rings with an outer steel jacket were used in the other.

The construction work was primarily focused on in the project with determination of the creep properties of the virgin rock as second issue. The latter were investigated by careful measurements of the change in diameter of boreholes of various size. A third issue was investigation of waste package components, particularly in-situ corrosion experiments, and the hydration of backfill material, and gas permeation of virgin rock. The combined effect of heat and gamma radiation on clay and backfill was investigated in one set of experiments and chemical and bacteriological phenomena in another. A full-scale, integrated in-situ test of the repository concept, i.e. the "PRACLAY" test, was in the design phase at the termination of CROP [1].

5.4.3.2 Mont Terri

The Mont Terri URL was constructed from an existing highway tunnel system (Fig. 5.4) and excavated by different techniques: (1) drill-and-blast, road header reaming, pneumatic hammering, and modified raise boring.

An important conclusion from the construction work was that support with steel fibre reinforced shotcrete of 200 mm thickness is sufficient for stabilising tunnels of 3–5 m diameter. In some cases, depending on the tunnel orientation with respect to the stress field and to the bedding, as well as to the existence of 3rd order discontinuities in the form of fracture zones, tunnels may need additional support with rock bolts. A horizontal tunnel with 1 m diameter excavated by raise-boring technique did not require any lining.

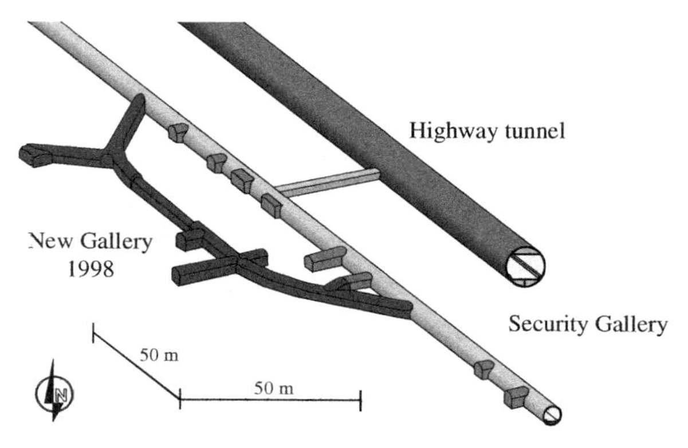

Fig. 5.4 Layout of the Mont Terri URL [1]

Several integrated experiments have been conducted [1]:

- Heater experiment in a vertical borehole in Opalinus Ton with the objective to investigate the response of the claystone to raised temperature and to study the interaction between the clay buffer and the host rock.
- Engineered Barrier Experiment (EB) comprising full-scale isothermal tests of the hydraulic behaviour of buffer consisting of compacted smectite-rich blocks, and granular bentonite.
- Ventilation Experiment for studying the influence of drying and wetting on the host rock focusing on the hydro-mechanical behaviour of the claystone.

5.4.3.3 Meuse/Haute Marne UR [1]

Two access shafts have been blasted in 2.4 m steps. A temporary support with grouted bolts and wire mesh was placed immediately after completing each of them and a permanent concrete lining finally installed, starting about 12 m from the open end. By the termination of the CROP project no testing of the virgin rock or buffers and backfills had yet been made.

The purpose of the URL research program is to provide reliable data on the system consisting of the repository integrated in the Meuse/Haute-Marne geological environment over a long period of time and to demonstrate that the repository performs acceptably from practical and safety points of view.

5.5 Study of Rock Properties in the URLs

5.5.1 Rock Structure, Geohydrology and Geochemistry

It was early realized that structural characterization of the host rock is necessary for modelling the hydraulic and geochemical performance of a repository and the first comprehensive investigation for this purpose was made in granitic rock at

Stripa in Sweden for identifying and characterizing important structural features. The initiative was taken by the Lawrence Berkeley Laboratories, USA, which performed, in co-operation with SKB, a field study in the late 80s that made this test site a real URL. The main objectives of this study were to investigate the hydraulic conductivity of crystalline rock on all scales and to determine the impact of high temperature on rock surrounding large boreholes with simulated HLW canisters. No complete structural model was derived at this time but later compilation of the outcome of topographic studies, deep drillings, geophysical measurements and piezometric data has shown that the granitic dome is a rather typical mass of "good" crystalline rock.

The granitic region where the Stripa mine is located is characterized by 1st and 2nd order discontinuities oriented and spaced as in Fig. 5.5. On this large scale the major low-order discontinuities make up two steep NW-SE and NE-SW oriented sets. Close examination of finer weaknesses has shown that those of 3rd and 4th orders also have these orientations and that there is one more set of such weaknesses that is more or less subhorizontal [4].

Figure 5.6 shows a generalized rock structure model of the Stripa rock mass derived for performing groundwater flow analyses and transport of contaminants from a central waste-containing room, appearing in Fig. 5.7.

BEM is a particularly suitable calculation method for dealing with complex systems since it requires significantly less computer power than FEM. The code

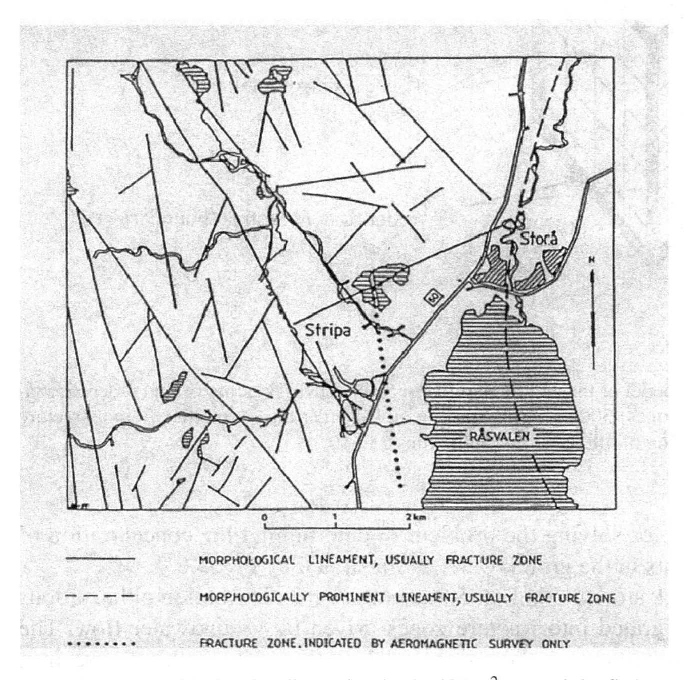

Fig. 5.5 First and 2nd order discontinuties in 48 km^2 around the Stripa mine [1,4].

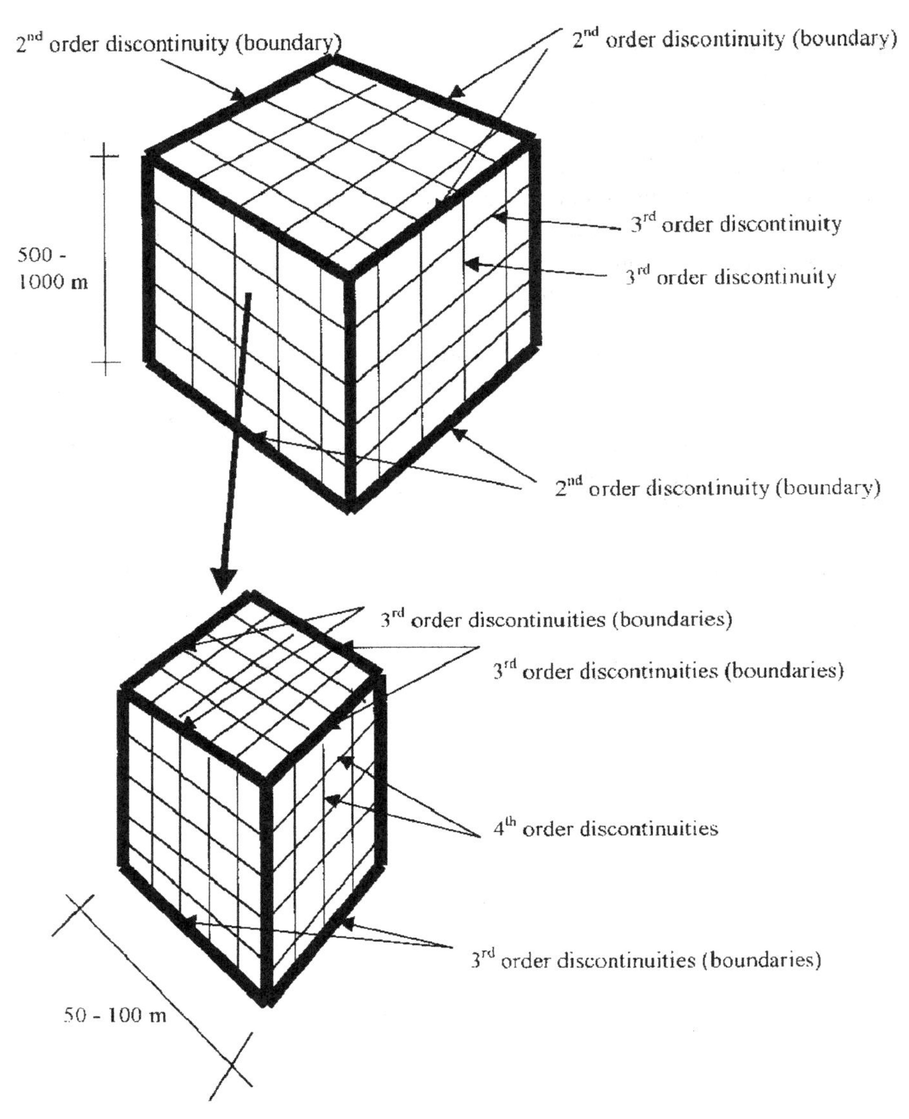

Fig. 5.6 Rock structure model of the central part of the Stripa URL. Spacing of 2nd order discontinuities (major fracture zones) 500–1000 m. Spacing of 3rd order discontinuities (minor fracture zones) 50 to 100 m. Spacing of 4th order discontinuities 5 to 10 m [5]

BEASY [6] was used for solving the problem of determining the concentration of hazardous contaminants in the groundwater as illustrated by Fig. 5.8.

The transparent rock structure in Fig. 5.8 illustrates the distribution of hazardous chemical elements migrated into fracture zones driven by groundwater flow. The concentration is at maximum in the room and less than 5% of this figure in a

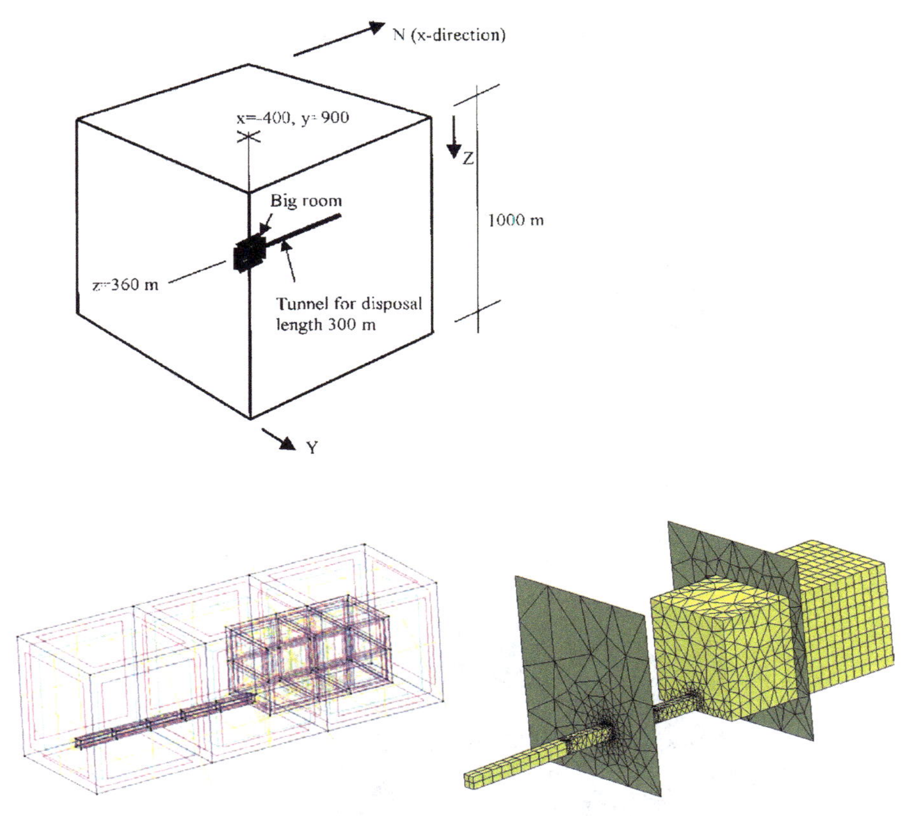

Fig. 5.7 The Stripa room and tunnel. The upper picture shows the room and tunnel viewed from east, and the lower picture the BEM model used for calculating percolation of them assuming a regional hydraulic gradient operating in the S/N direction. The rock structure model implies that each of the two spaces is intersected by a fracture zone of 3rd order [5]

fracture zone about 100 m from the room. The spectacular and apparently convincing presentation is of course correct on the premises given with respect to structural features, hydraulic gradient and conductivity but one realizes that the statistical variation of the individual parameters can combine to give quite different migration patterns.[2]

The Stripa Project comprised several attempts to test the validity of both large-scale and local structural models by conducting cross-hole measurement of water

[2] In this context it should be added that the uncertainty respecting the validity of models of the structure of crystalline rock for large-scale hydraulic calculations, led one of the Organizations to give up its attempt to consider crystalline rock for hosting a HLW repository and instead choose argillaceous rock in which diffusive migration of water and contaminants dominates.

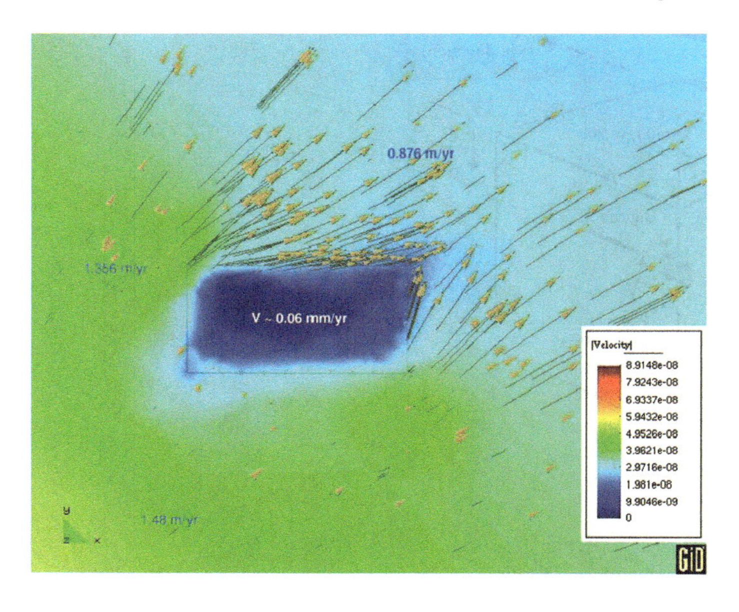

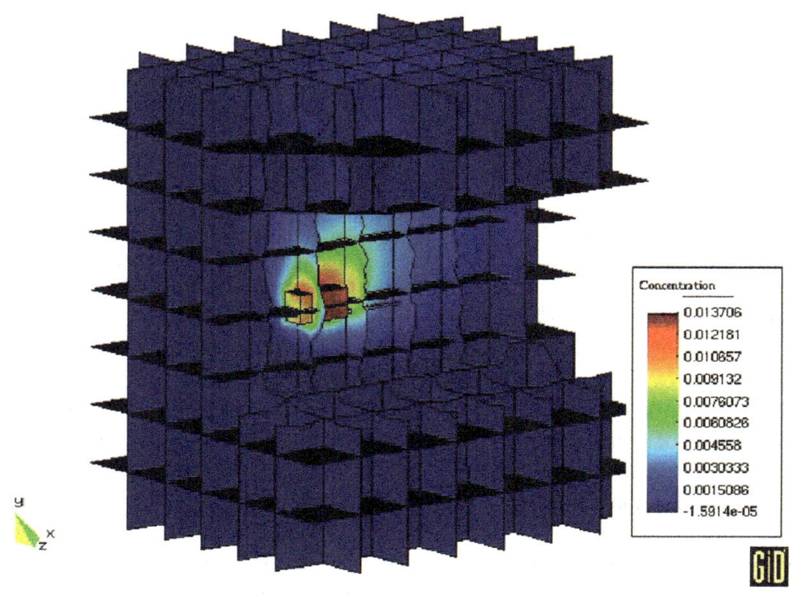

Fig. 5.8 The Stripa room and tunnel. The *upper* picture shows a side view of the hydraulic velocity field near the large room, which was assumed to be surrounded by a 1.5 m wide EDZ with 100 times higher hydraulic conductivity than virgin rock. The very low flow rate within the room is due to the clay-embedment of the waste and the much higher flow along the roof is caused by the interaction of the EDZ and the intersecting 3rd order fracture zone. The graphs show the calculated concentration distribution of contaminant released from the room after 127 000 years [5]

pressure and tracer tests with some success. For the near-field, i.e. the rock surrounding deposition holes and tunnels, prediction of groundwater flow largely failed, partly but not wholly because the existence of EDZ was not realized or accepted. Its existence was, however, verified by "ventilation" tests in the "BMT" drift already in the Lawrence Berkeley Laboratory's comprehensive field tests [7] and later confirmed by the Stripa Project [2].

The discouraging outcome of the flow modelling of the structurally relatively simple rock structure of the Stripa granite called for more systematic work, which continued in SKB's second URL at Äspö. At this time rock structure modelling had become an academic issue based on advanced statistics that is presently used extensively. The rock structure is taken to consist of statistically distributed fractures, and deterministic features representing major fracture zones. The fracture systems are defined by fracture intensity, orientation and properties assuming statistical distribution models [1]. The deterministic major features are site-specific and characterized by comprehensive flow measurements and in fact correspond to the categorized discontinuities of various orders used in this book.

What is then the reason for the difficulty in accurate prediction of groundwater flow in crystalline rock? Truly, the tools for identifying the deterministic, major facture zones, are there, but their complex, varying internal structure, intersected by boreholes for pressure and flow measurements, can only be partly identified. For the statistically distributed discontinuities, i.e. those of 4th order, the possibility to predict and describe their hydraulic performance is poorer than for the major discontinuities since percolation of them takes place in channels, which are often formed by intersection of factures. Considering the very strong structural variations in crystalline rock and, above all, the heat-induced changes in aperture and the clogging by migrating fragments of fracture fillings and erosion of such fillings that follows from the eternal massage that the earth crust undergoes by tide phenomena, it would be reasonable to spend only limited resources on hydraulic modelling. Logically, geochemical modelling involving prediction of changes in salt content and composition is also very uncertain.

Figure 5.9 illustrates the said: comparison of the results from pressure measurements and determination of salt content shows similar but not altogether the same patterns and for the near-field there is very poor agreement. And what will the conditions be in 1000 years from now? And after the first of the forthcoming glaciations? The answer is of course that hydraulic and geochemical modellings are too uncertain to be taken as a basis of assessing repository concepts and selecting sites in crystalline rock. The performance of such rock in other respects than large-scale rock mechanics is therefore uncertain and the principle should be to regard the rock as a mechanical protection of the "chemical apparatus", i.e. the canisters and the buffer.

For salt rock there are almost no problems related to water-bearing structural features as concluded from work in URLs in domal salt, while they can provide difficulties with stability and local flow in bedded salt. In argillaceous soft and very stiff rock there can be minor difficulties with inflowing water and moist air can

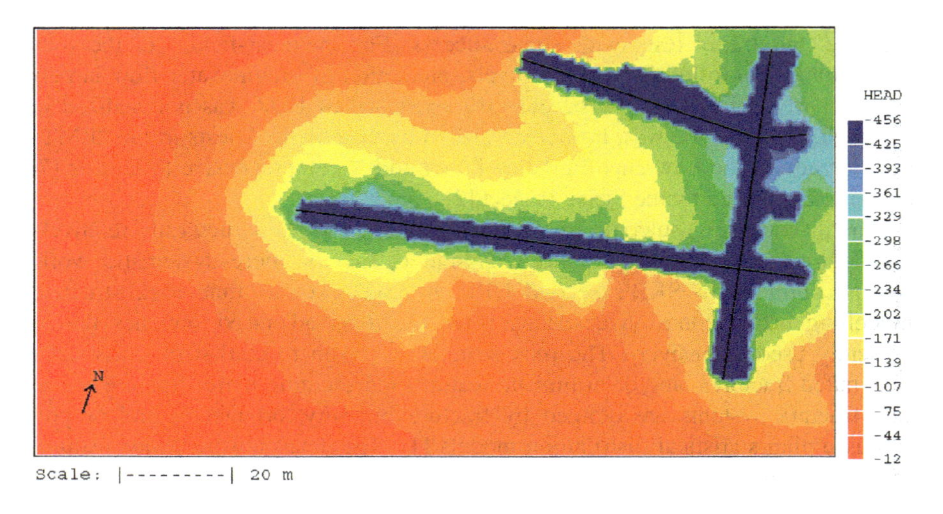

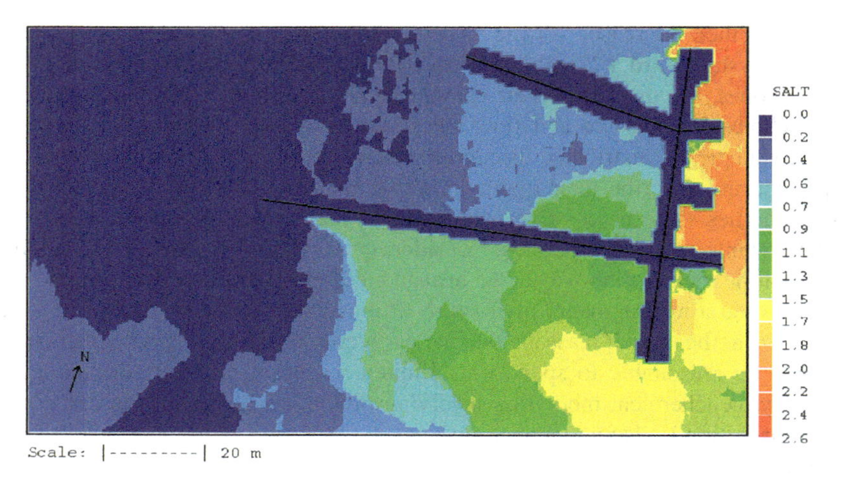

Fig. 5.9 Water pressure and salinity at 447 m depth at AEspoe URL in granitic rock. *Upper*: Pressure head (in meters). *Lower*: Salinity (in %). The water pressure at 2 m distance from the tunnel wall is between 100 kPa and 1.5 MPa (fits with measurements). The mean salinity of the water in boreholes is 5.5 to 8.7 g/l (no fit in the near-field) [1]

cause problems. Thus, wetting of Opalinus Ton caused by lack of ventilation can cause collapse of roof and walls of tunnels and drifts.

5.5.2 Stability

The high strength of crystalline rock is its most valuable property for hosting a repository since it provides excellent stability at reasonable depth. Slip of rock

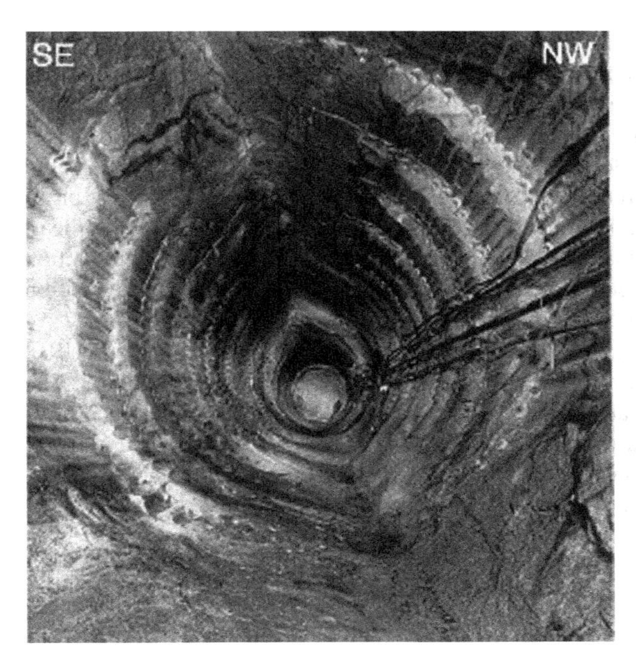

Fig. 5.10 Failed tunnel in the Canadian URL caused by high primary rock stresses [1]

wedges from roof and walls can occur but hardly collapse except where very high primary stresses prevail, like in the Canadian URL (Fig. 5.10) where horizontal stresses of more than 40 MPa prevail and caused rock burst that required elliptic cross section of blasted tunnels for minimizing stability problems.

The experience from this URL and from comprehensive theoretical analyses have indicated that problems with spalling and other stress-related unstable conditions are expected at primary stress levels of about 35 MPa and higher, unless drifts and tunnels are oriented favourably, i.e. parallel to the major principal stress. Horizontal deposition tunnels can then be constructed without great difficulties at depths up to about 400 m as indicated by ongoing investigations at Forsmark, Sweden. Concepts of the KBS-3 V type with vertical deposition holes will lead to critically high stresses in the surrounding rock and to breakage when heated by the HLW. Blasted tunnels can sustain higher primary stresses than TBM-bored ones since the EDZ causes stress relaxation and transfer of hoop stresses to a little bit outside the tunnel [4].

In summary, the URLs in crystalline rock have contributed very much to the knowledge by giving several opportunities to evaluate the stability of such rock but there are still uncertainties respecting time-dependent stress changes and the impact of tectonically and thermally induced bulk strain on the aperture of discontinuities that intersect deposition holes and tunnels.

The earlier mentioned problem with insufficient stability of openings in weak argillites and clay of softer type, like the normally or slightly overconsolidated Boom clay, has been investigated in detail in the Belgian and Swiss URLs. Comprehensive work with carefully measured stress and strain at and behind the advancing front of excavation in the Belgian URL has given detailed information on how and how fast supporting concrete liners must be placed for minimizing stability problems. Tests with heated holes and drifts for measuring thermal impact, like heat-induced creep, have been made (Fig. 5.11) but a remaining problem is the reduction in effective stress and strength in the natural Boom clay surrounding deposition tunnels by the heat-generated expansion of the porewater.

For overconsolidated argillaceous rock, like the Opalinus Ton in the Swiss URL, the stability conditions with respect to water have been investigated in detail. The experience is that water that starts flowing in tunnels will cause softening and initiation of disintegration and rock fall in a few days or weeks and that also moist air will cause such degradation with time, while tunnels ventilated with air with a low relative humidity can stay stable for many years.

5.5.3 EDZ

5.5.3.1 Crystalline Rock

The existence and role of the EDZ, appointed major hydraulic conductor in blasted rock as discussed earlier in the book, emanate from tests on different scales in the URLs. For crystalline rock, field experiments at Stripa and in AECL's URL have been of particular importance since they have given quantitative data. In summary, the existence of a "skin" zone with high hoop stresses and reduced hydraulic conductivity radially to blasted tunnels was proven as was also a hundred- to thousand-fold increase in conductivity of the blast-disturbed EDZ extending 0.3 to 0.8 m (in the floor) from the periphery. Similar results have been reported for crystalline rock

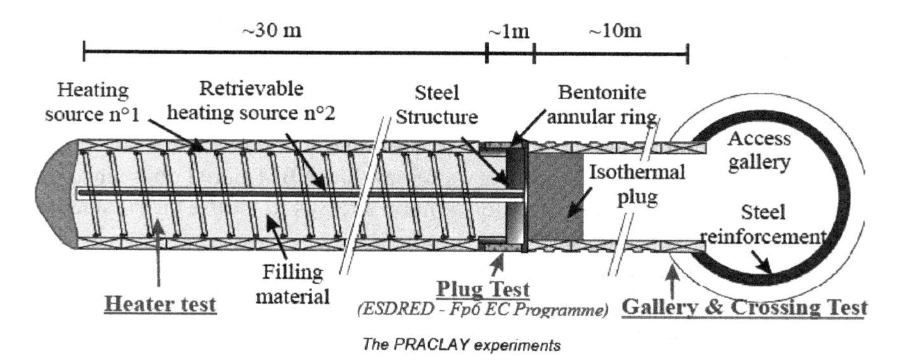

Fig. 5.11 Layout of the PRACLAY Experiments (After Bernier), [8]

in Germany [9]. The conductivity of the "skin" zone in axial direction of the tunnels was concluded to be about ten times higher than the average conductivity of the virgin rock.

5.5.3.2 Salt Rock

EDZ develops also in salt rock and extends up to some decimetres from large bored holes or excavated tunnels and drifts. However, it is of no importance as transport path for brine and possibly released radionuclides some tens of years after backfilling, since it self-seals by consolidation due to its very significant creep potential. In bedded salt, however, there are often thin layers of clay and silt that become embedded in the disintegrated material in the EDZ and counteract homogenization. Still, the average tightness of the EDZ is estimated to be fairly good since the non-salt components are not uniformly distributed after consolidation under confined conditions [1].

5.5.3.3 Argillaceous Rock

Strain measurements and comprehensive structural analyses in the Swiss URL have shown that the EDZ is at least 1000 times more conductive than virgin rock of this type and that it extends up to 2 meters from the tunnel walls (Fig. 5.12), [1]. This means that EDZ will represent effective flow paths and transport routes for radionuclides of at least the same importance as for crystalline rock, and that it should be cut off by constructing tight plugs that are keyed in the rock for creating closed hydraulic regimes.

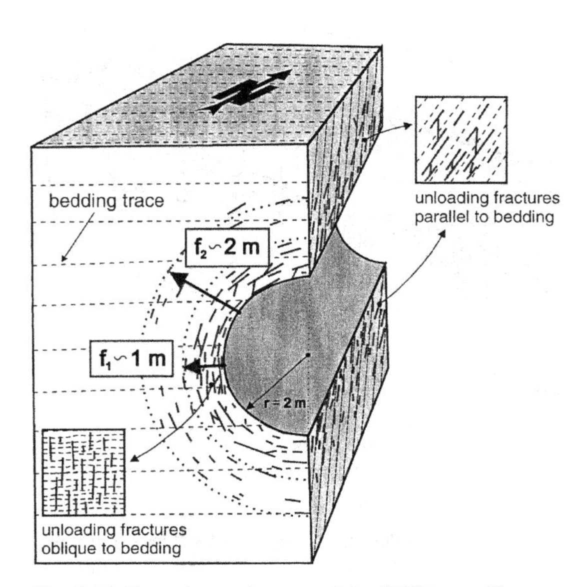

Fig. 5.12 Extension and nature of the EDZ in *argillaceous rock* [1]

5.6 Buffer and Backfill

5.6.1 Preparation and Manufacturing

The manufacturing and placement of the buffer have been in focus since the early eighties and have been investigated in several of the URLs. The technique to prepare dense blocks of smectite-rich clay powder was introduced and developed in SKB's R&D using isostatic compression (3D) under 100 MPa pressure. This gave cylindrical blocks that were sawed to fit in the deposition holes of the model repository drift B(uffer) M(ass) T(est) at Stripa (Fig. 5.13), [10]. Here, six holes with 760 mm diameter and about 3000 mm height were bored using coring technique in a blasted drift that was backfilled with a layerwise placed and vibrator-compacted mixtures of properly graded ballast (aggregate) and 10% MX-80 bentonite to 2/3 of the height of the drift and with a mixture with 20% MX-80 bentonite that was placed by shotcreting. The buffer blocks were heavily instrumented for recording the thermal evolution and maturation of the buffer mounted around 380 mm diameter electrical heaters of teflon-coated aluminium that were powered to different levels, i.e. 600 to 1300 W. The preparation of blocks from the columnar dense clay blocks gave the components in Fig. 5.14. Placing them in the 760 mm diameter holes gave tightly fitting cylindrical stacks of blocks with a 9 mm smaller gap to the rock wall.

In SKB's second URL, the Äspö site, blocks with larger dimensions (1650 mm diameter and about 300 mm height) were prepared for simulating the buffer in a real

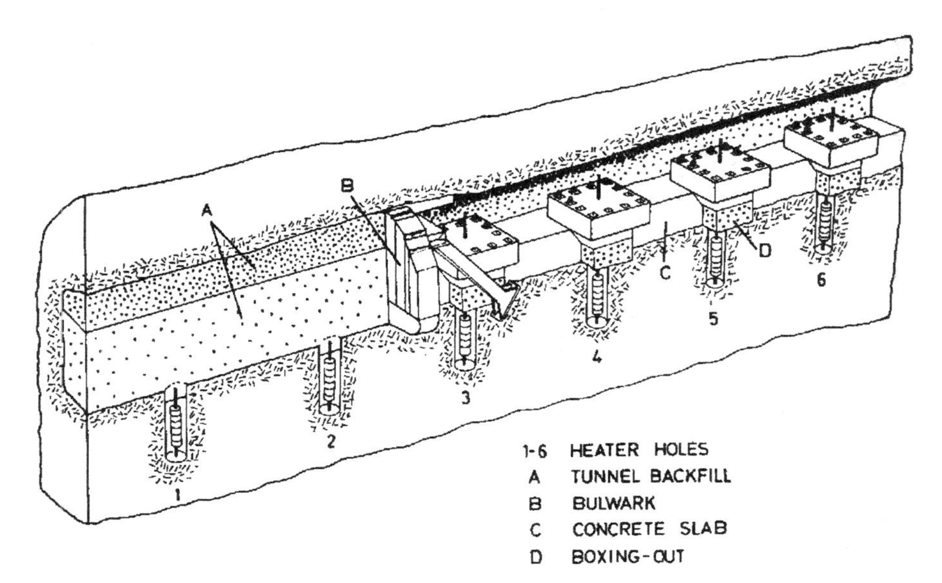

Fig. 5.13 View of the Stripa BMT test drift with six holes simulating canister deposition holes. The boxing-outs in the right part of the drift were backfilled as the full-sized part of the drift and could be opened at desired moments for excavation and sampling [10]

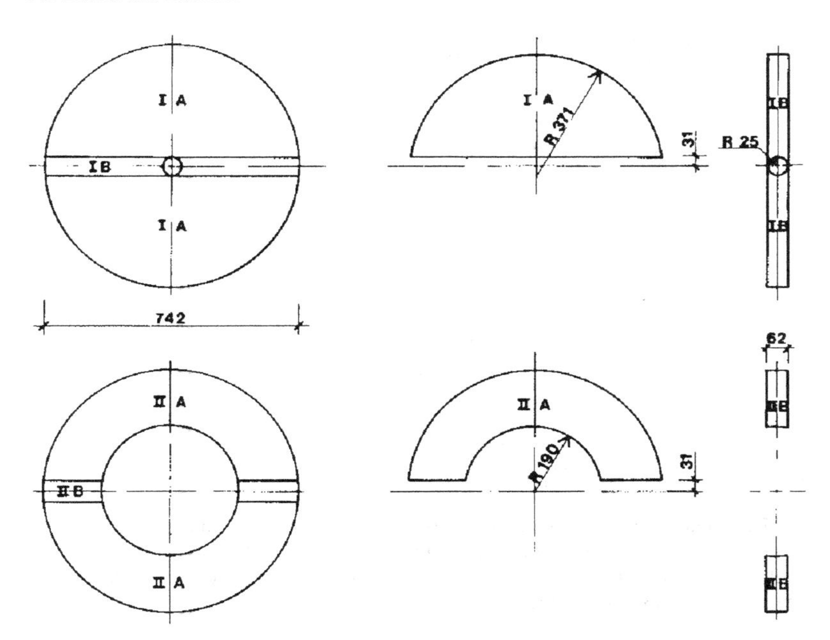

Fig. 5.14 Shape of buffer blocks. The straight pieces had to be made since the diameter of the columnar blocks was not sufficient for making full half-cylinders [10]

KBS-3 V repository with deposition holes with 1750 mm diameter and 8000 mm depth. Monolithic blocks were prepared by uniaxial compression of clay powder in rigid forms that were coated with molybden disulphide as lubricant. The shape was slightly conical so that the blocks could be pushed out without being significantly fractured. In most experiments the rather coarse-grained clay powder had a water content of 8–13% and was compressed under 100–150 MPa pressure yielding dry densities of 1850–2000 kg/m^3. Some tests were made with water contents up to about 18%, which gave blocks with a degree of water saturation of more than 80%, yielding a significantly higher thermal conductivity than the drier blocks. This property is valuable for transferring heat from the canisters to the rock in the repository [1].

There are several problems involved in preparing large blocks of clay powder, a major one being that too fine-grained material may not let air out sufficiently quickly at compression, by which problems arise like resistance to compression. Also, stress concentrations can cause fissures and weaknesses even if the shape of the form used for uniaxial compression is suitable, i.e. slightly conical (Fig. 5.15).

5.6.2 Handling and Placement

Handling and placement of buffer blocks of Stripa size is a simple task, while the big Äspö buffer blocks, weighing some 2000 kg or more, require a crane for putting

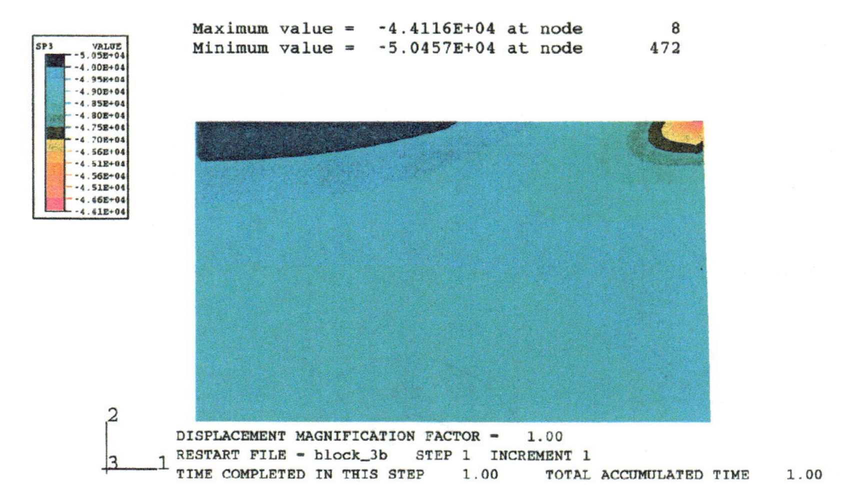

Fig. 5.15 FEM-calculated stress distribution of large block prepared by uniaxial compression (After Börgesson). The overstressing of the upper edge gives fissuring that can weaken this part and make it fall off in the handling of the block. The picture shows one half of the block with the left border representing the axis of symmetry

them on site to form the block column in a deposition hole. The canister will be lowered into the open space of the column according to the KBS-3 V concept and two techniques have been tested, one that utilizes straps as shown in Fig. 5.16, and the other based on vacuum technique. Both are expected to give difficulties, in the first case by fragments broken off from the blocks and hindering the block placement, and in the second malfunctioning of the vacuum technique in critical moments. One understands that preparation of supercontainers with well fitting blocks and canisters under favourable "laboratory-like" conditions on the ground

Fig. 5.16 Placement of a 2000 kg buffer block of MX-80 bentonite in a deposition hole [1]

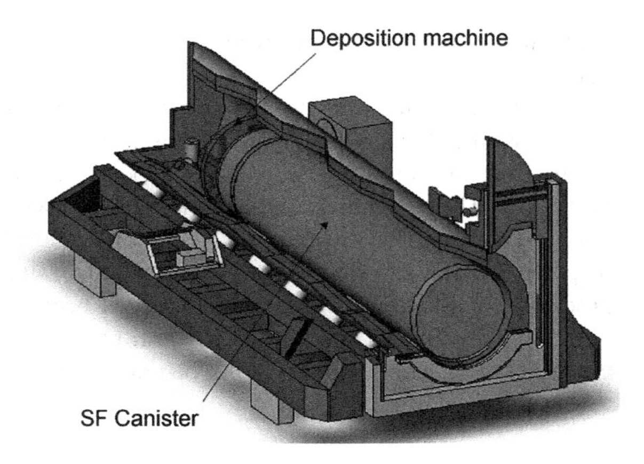

Fig. 5.17 Deposition machine loaded with the canister with spent fuel (SF) inside the shielding cask (After SKB)

level, and bringing them down in the deposition holes as discussed in Chap. 4 is a much safer way of placing the engineered barriers.

For horizontal deposition tunnels one possibility would be to put in hollow buffer blocks first and then push in the canisters with tools of the type shown in Fig. 5.17. The process is simple on the drawing table while a number of problems appear in practice. Thus, breakage of buffer blocks gives fragments that fall and can make it difficult or impossible to move the canisters, which can be stuck several tens of meters from the outer end of the tunnels. The most critical condition would be caused by unforeseen rock burst or fall of rock wedges in the canister placement phase, making it impossible to move the blocks in or out. Another difficulty, which

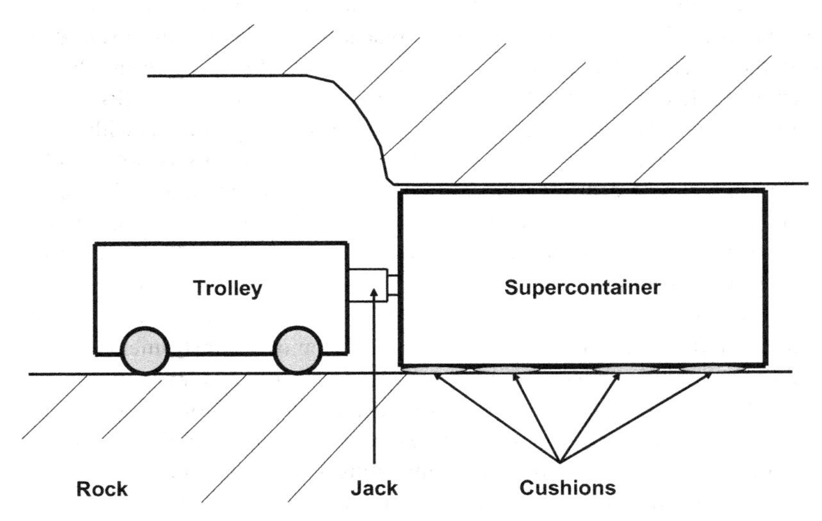

Fig. 5.18 Functional sketch of the deposition machine moving a supercontainer with spent fuel into a horizontal deposition tunnel

is inevitable, is water inflow even if comprehensive grouting of the rock has been made.

For avoiding the problem of fitting the canisters into hollow buffer blocks in the tunnels with a diameter of about 1750 mm with water dripping and spraying from the roof making the floor slippery by eroded clay, the supercontainer principle is preferable. These containers can be prepared elsewhere in the repository where the humidity is low or on the ground level, in dry atmosphere, and brought down to the respective tunnel into which they can be moved on site by use of hydraulic jacks. The procedure, indicated in Fig. 5.18, has not yet been tested on a full scale. Problems with falling rock and inflow of water will naturally appear also if supercontainers are used but the clay blocks will not be readily dispersed.

5.7 Instrumentation and Data Acquisition

5.7.1 What Shall be Measured?

One of the most important objectives of the URLs has been to test buffers and their interaction with near-field rock under repository-like conditions and a number of important conclusions have been drawn from such tests in SKB's Stripa and Äspö URLs as well as in the Belgian, Swiss and Canadian equivalents. Instruments can be inserted in the buffer at room temperature and normal RH in an air-conditioned part of the URL or in a suitable building on the ground surface. Instruments for measuring rock strain, like extensometers, temperature and pressure are also placed without difficulty and hurry in crystalline rock, while in salt and argillaceous rock, as well as in clastic clay, instruments must be placed immediately after excavation to avoid problems associated with convergence or failure. Figure 3.36 in Chap. 3 serves as an example of the general principle of instrumentation of the EBS.

There are two important characteristics of demonstration experiments with buffer materials, one being that access to or replacement of instruments is not possible after installation without perturbing the test, and the other that the planned duration of the experiments may be far beyond the guaranteed lifetime of the instruments. While instrumentation of ordinary underground constructions like railways and road tunnels is not exposed to particularly severe conditions, the sensors and gauges in URLs and real repositories are affected by high pressure, temperature, corrosive atmosphere and radiation. This requires careful selection of the instruments, which must be reliable, durable and mechanically strong. Ideally, only well proven sensors with acceptable long-term behaviour should be used and testing of different types of instruments is in fact an important objective of the experiments in the URLs. The risk of loosing data can be minimized by duplicating with other types of measurement but the survival potential of the back-up instruments must of course also be proven. The installation procedure must be made with care in order to minimize perturbances, which can directly affect the validity of the data obtained. Mechanical couplings, fittings, etc. must be reliable and not introduce additional noise into the data acquisition system (DAS).

One of the most important objectives of the CROP project was to compile information on instruments used or planned to be used in the various national URLs and repositories. This led to comprehensive lists of sensors and gauges for recording temperature, pressure and moisture under the demanding conditions provided by the buffer. Among the large number of observations was the finding that ordinary steel-coated thermocouples corrode rapidly at ordinary buffer temperatures even for rather low salt contents in the groundwater. Another experience was that there is presently no technique for measuring moisture in buffers and backfills that works for both low and high water contents. As to pressure cells, a first major problem is that the accuracy is low at low pressures for cells designed for sustaining pressures of 10 MPa or more, and a second that many cell types are sensitive to high temperatures and tend to fail or suffer from zero-point shifting after a few years. If one wants to record pressures at less than about 100°C for decades it seems that active fluid pressure cells of Glötzl type are the best. Piezometers of vibrating string type tend to age and be less reliable after a few years at repository temperatures, and also for this purpose Glötzl gauges seem to be most reliable. A problem common to all instruments is that the substantial strain in the buffer can tear off cables or tubings. Instruments for use in URLs have undergone much development but soon tend to be outdated and new types currently appear on the market.

5.7.2 Practicalities

One naturally would like to have all important processes recorded by numerous sensors but the possibility to find sufficient space for them and for the cables and tubings is a limiting factor. Monitoring and control systems shall be designed to include:

- Power supply and signal conversion.
- On- and off-site control and monitoring of special processes.
- On- and off-site steering of the experiment, like changing the heater power.
- On-site data collection, archieving and display of important variables:

  - temperature, gauges must be placed so in the buffer and backfill that temperature gradients can be evaluated.
  - swelling pressure.
  - porewater pressure.
  - wetting and drying of the buffer and backfill within the buffer and at the buffer/rock contact.
  - change in chemical conditions.

- Post-processing, capabilities for reporting.
- Data base management.

5.7.3 Selection of Instruments

A major objective of the buffer and backfill experiments in URLs is to obtain data on the evolution of the different processes that take place in the EBS and in the host rock under conditions similar to those expected in a real repository. Therefore, instrumentation is a major issue and so is the acquisition of data. As shown in the subsequent text sensors and cables connecting them to the recording units can give false information by serving as water migration paths and great care has to be taken in selecting positions for them. Scoping calculations with parameter variation have to be carried out in the planning of all experiments in order to select sensors with suitable measuring intervals. Redundance by installing back-up sensors of a second type is recommended and was consistently implemented in the Stripa URL, where simple but very reliable gauges like manometers were used for pieziometric measurements. It is essential to answer the following questions with respect to the test objectives:

- What resolution is really needed?
- What measurement intervals are needed?
- Can non-destructive measurement methods be used?
- What is the natural variability of the measured parameters?
- How reliable are sensors and what is their longevity?

Some of these questions can be answered by past experience, while others require pre-testing. In any case, at the end of each experiment evaluation of the performance of the individual components and re-calibration has to be made.

5.7.4 A Real Problem

The matter of installing gauges in buffer and backfill was considered a routine and simple matter in early field tests but has turned out to be both difficult and very important since water from the surrounding rock can flow along pipes and cables to the sensors in the clay barriers and make them react much sooner than barriers with no instruments.

The diagrams in Fig. 5.19 represent the measured swelling pressure of the buffer 165, 270 and 370 days after starting the so-called "Retrieval Test", which represents a full-scale KBS-3 V case in SKB's Äspö test site, with the buffer/rock contact being a constantly water-filled filter. Theoretically, the rate of hydration implies initiation of the wetting of the clay close to the canister after a few months according to models developed and used in the Prototype Repository Project [1], an example being given in Fig. 5.20. The field observation that the wetting was quicker at the hot canister periphery than 5 cm away from it is a definite proof that the recordings did not accurately depict the true wetting rate. The water pressure was 100 kPa, i.e. a very small fraction of the measured total pressure.

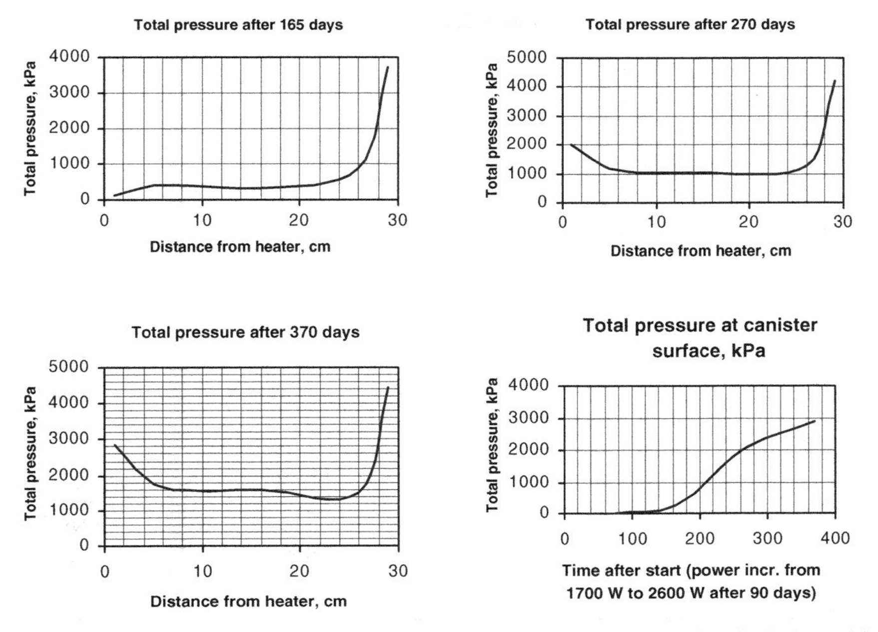

Fig. 5.19 Recorded total pressures in SKB's Retrieval Test at AEspoe. One finds that, with the exception of the nearly constant 4.5 MPa pressure 20 mm from the wet boundary, the highest pressure developed at the hot canister surface in all but one test. The anomaly was caused by water migrating along the cables by which the pressure cell reacted much earlier than where there were no instruments

Comparing the recorded wetting rate, manifested by the growth in swelling pressure, with theoretical predictions of the degree of water saturation in the diagram in Fig. 5.20 shows very poor agreement. This and all other predictions using models discussed in Chap. 3 gave complete saturation after 140 days, while the clay was in fact only partly saturated after one year. The diagram shows that the degree of water saturation close to the canister ("near heater") would drop in the first 80 days and then increase to 100% some 60 days later. This is in complete contrast to the increased degree of saturation recorded in the experiment.

5.7.5 Data Acquisition Principles

5.7.5.1 Data Collection, Storage and Handling

The data acquisition system (DAS), used in a URL, must be designed according to the needs and requirements of the specific sensors. A general centralised DAS is preferable but sensor-specific systems may have to be used. For example, high frequency seismic measurements with ten thousands data points per second have to be sampled, while pore pressure measurements only require a few recordings per

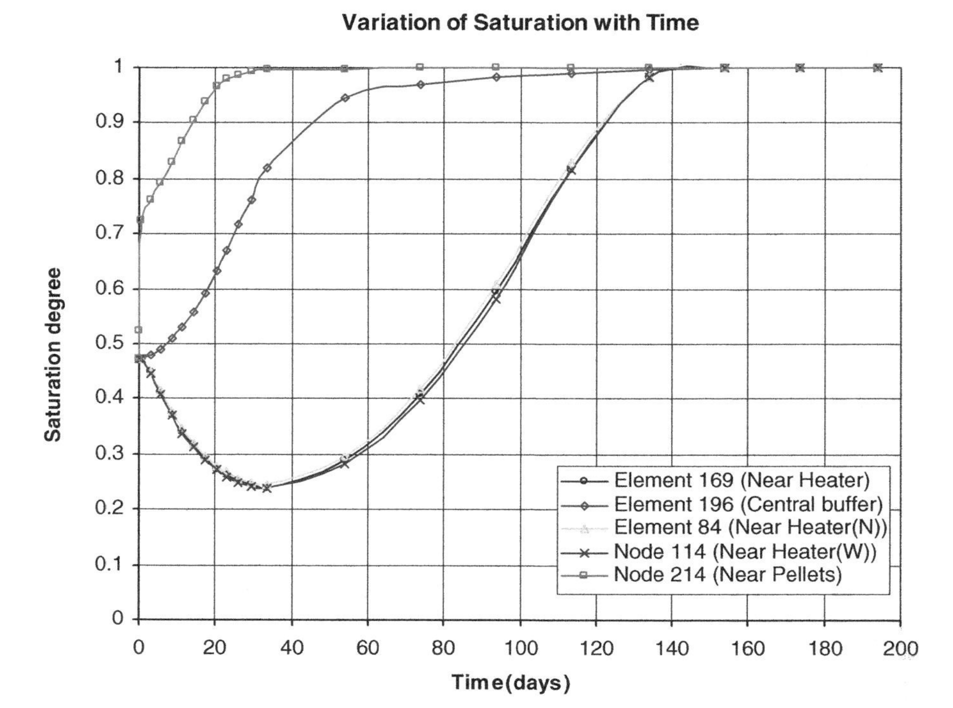

Fig. 5.20 Predicted evolution of the degree of water saturation in the KBS-3 buffer consisting of dense blocks and a loose filling of pellets at the outer boundary. Both the blocks and the pellets consist of MX-80 clay [1]

hour or day. If measurements with such different requirements have to be combined, centralised DAS should naturally not be used.

Most measurements are performed by automated data-acquisition. The database management system must have facilities for the storage of raw data, data reduction, and organization and visualization of the results. Pioneer work was made for data collection at the Stripa URL already back in the early eighties and more advanced acquisition systems have been developed and used in all subsequently established URLs. An example of a Local Monitoring System (LMS), designed and implemented by Aitemin, Spain, and used in the Mont Terri URL, is shown in Fig. 5.21.

5.8 Testing

5.8.1 Principles

Field testing of buffers and backfills is made for getting accurate data on their performance under repository-like conditions, which requires that the other engineered

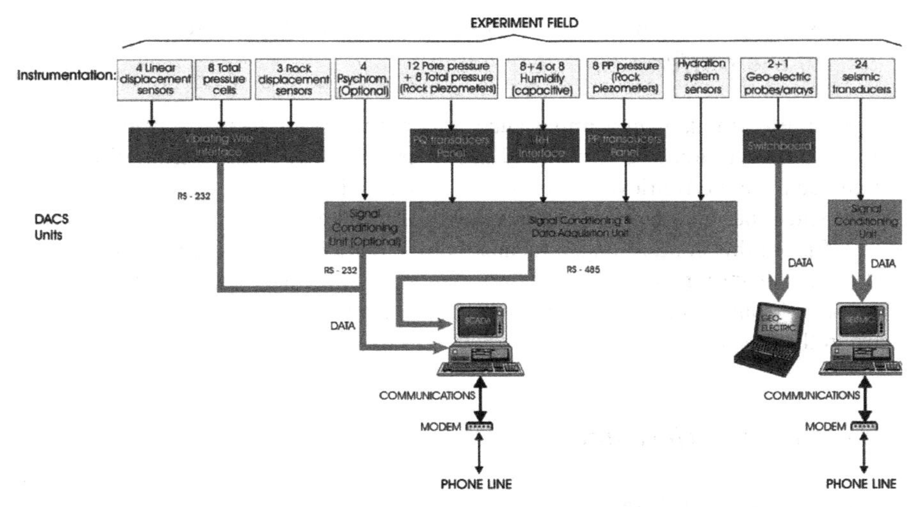

Fig. 5.21 Local Monitoring System (LAS) installed by Aitemin, Spain, for buffer experiments [1]

barriers are also on site and that they are all installed in rock of relevant type and of sufficient size. The main processes to be recorded is water uptake by the buffer and backfills, expansion of the buffer, and thermal evolution of the entire system.

5.8.2 Powering

Experiments with buffer clay in deeply located URLs in crystalline rock usually require use of electrically heated canisters and this is known to cause problems with short-circuiting of electrical cables and problems with deformation of cables and malfunctioning sensors. Ideally performing installations are represented by the systems used in SKB's Stripa mine [11] while considerable problems were met in other URLs like short-circuiting and loss of power control leading to strong over-heating that jeopardized and nearly ruined a buffer experiment. At Stripa the power system was designed and constructed by ASEA Atom.[3] Each of the heaters contained three resistors that provided heat and each of them was supplied with power through its own cable. The resistors gave up to 1000 W power and since most of the experiments required 600 W for resembling the function of HLW canisters there was sufficient redundance. The key component in the system was a transducer that measured the power and discharged a current that was proportional to the power. Stable pulsed voltage was provided by thyristor technique.

[3]Supported by Gunnar Ramqvist, El-tekno AB.

5.8.3 The Role of Rock as Supplier of Water for Wetting of the Buffer

A question of fundamental importance is how the access to water from the rock affects the saturation rate. Unlimited access provided by richly fractured rock, or simulated as in the Retrieval Test (Sect. 5.7.4), is valuable for the maturation of the buffer since the risk of permanent microstructural alteration by desiccation is limited [12], but this condition also means that the migration rate of possibly released radionuclides into the rock will be high. We recognize this reasoning from the discussion in Chap. 3 and return to it here to see what sort of information the URL tests gave in this respect.

5.8.4 Buffer Performance

5.8.4.1 Maturation Rate

We have seen in Chap. 3 that the development of the swelling pressure is directly related to the water uptake. When work started on developing the Swedish concept for disposal of HLW (spent fuel) in the period 1970–1980 smectite clay was appointed main barrier around the canisters on the basis of available scientific knowledge and some exploratory small-scale experiments. The need for experiments with highly compacted smectite clay surrounding heaters in large boreholes led to the BMT test of the Stripa project [1,10]. It required prediction of the maturation rate, i.e. the time-dependent water uptake and homogenization, for selecting suitable instrumentation and comprehensive laboratory tests were performed in the planning stage. Parallel to the field tests, which ended in 1992, intense work on development of theoretical models started as described in Chap. 3 and there was great satisfaction when it was found that the evolution of the buffer of dense clay (MX-80) agreed in principle with the predictions for buffer in "deposition holes" located in richly water-bearing rock (Fig. 5.22). However, for holes in less wet holes the evolution of the swelling pressure, being the best indicator of the rate of water saturation, remained largely unknown and the way of predicting the rate of maturation without validation (Fig. 5.23). This condition still remains as we saw in the description of the modelling attempts in Chap. 3.

The evolution of the swelling pressure in holes located in rock with intermediate ability to give off water to the clay is illustrated by Fig. 5.24.

These URL experiments were the first to show the importance of the rock's hydraulic performance to the maturation of smectite-rich buffer. They not only demonstrated that the rate of water flow through the rock to the deposition holes controls the rate of maturation of the buffer but also that the flow, which is determined by the hydraulic conductivity and the pressure gradient, needs to be sufficiently high for maintaining overpressure or at least zero water pressure in the fractures. We see that this in complete agreement with the theoretical modelling that gave the graphs in Fig. 3.19.

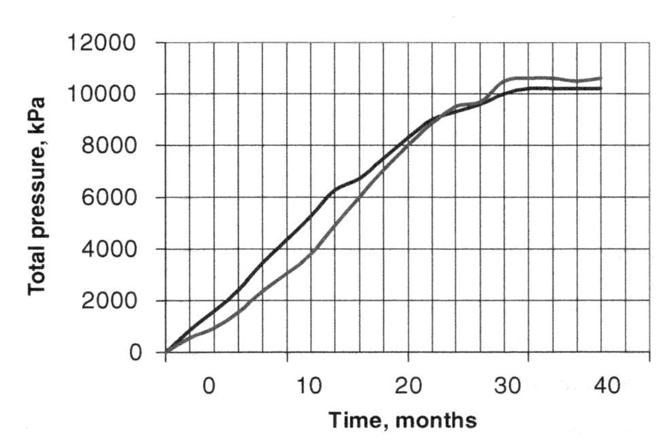

Fig. 5.22 Swelling pressure recorded by Glötzl cells at mid-height of the 600 W heater in the "wet" 760 mm diameter hole. The radial thickness of the clay was 190 mm. Serie 1 represents the *lower* and Serie 2 the *upper* part of the buffer

At termination of the respective tests, excavation and sampling were made for determining the water content distribution. Figure 5.25 illustrates that the water uptake of the buffer from the rock was far from complete after 15 months even in fairly "wet" holes. Within 10 cm distance from the heater the buffer with 19 cm radial thickness had in fact undergone desiccation and drop in water content from initially 10% to about 8% adjacent to the heater surface. This process, which has later been confirmed by field tests in other URLs, is not properly treated by the theoretical models described in Chap. 3. This is because the desiccation is associated

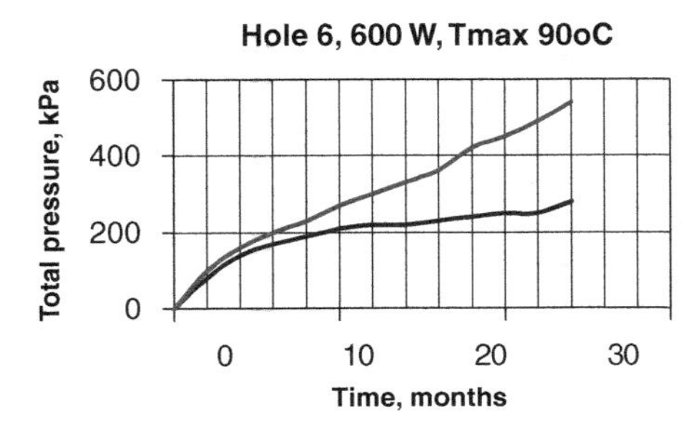

Fig. 5.23 Swelling pressure recorded by Glötzl cells at mid-height of the 600 W heater in a hole with an inflow of water that was very much lower than in the wet Hole 2 (BMT project). The upper curve represents the lower part of the buffer and the lower curve represents the central, most heated part. For the case of unlimited access to water the pressure would have been more than 10 times higher (cf. Fig. 5.22)

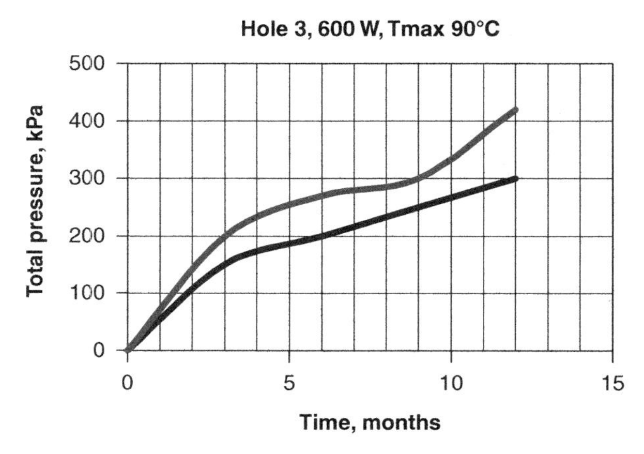

Fig. 5.24 Evolution of swelling pressure in Hole 3 in the BMT project with an inflow of water that was intermediate to that of Holes 1 and 6. The lower part of the buffer (*upper curve*) matured more rapidly than the mid part (*lower curve*)

with cracking that greatly influences the hydraulic and mechanical performance of the buffer.

There is another important observation from examining Fig. 5.25 and other diagrams of the water content distribution in such tests, namely that the curve shape is typical of diffusion-controlled migration. This has been repeatedly found to be the case in one-dimensional [13] and axisymmetric tests [14] with smectite-rich buffer tested under isothermal conditions as well as exposed to thermal gradients, and has led to development of conceptual and theoretical models of buffer evolution based on diffusive uptake of water as discussed in Chap. 3. The diffusion coefficient has consistently been found to be around 3E-10 m^2/s for dense smectite-rich clay.

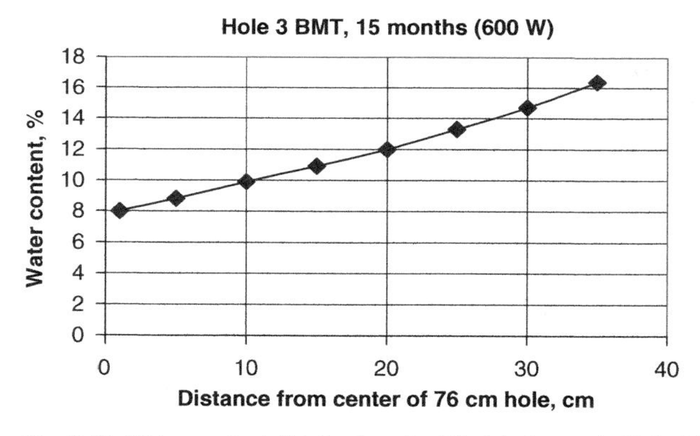

Fig. 5.25 Water content distribution at midheight heater in Hole 3 in the BMT project at termination after 15 months. The initial water content was 10% and desiccation, that remained after this period of time, took place within about 100 mm distance from the heater

5.8.5 Conceptual and Theoretical Models

5.8.5.1 Selection of Processes to be Modelled

The evolution of the engineered barriers in URLs is described by models that are designed to be used for investigation and understanding of processes like the temperature increase in buffer clay and surrounding rock, and water uptake and redistribution in buffer clay, as discussed in Chaps. 3 and 4. The following features or processes have to be modelled for predicting the evolution of buffer clay:

1. Redistribution of stresses by excavation of deposition holes or tunnels, particularly respecting the formation of the EDZ (all rock types).
2. Development of water pressure and degree of water saturation of the host rock surrounding the tunnels (crystalline and argillaceous rock as well as clastic clay).
3. Heat-induced redistribution leading to desiccation of the buffer close to HLW canisters and to wetting of the rest of the buffer and of the EDZ (all rock types).
4. Convergence of the tunnel that can compact the buffer (salt and argillaceous rock).
5. Generation and migration of gas produced by metal corrosion, including displacement of water in the host rock, buffer and backfill (all rock types).
6. Pressurizing of the buffer by the rise in volume of corroding canisters (all rock types).

Some of these processes take place in the operational or monitoring phase extending over as much as 100 years or possibly even more (Fig. 5.26), while others last

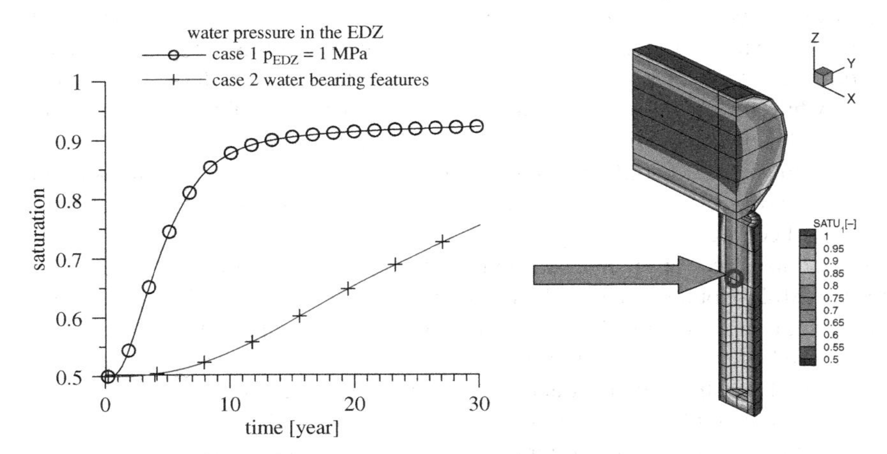

Fig. 5.26 Water saturation rate of buffer in KBS-3 V deposition hole (Calculation by Lutz Liedtke, BGR, Germany). The case without discrete fractures and water given off only from the EDZ to the buffer, representing very tight rock, means that complete water saturation may take many hundreds or even thousands of years, while in "wet" rock, representing ordinary crystalline rock, a high degree of water saturation may be reached already after one or a few decades

for much longer time. Scoping calculations using derived models indicate where sensors should be placed and how long they need to work for providing data of interest.

The problem is that several of the processes are strongly non-linear and coupled to each other, which makes assessment quite difficult. In addition, the boundary conditions as well as the most important parameters are often not adequately known in the test in the URLs. Therefore, separate mock-up tests on the ground surface have been conducted by several of the Organizations to verify codes and models on simplified test set-ups with well-defined boundary conditions.

5.8.5.2 Comparison of Theoretical Predictions and Experimental Data Concerning the Interaction of Rock and Buffer

We have seen in Chaps. 3 and 4 that the predicted and recorded physical evolution of the buffer in crystalline rock can be different even under well defined boundary conditions with respect to the ability of the surrounding rock to give off water. For unlimited access to water the predicted and actual maturation rate of buffer consisting of highly compacted smectite is still not too different while for "dry" rock they may differ significantly. For argillaceous rock, which is even tighter than "dry" crystalline rock, the corresponding difference has also been investigated, the predictions being based on the use of a number of techniques including analytical methods, finite-element or finite-difference codes e.g. FLAC 2D/3D, and boundary element codes like BEASY.

A vital issue is to predict the stress situation in the rock in the construction phase and a first step is to estimate whether the hoop stress will be lower than the compressive stress generated by the excavation of tunnels and rooms. For normally or very weakly overconsolidated clay, the unconfined compressive strength is usually much lower than the hoop stress, which means that the safety factor is low and immediate support in the form of a lining is required. This is the case for the Boom Clay at Hades, while for the cemented Opalinus Clay the factor of safety ranges between unity and 3, hence requiring immediate stabilization only locally.

Due to the nature of argillaceous rock, it is necessary to use full coupling between hydraulic and mechanical models (HM coupling). In most cases, hydraulic modelling has been performed using Darcy's law, while various different constitutive laws are considered for the mechanical part. Models used so far at Hades and Mont Terri are: Mohr-Coulomb, Hoek-Brown, Drucker Prager and Cam Clay models. For certain purposes these models have been modified to account for special conditions like creep taken to be a form of viscous flow. The agreement between the predicted and recorded redistribution of water in the nearfield rock has been found to be poor. Thus, the predictions indicated a desaturated zone in the Boom Clay at distances up to 1.5 m from the tunnel periphery while the experiments indicated saturated conditions. A further discrepancy was that the actual water flow rate was only about 7% of the predicted value.

The overall performance of the models has been assessed for parameters like the pore pressure and the effective stress in different phases of the field experiments, e.g., buffer hydration by uptake of water from the rock, artificial hydration by injecting water in experiments and a few more, and the conclusion has been drawn that no single model is able to reproduce faithfully the complete evolution of stress and strain. It has been pointed out by several investigators that model predictions are extremely sensitive to certain material properties, in particular to the water retention capability and the hydraulic conductivity of the rock surrounding buffer clay.

5.9 Plug Construction

Construction of plugs, which we saw examples of in Chap. 4, has been tried in all the URLs and considerable experience has been gained concerning constructability and strategies of locating them with respect to the rock structure and flow paths. One experience is that the heat produced by casting large volumes of concrete causes defects in the plugs and that it can not be fully compensated by cooling pipes, and another that casting big plugs is very tedious and expensive. This is not encouraging especially since most of the concrete plugs, even those equipped with O-ring-type seals of dense clay, have turned out to leak right through the concrete. Also, there is still uncertainty respecting the long term performance of plugs made of concrete, a matter that has not been and can hardly be investigated in URLs. It seems, however, that the most recently developed concrete types, i.e. silica concrete with only little cement (of low pH-type), can serve well much longer than ordinary concrete.

An example of a plug design for argillaceous rock that relies on smectite-rich clay and implies construction of a big masonry of highly compacted blocks is shown in Fig. 5.27. This plug type represents a very tight connection of different parts of a tunnel or drift and cuts off the EDZ by extending into it. The chemical interaction between the concrete elements and the clay can be limited by using low-pH cement as discussed in Chap. 4. If the clay is given high density the end supports need to

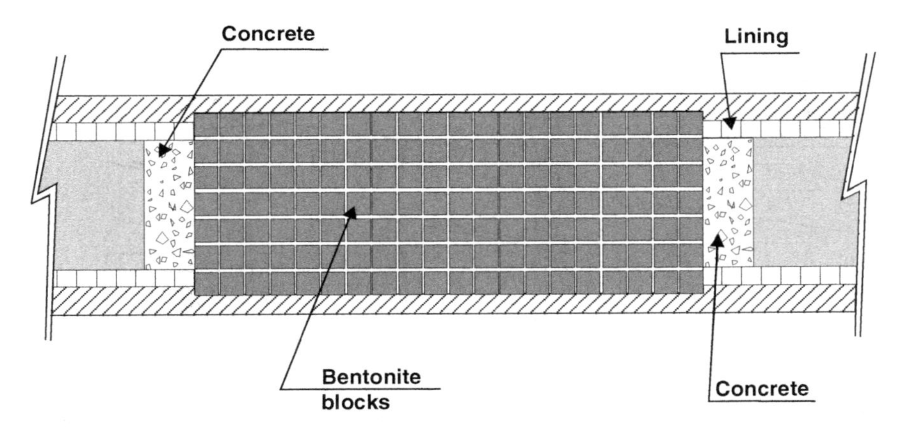

Fig. 5.27 Plug design proposed by ENRESA [1].

be designed to take the pressure and be keyed in the rock for avoiding displacement of the plug ends. For a 20 m^2 cross section of a smectite-rich plug with a density of 2000 kg/m^3 the force acting on each concrete support is about 20 000 t (200 MN).

5.10 Borehole Sealing

Potentially unstable fracture zones intersected by boreholes have to be stabilized for avoiding rock fall in conjunction with subsequent placement of clay plugs. Stabilization is best made by reaming the hole and filling it with cement-stabilized quartz concrete followed by re-boring. The sealing function is provided by clay-based plugs in the parts of the hole that are located in normally fractured rock, while the parts that have been stabilized are plugged by filling them with cement-stabilized quartz concrete similar to what is used for the stabilization. The quartz component is chemically stable while the cement, which is of low-pH type, can possibly be dissolved and lost. This is not deemed to be critical to the performance of the fill since its gradation can easily be selected such that it should not be eroded by groundwater flow.

Deep boreholes intersecting a repository may serve as short-circuits and need to be sealed as described in Chap. 4 (Fig. 5.28). Successful exploratory experiments have been made in the Finnish URL at Olkiluoto as summarized here (Fig. 5.29),

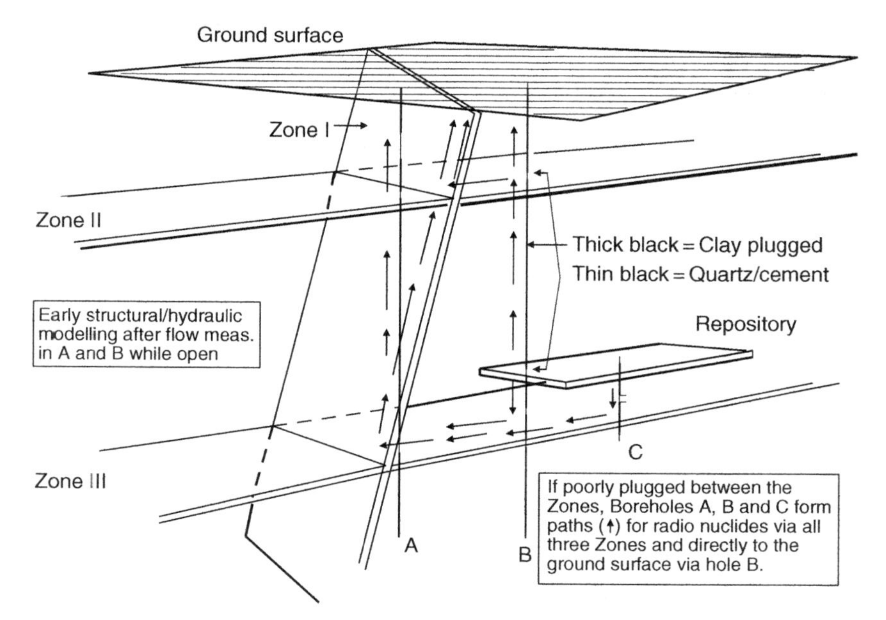

Fig. 5.28 Schematic picture of effect of poor sealing of steep boreholes that interact with a repository

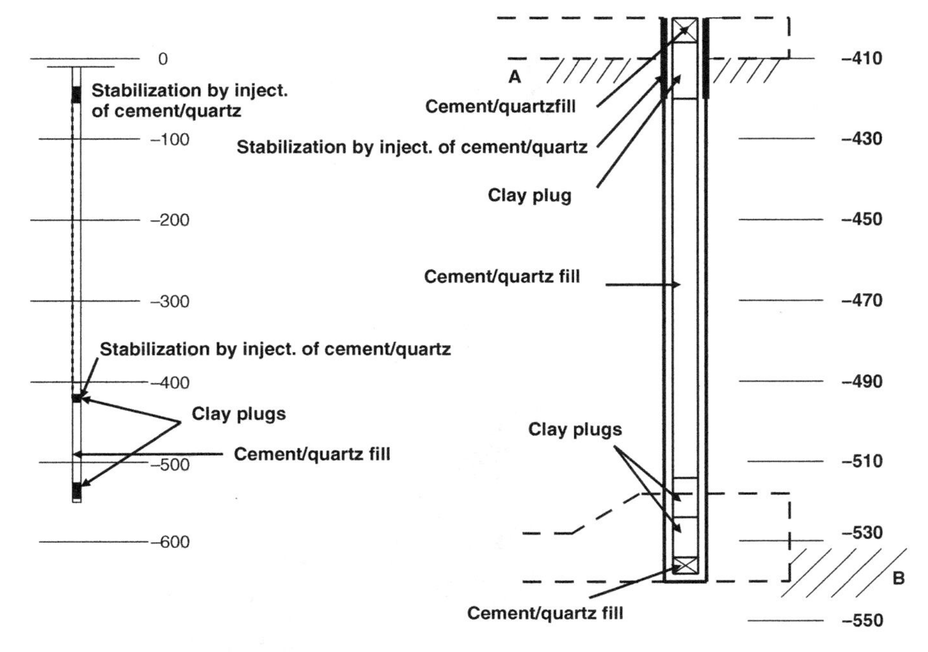

Fig. 5.29 Schematic illustration of the planned plugging of a 535 m deep borehole at Olkiluoto [15]

[15]. The 12 m long clay plug at the bottom was of "Basic" type, i.e. perforated tubes with highly compacted smectite clay (Fig. 5.30), brought down in a borehole of about 525 m length by use of drilling rods. It was left for a day for initial maturation and then covered with concrete. The maturation of the plug is predicted to lead to complete saturation and nearly complete homogenization in a few months. Figure 5.30 shows details of the operation, indicating the need for drill rigs, cranes etc for placement and lifting of heavy tools.

Clay plugs must be prevented from expanding in the axial direction of the boreholes, which requires some type of mechanical lock at the free ends of the holes. The borehole sealing project conducted by SKB and POSIVA has comprised development of two possibilities, one implying casting of silica concrete, based on low-pH cement and quartz ballast (aggregate) including quartzite fragments, and the other based on placement of an expander of copper. Both are keyed into the boreholes walls by extending into reamed parts of the borehole as indicated in Fig. 5.31, which shows an overcored rock column sectioned by sawing for exposing the interior of the copper plug. A corresponding picture of the concrete plug was shown in Chap. 4, Fig. 4.66. Loading tests of concrete plugs cast in 200 mm boreholes with 50 mm deep and long slots demonstrated that such plugs can sustain an axial force of more than 100 t (1 MN). An appreciable load can be taken also after complete loss of the cement component that may take place in a very long period of time.

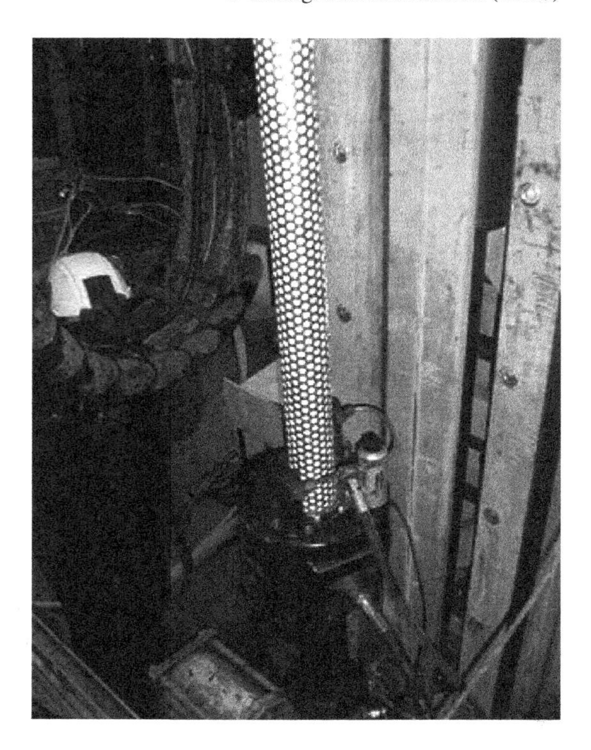

Fig. 5.30 Jointing of two 2.5 m long units before placement at a about 500 m depth in an 80 mm borehole at Olkiluto [15]

5.11 General Conclusions from URL Activities

A final objective of the CROP project was to find out to what extent the national URLs have supported or changed current ideas of how repositories of the respective concept types perform, and to identify possible shortcomings and agreements between theoretically derived and measured data, primarily respecting groundwater flow and its relation to rock structure, and secondarily how the evolution of buffers depends on the hydraulic and mechanic performances of the surrounding rock.

5.11.1 Rock Structure, Rock Mechanics, and Groundwater Flow

Working out relevant and reliable rock structure models on large and small scales is extremely difficult or impossible for crystalline rock and the impact of the thermal pulse caused by the radioactive decay in the HLW, tectonical events and glaciations may change the structural constitution so much that the mechanical and hydraulic performances of the host rock can not be accurately modelled. In these respects argillaceous rock is less difficult but also for such rock it has been found that identification of water flow paths in the near vicinity of deposition tunnels is uncertain and that neither inflow into them or the piezometric conditions around them

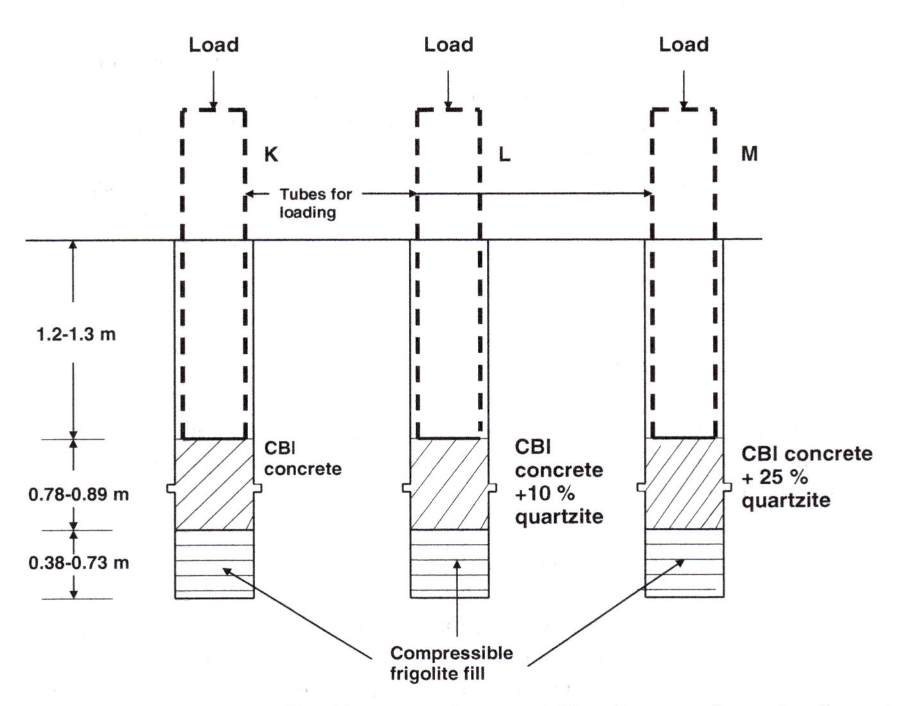

Fig. 5.31 Plug tests. *Upper*: The shiny copper plug extruded into the recess. *Lower*: Loading test of concrete plugs with different amounts of quartzite fragments for increasing the shear strength [15]

can be reliably predicted. It has turned out that for both these rock types water flow is commonly strongly overrated but that there are examples, especially for crystalline rock, where inflow into tunnels has been much higher than predicted, causing great difficulties in backfilling of tunnels and shafts. Water flow in these types of rock commonly takes place in channels, especially where water-bearing fractures are intersected, which makes identification of such flow paths virtually impossible.

The block structure of crystalline rock means that the magnitude and orientation of the major and minor principal stresses vary significantly and that averaging larger numbers of measured pressures may lead to considerable under- or over-estimation. Such rock is very stable provided that the rock stresses are not very high. In salt and argillaceous rock the greater structural homogeneity is much better and causes less variation in measured pressure. For bedded salt and argillaceous rock the major problem is related to stability and conceptual models of degradation leading to rock fall in conjunction with and immediately after excavation and long thereafter, have been worked out. For domal salt the only practically important process that can be investigated in URLs, except for determination of thermal impact, is the creep potential, for which practically applicable empirical rheological models have been derived.

5.11.2 Buffer and Backfills

In most cases the rate of maturation of clay buffer in crystalline rock has been found to be significantly lower than predicted and the greatest risk is that the buffer will desiccate near hot canisters and undergo microstructural changes, like cementation, that can be permanent. The poor agreement between theoretically predicted water saturation and actually measured wetting is both because the hydraulic properties of the near-field rock can not be accurately determined, and because the mechanisms of wetting of dense smectitic clay under thermal gradients and limited access to water from the rock are not well known.

5.11.3 Practical Handling

Full-scale practical handling of canisters and buffer can be investigated in URLs, but no equipments, adaptated or prepared for use in real repositories, were developed at the time of the CROP project.

5.12 Monitoring of Real Repositories

The experience from the URLs tells that cables can serve as transport paths for water and hence of possibly release radionuclides. Even if they can be tightly embedded in clay the cable material can corrode or degrade, which can result in effective flow paths, and it has therefore been proposed to leave out instrumentation in constructing repositories, at least in the near-field. Attempts have been made to use remote, indirect techniques for measuring processes like buffer wetting but the resolution power of currently available systems is not sufficiently good. However, at distance from deposition tunnel systems watchdog instruments may well be installed for recording water pressure and for sampling groundwater for analysis on the ground surface. However, they should be removed in the course of the closure of the repositories and leave just one important way of assessing the function of the entire repository: sampling in boreholes located at reasonable distance from the repository, preferably short ones extending from the ground surface.

References

1. Svemar Ch, 2005. Cluster Repository Project (CROP). Final Report of European Commission Contract FIR1-CT-2000-2003, Brussels, Belgium.
2. Gray MN, 1993. OECD/NEA International Stripa Project Overview Volume III: Engineered Barriers, Published by Swedish Nuclear Fuel and Waste Management Co. (SKB), ISBN 91-971906-4-0, Stockholm.
3. Huertas F, Fuentes-Cantillana JL, Jullien F, Rivas P, Linares J, Farina P, Ghoreychi M, Jockwer N, Kickmaier W, Martinez MA, Samper J, Alonso E, Elorza FJ, 2000. Full-scale engineered barriers experiment for a deep geological repository for high-level radioactive waste in crystalline host rock (FEBEX project): Final Report. European Commission, Nuclear Science and Technology, R and T specific programme 'Nuclear fission safety 1994–98', EUR 19147 EN, Brussels.
4. Pusch R, 1994. Rock mechanics on a geological base. Developments in Geotechnical Engineering, 767 Elsevier Publ. Co. ISBN: 0-444-89613-9.
5. Popov V, Pusch R, (Eds), 2006. Disposal of hazardous waste in underground mines. WIT Press, Southampton, ISBN: 1-85312-750-7, ISSN: 1476-9581Popov LRDT.
6. BEASY User Guide, 1996. Computational Mechanics BESY Ltd, Southampton, UK.
7. Gale J, Witherspoon PA, Wilson CR, Roleau A, 1982. The "Macropermeability Experiment". Geological Disposal of Radioactive Waste. In-situ experiments in granite. Proc. of the OECD/NEA Workshop, Stockholm October 25–27 1982.
8. Bernier F, 2005. A repository scale in-situ programme: The PRACLAY experiments, SEK-CEN website: sckcen.be/sckcen/ScientificReports/2005/.
9. Liedtke L, Shao H, Alheid HJ, Sönnke J, 1999. Material Transport in Fractured Rock – Rock Characterisation in the Proximal Tunnel Zone. Federal Institute for Geosciences and Natural Resources, Hannover, Germany.
10. Pusch R, 1994. Waste disposal in rock. Development in Geotechnical Engineering, 76. Elsevier Publ. Co. ISBN: 0-444-89449-7.
11. Nilsson J, Ramqvist G, Pusch R, 1984. Buffer Mass Test – Heater design and operation. Stripa Project, Techn. Report 84-02. SKB, Stockholm.
12. Pusch R, Yong RN, 2006. Microstructure of smectite clays and engineering performance. Taylor & Francis, London and New York, ISBN 10: 0-415-36863-4.
13. Kröhn K-P, 2003. New conceptual models fro resaturation of bentonite. Applied Clay Science, Vol. 23 (pp. 25–33).
14. Pusch R, Kasbohm J, Pacovsky J, Cechova Z, 2007. Are all smectite clays suitable as "buffers"? Physics and Chemistry of the Earth, Elsevier Publ. Co., Vol. 32 (pp. 116–122).
15. Pusch R, Ramqvist G, 2006. Cleaning and sealing of boreholes. Report of Sub-project 3 on plugging of borehole OL-KR-24 at Olkiluoto and reference borehols at Äspö. SKB Int. Progress Report IPR-06-30. SKB, Stockholm.

Chapter 6
Site Selection

6.1 Deep or Shallow?

Is it possible to locate a HLW repository at shallow depth? The question is certainly justified since it would be much cheaper than deep disposal. The answer depends on the life length of the radioactive content and on its composition, but primarily on the isolating ability of the engineered barriers. Technically, the solution can be mausoleum-type repositories or soil-covered storages on the ground surface, or repositories with lined or unlined rooms in rock at a depth of a few tens to a hundred meters. Deep disposal is, however, generally preferred.

6.2 Criteria for Locating Repositories at Depth Crystalline Rock

In addition to national, regional and local socio/economic/environmental factors that can rule out a number of proposed sites for a HLW repository several criteria have been defined for acceptance. The major one is that the geological conditions must offer sufficient isolation of the HLW from the biosphere over a defined period of time, considering all known physical and chemical degrading processes including exogenic impact. Since economy has become an important factor in energy production and hence makes the cost for establishing HLW repositories a very important issue, it is clear that a further requirement is that the rock should not contain exploitable amounts of useful ore. In recent time the matter of electrical potentials implying possible severe corrosion of canisters has become an additional matter. Such potentials are known to cause quick corrosion of instruments and tools deep down in the rock but the impact on canisters is not yet known. However, for the time being it should be taken into consideration in the site selection process and sites with such potentials avoided. We will briefly comment on some of these matters in this chapter.

R. Pusch, *Geological Storage of Highly Radioactive Waste*,
DOI: 10.1007/978-3-540-77333-7_7, © Springer-Verlag Berlin Heidelberg 2008

6.3 Content of Valuables

Rock containing valuables in the form of precious metals like gold, silver, palladium and platinum, and metals of interest to the industry, like titanium, chromium, nickel and uranium, as well as iron-rich ore and high-class rock for the stone industry, should not host a repository. One needs to keep in mind that presently used methods for extracting and refining such valuables may be replaced by new techniques that can make also low-concentration resources interesting in the future. For argillaceous rock, corresponding valuables are represented by oil and gas. It should be understood that careful prospection and analysis of rock samples are required before one can think of assessing a rock region as a potential repository site.

6.4 Mechanical Stability

6.4.1 General

Mechanical stability of the rock hosting a repository is definitely needed in the construction phase and also subsequently although moderate degradation has to be accepted in the very long period of time of required isolation as long as it does not significantly perturb the engineered barriers. This is in fact a most difficult assessment since the ongoing evolution of the bedrock is not well understood [1]. Most people engaged in predicting the performance of the various components in a repository take the presently defined rock structure to be the same for hundreds of thousands of years and do all sorts of sophisticated calculations assuming the rock to maintain its present constitution. Thus, the impact of the neverending massage of the earth crust and change in shape and size of the globe is usually disregarded in extrapolating the interaction between rock and engineered barriers over any period of time. This is kept in mind in the examination we will make here of the basic principles of selecting sites for repositories and see how reasonable the presumptions are in predicting the long-term performance of a HLW repository.

A first, basic principle is that a repository should not be located in a 1st order discontinuity because of its content of very permeable parts and clay-rich lenses, which make construction difficult as illustrated by the tremendous problems met with in the ongoing construction of two 8.6 km long railway tunnels in the Hallandsås horst in Sweden. This continuous project started in 1992 and is not expected to be completed until 2015 at the earliest, requiring comprehensive stabilization by grouting and freezing. While repositories should be located between such weaknesses, they will be intersected by weak zones of the second largest type, represented by 2nd order discontinuities. These can also cause difficulties in the construction phase and serve as major paths for groundwater flow, and should hence not be accepted in the HLW disposal area but be allowed to intersect ramps and shafts leading down to the disposal area. Minor fracture zones, i.e. those termed 3rd order discontinuities, have to be accepted in the disposal area but should not intersect deposition holes and not cross deposition tunnels where canisters are planned to be located.

The problem with all these low-order discontinuities is not only that they provide quick transport of possibly released radionuclides but also currently undergo shear strain that can affect the canisters and their clay embedment. Such strain can occur at a more or less constant rate or have the form of intermittent movement associated with seismic events or generated by high deviatoric stresses in the rock mass. This is a matter that needs consideration of the regional structural geology in general and of ancient and possible future tectonics in the site selection process. Unfortunately, most geologists and rock modellers appear to take the present rock structural constitution as permanent while the long perspective to be considered should force them to develop models for predicting the forthcoming evolution of the earth crust. The basis of such work was published half a century ago [2,3] and a lot more in the latest 20 years [4] but has hardly been referred to in the site selection process. We will have a look here at some modern theories of crustal evolution for getting an idea of how large-scale changes in rock structure can affect repositories in crystalline and argillaceous rock.

6.4.2 Impact of Tectonics

A prerequisite for rock displacements that can cause significant changes in the hydraulic performance of large rock units and generate damaging strain is that critically high stresses prevail. It is recognised that large vertical and subhorizontal movements can occur in the earth crust by slip over faults and potentially lead to destruction of underground constructions and even generate tsunamis. The issue is to identify and avoid such regions in the site selection process.

Vertical displacements are associated with certain types of stress fields, so called, stress regimes (Fig. 6.1).

The aim of a study of the risk of host rock undergoing critically large strain would be to develop an accurate mathematical model for identification of stress regimes in different parts of the region in question. It should be based on local stress measurements accumulated within the World Stress Map Project (WSMP), [5]. The WSMP data mostly contain stress orientations obtained by different geophysical methods representing earthquake focal mechanisms. These measurements indicate stresses at different depths through the lithosphere. One needs to remember, however, that

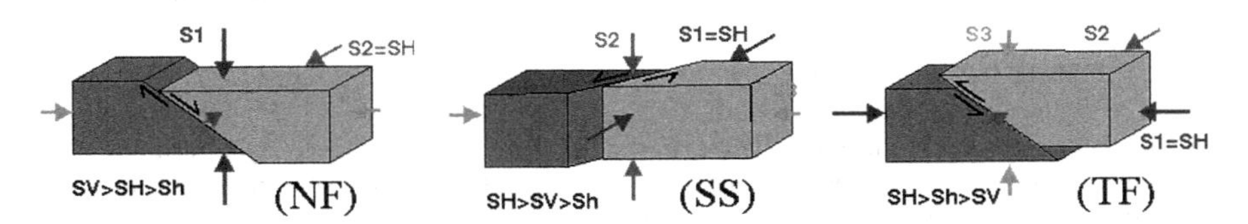

Fig. 6.1 Stress regimes [5]: NF=Normal faulting, SS=Strike-slip faulting, TF=Thrust faulting

discrete measurements do not allow high accuracy in the prediction of mechanical behaviour and failure of rocks, while the knowledge of stress fields does.

The problem is that the theory for the transition of discrete data into continuous stress fields is not yet completed and further investigations are necessary to provide mechanically consistent models of the tectonic plates. This requires the development of novel approaches since the existing ones based on inverse analysis [6,7] are not capable to single out stress characteristics from modelling. At the moment an alternative approach have been developed for elastic regions [8,9]. The approach makes use of stress trajectory concepts to reconstruct mechanically consistent 2D stress fields with minimum number of arbitrary parameters, which is important for parametric analysis. For instance, such important characteristics as stress trajectories are reconstructed uniquely and the maximum shear stress include an arbitrary multiplier.

Further development by introducing the curvature of lithosphere, complex geometries, and different rheologies is ongoing and expected to yield a valuable instrument for identification of potentially unstable regions and for quantifying the fate of any region with respect to tectonic impact. As an example, preliminary analysis of the Sumatra region shows that the principal stresses acting in-plane are close to each other in an elongated zone near Sumatra. Figure 6.2 presents stress trajectories and a contour map of maximum shear stresses Tmax = (S1–S2)/2 in normalised form, i.e. a(S1–S2), where a is positive parameter, $a = 1$ in the figure. Light colour fills the areas where horizontal principal stresses are close to each other. Asterisks located represent the epicentre of the earthquake happened on the 26 of December 2004, and 28 of March 2005; other circles representing earthquakes with magnitudes

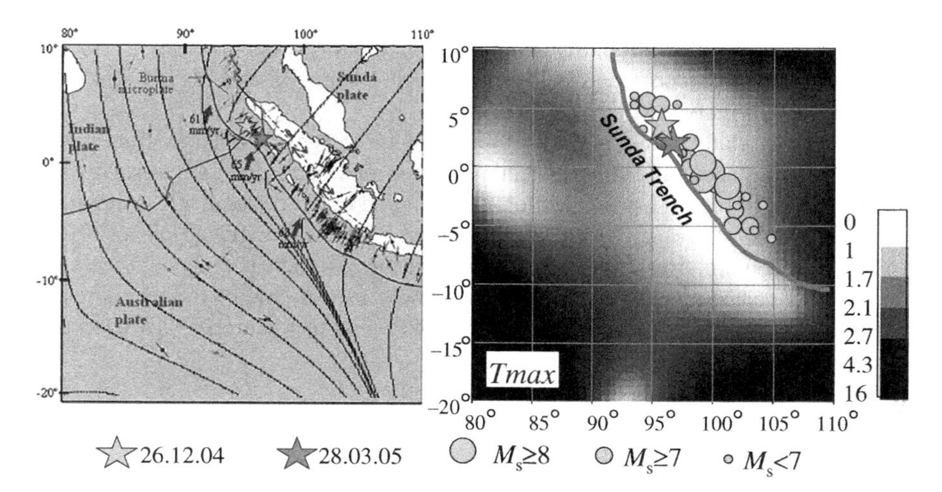

Fig. 6.2 Pattern of stress trajectories (*left*) and map of maximum shear stress (*right*) near Sumatra. Data on the left have been supplied by the WSM project, calculations have been produced by the Australian Computational Earth Systems Simulator, ACcESS, within the framework of the P10 project [10]

higher than 6 that occurred more recently. It is evident that for these two cases the difference between the principal stresses acting in plane is relatively small and it is therefore expected that the vertical principal stress is not the intermediate one. The same type of assessing the potential for unstable conditions in Europe leads to the conclusion that Sweden and Finland are fairly safe but that countries like Portugal, where severe tsunamis have in fact taken place, contain regions that are unsuitable for hosting repositories.

Naturally, energy issues are in focus in predicting both micro- and macroscopic strain and for the far-field one needs to refer to the structural constitution of the host rock using current categorization schemes of scale-dependent discontinuities of different orders [11,12,13,14]. Those of 1st order, which persist for tens to thousands of kilometres and contain strength-reducing gouge, undergo current creep-related shear strain. The 2nd order discontinuities are shorter and stronger but may undergo significant strain if the stress field is critically oriented and escalated. The 3rd order discontinuities, which are numerous in crystalline rock and persist for a few hundred meters, have a significant strength, expressed in terms of the friction angle, and may be triggers of large instantaneous shear strain in a rock mass.

6.4.3 Structural Implications

The structural constitution of the earth crust where large strain can take place can be generalized as in Fig. 6.3. The "virgin rock" has a structural constitution characterized by the ordinary spectrum of discontinuities of different orders. The regional

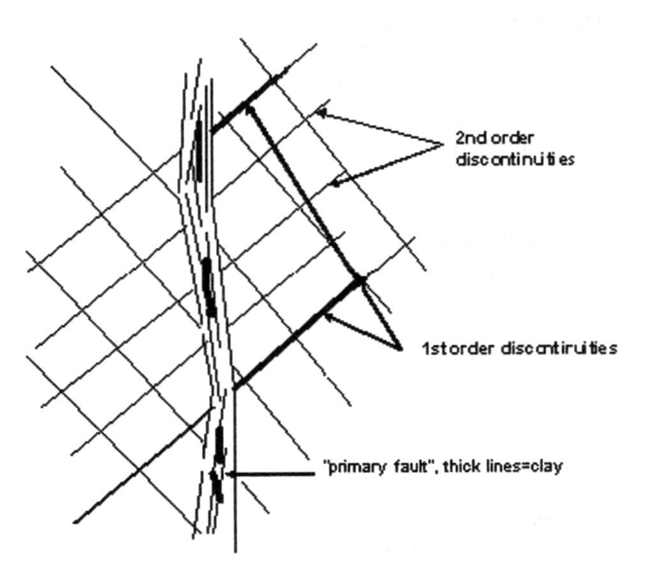

Fig. 6.3 Proposed structural constitution of the seabed where tsunamis can develop. The rock matrix between the discontinuities has high rigidity and strength

strain is controlled by the "mother fault" of 0th order represented by the San Andreas fault or its megascopic relatives that persist for thousands of kilometres, while 1st order discontinuities make up a series of more or less parallel major weaknesses aligned with or crossing the fault. Discontinuities of 2nd order are comparably weak and form parallel or oblique sets with a persistence of some kilometres. Those of 3rd order persist for a few hundred meters and their strength is comparatively high.

Several mechanisms combine to yield the large-scale quick breakage that is required for generating high-magnitude earthquakes. A prerequisite is the accumulation of stress by creep strain in the "mother fault", the first structural members of the "virgin rock" to react being the weakest ones, i.e. the 1st order discontinuities. They contain brittle parts serving as asperities that break when the strain has become critically large. The breakage is associated with release of high energies, yielding vibrations, and shearing that causes additional fracturing and reduction in strength and transfer of stresses to less stressed members of the same category of weaknesses and to the network of 2nd order discontinuities. The latter will thereby undergo shear strain that will in turn activate those of 3rd order etc. The shear strength of 4th and higher order discontinuities, i.e. the discrete fractures that are water-bearing and inevitably intersecting deposition holes and tunnels, is judged to be sufficiently high to reduce their deformation to less than a millimetre by earthquakes of magnitude M5 [13]. Instantaneous shearing of 4th order discontinuities by 100 mm would correspond to earthquake magnitudes exceeding M7.

The primary driving force yielding large, instantaneous rock strain is the shearing of very large faults, i.e. those of what we might call 0th order that are typically related to orogenesis. Their long strain history has softened them and disintegrated them to become clayey by mechanical disintegration in conjunction with percolation of hydrothermal solutions. The degrading processes have proceeded far in some parts and less in others depending on their strength and the shear resistance hence varies along the fault (Fig. 6.3). The high-resistance spots are the ones that create the typical "stick-slip" performance but since the weak parts dominate in ancient faults the energy-release in the form of earthquakes may not be of very high magnitude. Instead, earthquakes of magnitudes higher than M8, which generated the tsunami in December 2004, imply release of energy that is thought to be related to slip and propagation of 1st and 2nd order discontinuities or to propagation of freshly formed faults. It is believed that 1st and 2nd order discontinuities may well undergo catastrophic sliding [15].

6.4.4 Block Movements

One commonly assumes that large movements in the earth crust has the form of horizontal displacement along steep discontinuities as inferred from Fig. 6.3. Such movements may be disturbed by obstacles as in Fig. 6.4 where the lineation follows the contours of the granite dome. Vertical or steep movement can have the form in Fig. 6.5, which has been proposed as a tsunami-generating process. Naturally, such

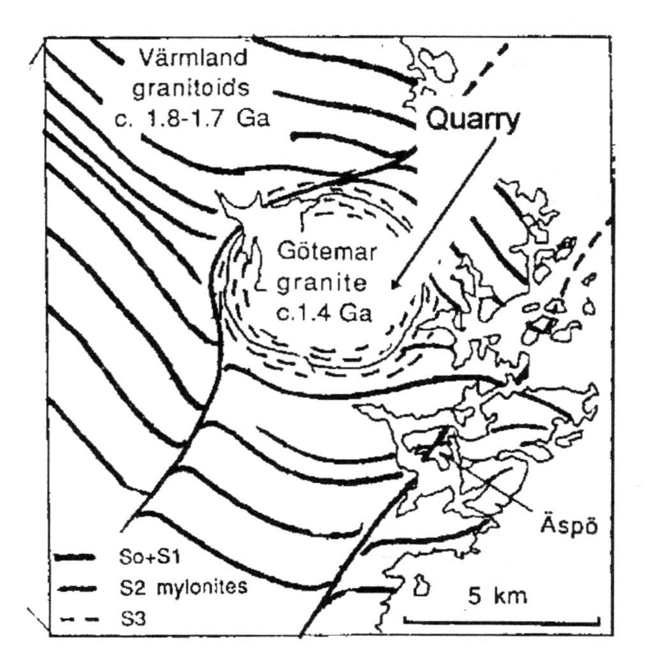

Fig. 6.4 Lineament pattern around strong dome in crystalline rock [16]

events can take place where similar structural conditions prevail in areas of potential interest for locating a repository.

There are several possible mechanisms that can yield the large-scale instantaneous breakage that is required for generating high-magnitude earthquakes yielding both upward/downward and lateral displacement of mega-sized rock blocks. One possibility is release of high energies associated with formation and propagation of new major discontinuities in "virgin" rock when shearing of the "primary fault" is hindered by interaction with large obstacles in the form of strong rock units like domes and "microcontinents" shown in Figs. 6.4 and 6.5.

6.4.5 Evolution of Low-Order Discontinuities

The older and larger the discontinuities, the more have they been affected by thermo/mechanical/chemical processes. This has caused the well known mechanical disintegration and conversion of rock-forming minerals – primarily feldspars and heavy minerals – to clays that can make tunnel construction very difficult. Shearing of the rather ductile large faults does usually not lead to earthquakes of magnitudes higher than 7–8. The extreme magnitudes (M8-M9) of the event on 24th December, 2004, and 25th March 2005, in southeastern Asia, and also in mid December 2004 in the Antarctic area, are therefore believed to be associated with the formation and

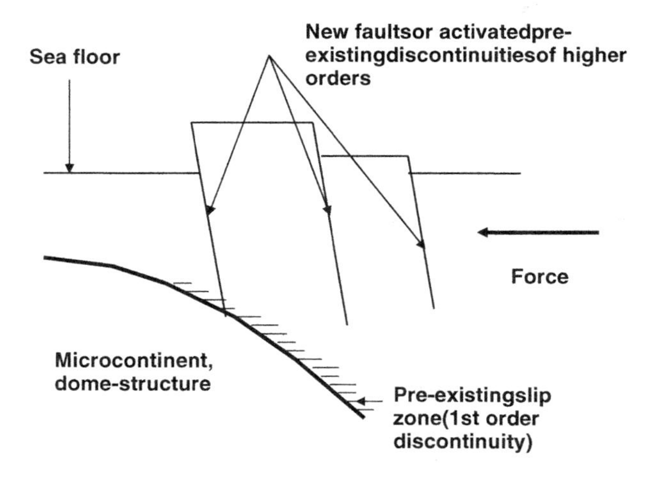

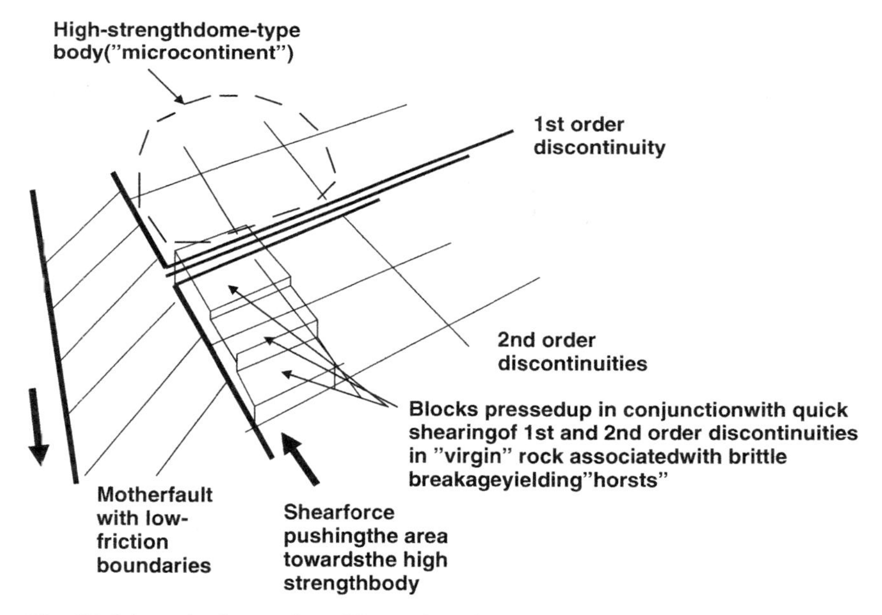

Fig. 6.5 Schematic picture of possible erection of large rock units that can generate a tsunami if it takes place at sea. *Upper*: Section parallel to the forced movement. *Lower*: Perspective view

propagation of new discontinuities. An early attempt to work out conceptual models for this is illustrated by Fig. 6.6 but the present understanding is that "stress due to dislocation" is not relevant because the stress field is singular at the dislocation tip, and slip is impossible along stress trajectories because the shear stress vanishes there.

Much effort has been exercized in investigating how secondary faulting can lead to displacement of large rock block units, the boundary conditions of which can be defined by using suitable categorization schemes for rock discontinuities, like the one used in this book. Comprehensive rock-mechanical analyses for investigating the response to tectonic events have been made and they have indicated ways of modelling strain on all scales. Attempts have also been made to use this knowledge in working out models of uplift of large rock units ("micro-horsts") causing tsunamis.

Among existing stress/strain models the one in Fig. 6.7, implying rock dilatancy under uniform compression (Fairhurst-Cook 2D model), has been considered. It seems, however, that the pattern shown in Fig. 6.7b might not be mechanically

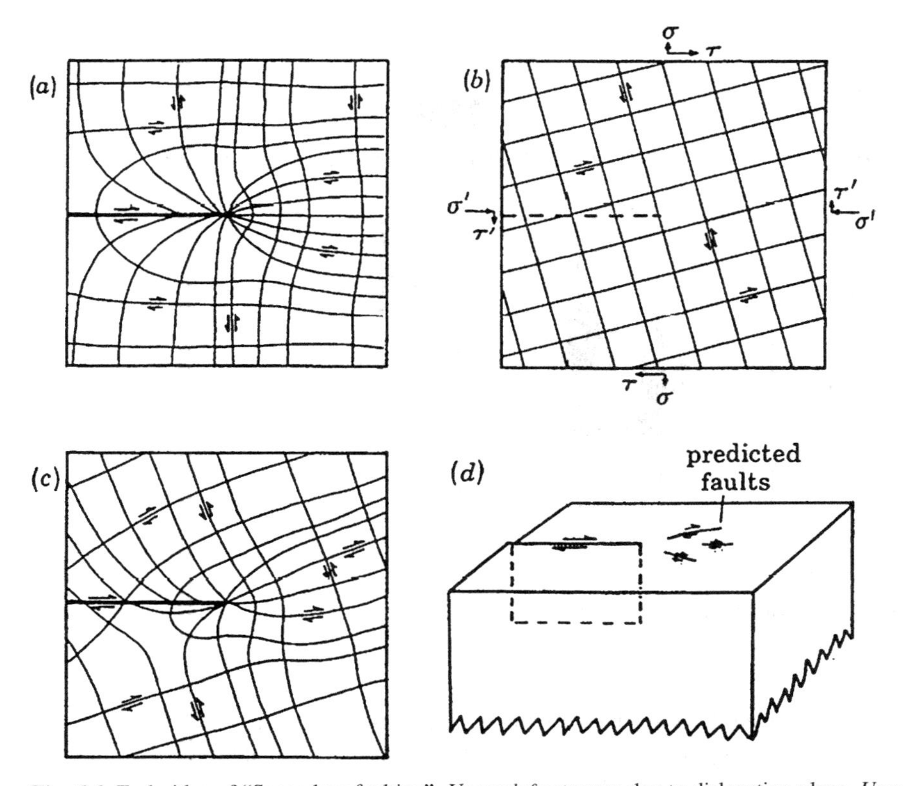

Fig. 6.6 Early idea of "Secondary faulting". *Upper left*: stresses due to dislocation plane. *Upper right*: Stress trajectories of an external field. *Lower left*: Combined stress conditions of (a) and (b) yield the predicted secondary faulting in the *lower right* figure [3]

a) displacement on a single fault (simple fracture)

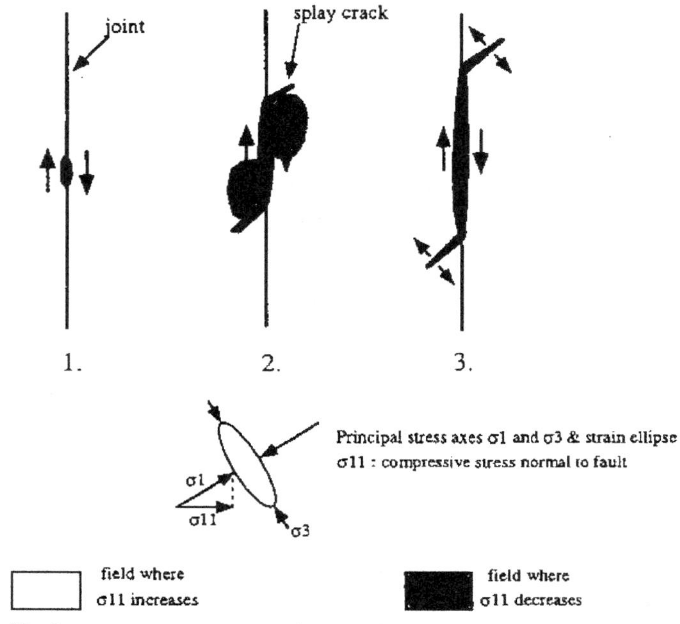

b) displacement on a pair of faults (complex fractures, fault zone)

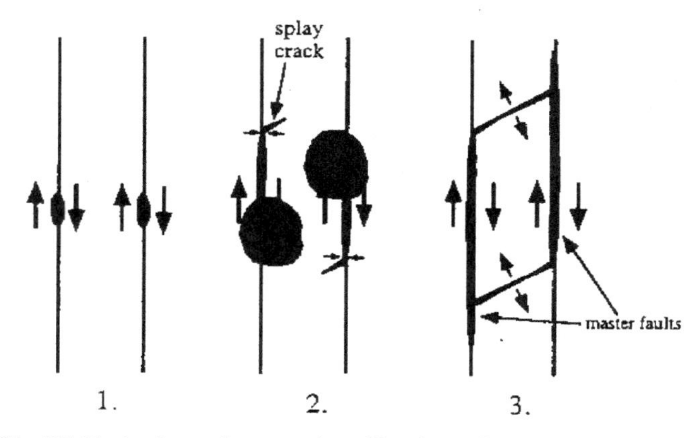

Fig. 6.7 Nucleation and propagation of breakage along preexisting single and pairs of discontinuities [4]

realized in the earth's crust because the wing cracks may not reach the neighbouring major discontinuity. Models of this type work on a small scale but not in the real world since they lack impact of the variation in structure and strength of the rock matrix. The strength distribution, which is particularly important, is strongly related to the varying strength of the major discontinuities.

A case of this type, exemplified by the photo in Fig. 6.8, is the Pärvie fault along which vertical displacement of up to 25 m occurred in conjunction with deglaciation of northernmost Sweden. It is believed that vertical movement of this discontinuity took place quickly in conjunction with earthquakes of magnitudes M5 to M9.

Breakage of large blocks, like the ones moved along slip surfaces like the ones indicated in Figs. 6.4 and 6.5, is controlled by their internal structure. Simplifying the structural constitution of such units one can illustrate the breakage at critical strain as in Fig. 6.9. Recognition of potentially unstable regions where large block movements can take place and where failure of the type shown in this figure can take place requires that the rock structure has been defined and that sufficiently comprehensive rock stress measurements are made.

Fig. 6.8 Aerial photo of major steep displacement along a 1st order discontinuity ("Pärvie fault") in conjunction with deglaciation causing critical stress conditions. The height of the steep slope is up to 25 m. (Photo by Talbot)

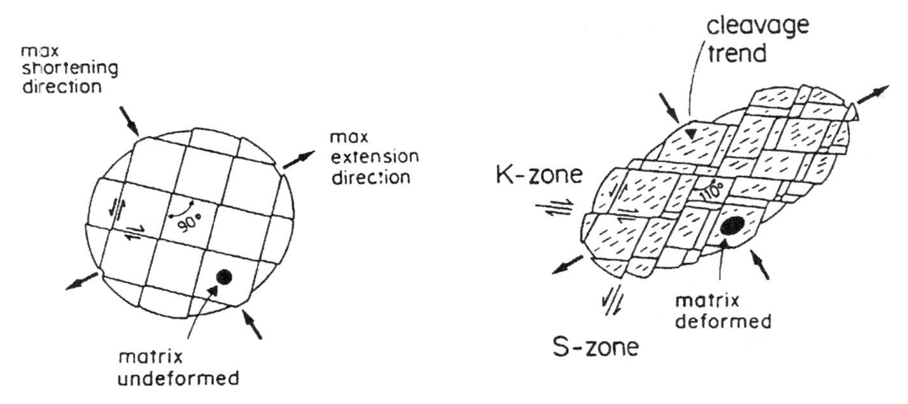

Fig. 6.9 Breakage and internal deformation of rock block that is displaced [4,17]

The stress field and displacements generated by rock blocks that are pressed against a steep boundary represented by a high-friction discontinuity can be examined by numerical methods like FEM and BEM techniques. The latter has been used in pilot studies showing that very high shear stresses are built up near the boundaries where brittle breakage leading to erection of large block fragments can occur. This case is relevant for high-strength boundaries of blocks with no slip along the steep plane against which the overriding rock is pressed, and represents the embryonic stage of folding (Fig. 6.10). At critical strain of the assumed brittle body, shear planes are formed and continued lateral compression can cause sudden upward displacement of a segment separated by a major axial shear plane.

6.4.6 Energy Issues

The primary driving force of large rock strain is the shearing of a steep, very large 0th or 1st order discontinuity, a "fault". The long strain history of such faults has

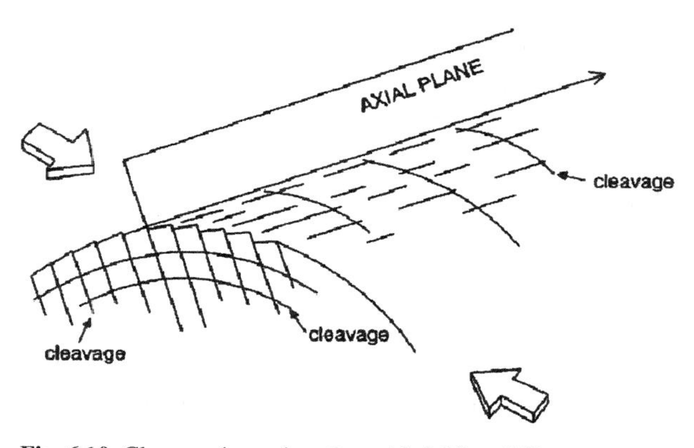

Fig. 6.10 Cleavage in conjunction with folding [18]

caused very rich fracturing and percolation of hydrothermal solutions have transformed parts of them to clay, often with significant content of smectites. The degrading processes have proceeded far in some parts and less far in others depending on the strength and hydraulic conductivity of the rock that is intersected by the fault and the shear resistance hence varies along the fault. The high-resistance spots are the ones that create the typical "stick-slip" performance but since the weak parts dominate in ancient faults the energy-release in the form of earthquakes is not of very high magnitude. Earthquakes of magnitudes higher than 8, which gave the tsunami in December 2004, implied release of energy that is thought to be related to propagation of freshly formed, major discontinuities [8].

6.4.7 Numerical Modelling of Large-Scale Strain

Naturally, prediction of strain by use of mathematical models requires that the conceptual deformation model is reasonable and that representative material parameter values can be identified. Likewise, the conceptual model of the rock structure must resemble the true one, which is the most challenging issue. We will see here what the magnitude can be of the shear strain distributed in a simple 2D rock structure containing relatively large discontinuities of 2nd order and integrated regular systems of 3rd and 4th orders in fractal-like patterns, assuming that the primary stress field with the principal stresses 30 and 10 MPa is rotated by 45° (Fig. 6.11), [19]. This study was made by analytical technique assuming the friction angle of the largest weaknesses to be 10°, 15° of the 3rd order fracture zones, and 25° of the discrete 4th order fractures. The latter mobilize a significant resistance to deformation. Shear strain takes place in the weakest elements, i.e. the 2nd order zones being rich in fractures and containing some clay. The rotation of the stress field generates shearing on the millimetre scale of the discrete fractures of 4th order, while the 3rd order fracture zones are sheared by a decimetre and the 2nd order zones by about one meter. First order discontinuities, i.e. very large weak zones, would deform by tens to hundred meters. Rotations of this magnitude do not take place instantaneously but can be assumed to occur stepwise triggered by large seismic events or intercontinental relative displacements, and retarded by shear resistance to be fully developed in hundreds to hundreds of thousands of years.

This sort of displacement of block units along low-order discontinuities is associated with breakage causing activation of higher-order discontinuities and creation of new ones. This is naturally of fundamental importance in the case of repository rock, since the direction and intensity of groundwater flow can change totally in a limited period of time. The risk of such events is significant because of the generation of stresses caused by the excavation of rooms, tunnels and drifts, and because the thermal pulse will further raise the stress level. Forthcoming glaciations will increase the vertical pressure very significantly, i.e. by 30 MPa as in the latest glaciation period, and thereby also the lateral stress in the rock mass, generating additional shear stresses and slip along steeply as well as subhorizontal and graded major discontinuities.

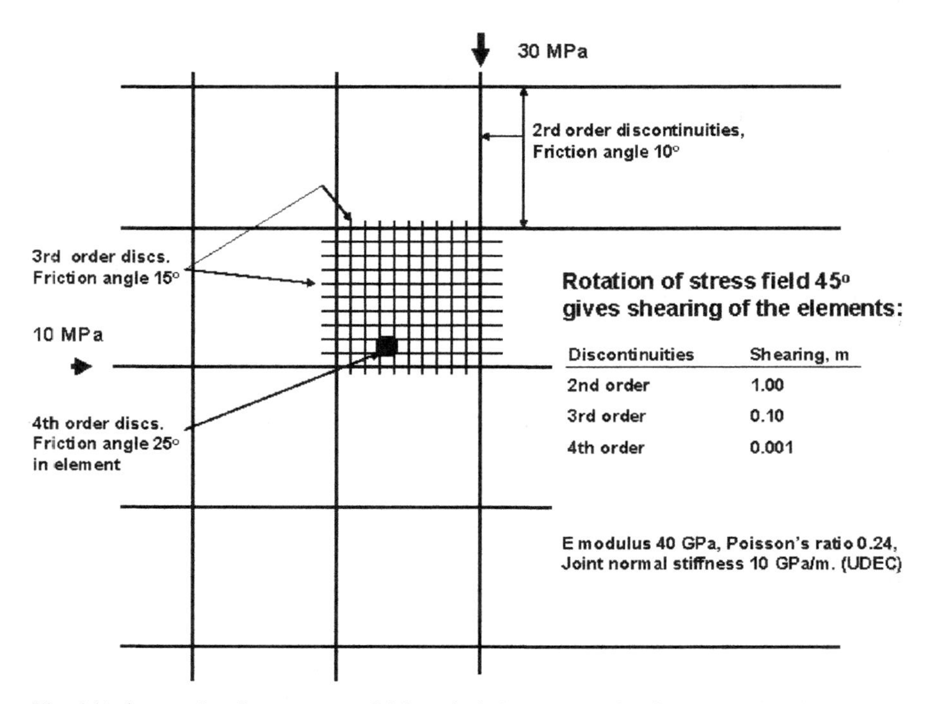

Fig. 6.11 Assumed rock structure model for calculating shear strain of the respective elements. The spacing and persistence of the 3rd order discontinuities is 100 m while the 4th order discontinuities have a spacing of 5 m and a persistence of 25 m

In practice, shearing of low-order discontinuities is associated not only with deformation and fracturing of blocks located between them but also with a change in interaction of adjacent blocks by formation of local gaps and breakage at over-stressed contacts (Fig. 6.12).

Pressure measurements near unloaded joints in a block system of this sort will naturally give much lower values than where the blocks are in tight contact. One would therefore expect rather strong variations in primary pressures in the rock hosting a repository, while for relatively small rock masses the variations should not be very significant. This is illustrated by the Stripa URL, where a rather uniform stress pattern have been found as indicated by Fig. 6.13. The figure shows the orientation of the major horizontal stress and of 4th order discontinuities in the crystalline rock in the northern part of the URL. The orientation of the major principal plane follows approximately the orientation of one of the major sets of this order.

The conclusion from various other attempts to determine deformations and critical stress regions in large models by performing numerical calculations is that the results are very uncertain because of the limited information on the true rock structure. The smaller the rock volume the more reliable are the results and one has to reduce the volume to contain one or two deposition holes in a repository for getting reasonably representative data. Provided that the rock contains only a few 4th order

Fig. 6.12 Block assemblages exposed to external stresses (After Stephansson)

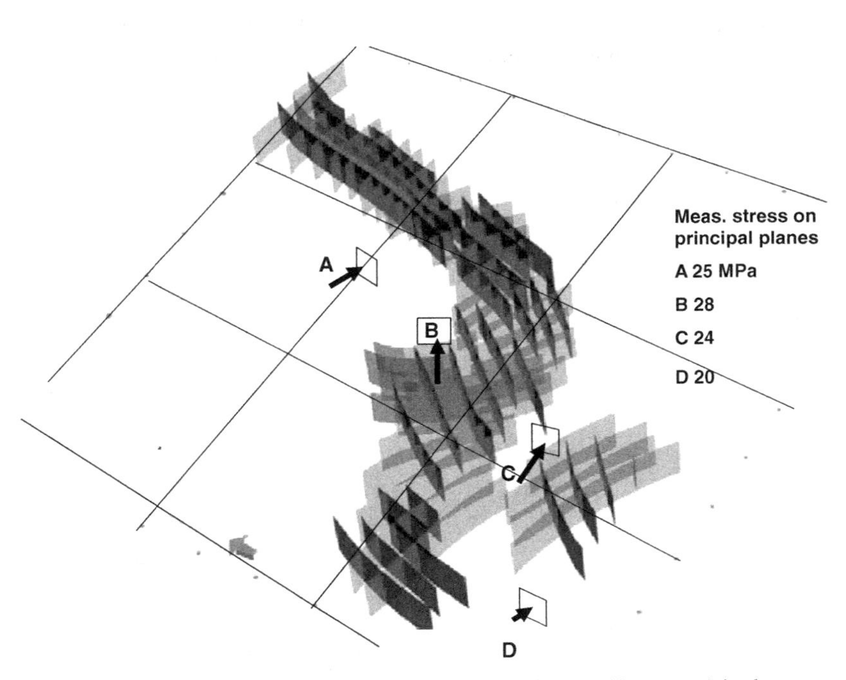

Fig. 6.13 Change in orientation of the major horizontal stress (*fat arrows*) in the system of 4th order discontinuities in the Stripa URL at 360 m depth. This stress component is oriented more or less parallel to the set of water-bearing fractures (*dark*). The spacing of the lines is 30 m [20]

discontinutities and that all discontinuities of higher orders are neglected one can derive an approximate stress distribution in the near-field rock [14,21].

6.4.8 Impact of Glaciation and Deglaciation

There are good reasons to believe that one or several glaciation cycles will occur in the north and hit Scandinavia, large parts of Asia and North America and areas earlier glaciated in Quaternary time. A probable evolution of glaciers is illustrated in Fig. 6.14.

The matter of glacier-induced stresses and associated strain in the earth crust have been discussed for decades and so have the involved hydraulic processes, i.e. the water pressure and flow [22]. There is no doubt that a repository of the type presently favoured by SKB will be affected in various ways by one or several glaciations, the main processes and factors being the following:

- The moving ice front will erode the landscape and scrape off soil and shallow rock. This means that the backfills in adits, like ramps and shafts, leading down to the repository will be removed and deeper parts of them exposed. This reduces the distance to the waste and thereby increases the risk of radioactive contamination of the biosphere.
- Glaciers formed in elevated areas will flow towards deeper ones and erode river valleys. In conjunction with the retreat of the ice some fifty to hundred thousand

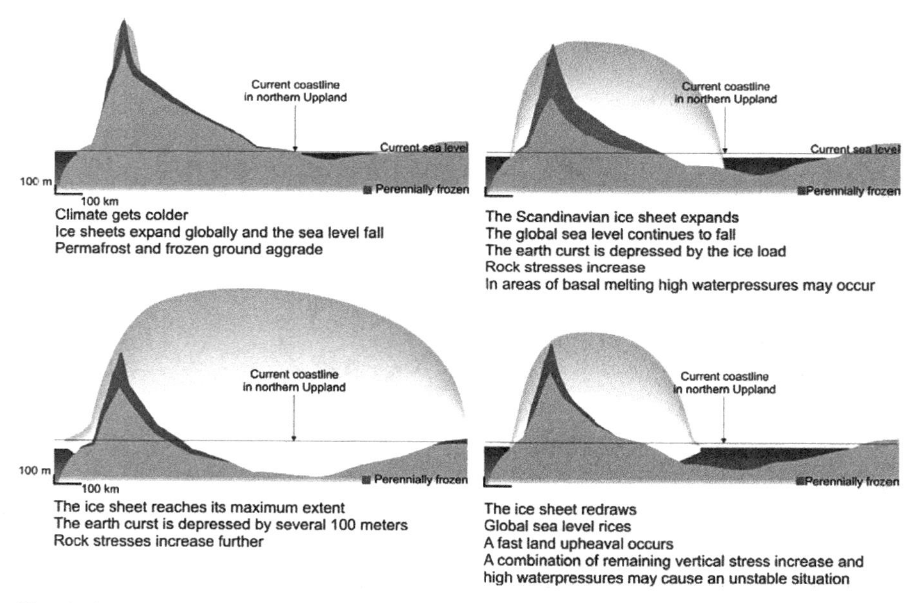

Fig. 6.14 Expected stages in forthcoming glaciation and subsequent early deglaciation (By courtesy of SKB)

years later the deep scars will be filled with debris emanating from the lower part of the ice making the river valleys serve as major conductive zones.
- The overburden pressure at 400–500 m depth will increase from presently 10–15 MPa to around 40 MPa and lateral pressures will be only slightly lower. On the retreat of the ice the vertical pressure will approach the preglaciation pressure while the lateral will only be slightly reduced at this depth, at least for some thousands of years.
- On the retreat of the ice high water pressures will prevail in steep fractures in the melting glacier causing high water flow at the contact with the underlying rock and in the rock. This can drive low-electrolyte meltwater into fracture systems, possibly even to the repository level.

These matters have been dealt with in comprehensive studies from which a few major results will be used to see whether they can reduce the isolation potential of HLW repositories.

6.4.8.1 Loading

A first issue is the stability of the repository rock if the vertical and lateral principal stresses rise by about 30 MPa, which corresponds to the around 3 km thick Pleistocene glaciers, causing a net vertical pressure of 50–75 MPa. Using basic rock mechanics one finds that the hoop stress at the periphery of openings, like deposition tunnels and deposition holes, will be on the order of 100–150 MPa. The data in Table 2.1 indicate that there should still be at least 1.5-fold safety for granite while for argillaceous rock one would expect comprehensive failure. However, critical conditions will in fact be reached also of at least 20% of the deposition holes in granite because of the statistical variation in strength. Where the present primary rock stress is 40–50 MPa the superimposed ice load will cause comprehensive breakage and collapse of the repository rooms even if they are backfilled with dense soil material. In salt, the load would cause considerable strain and healing of those parts of tunnels and holes that may remain after creep-generated convergence.

6.4.8.2 Glacial Erosion

A second issue is the erosion of repository rock caused by the movement of the glaciers. If the topography is suitable, i.e. convex upwards over the repository area, the erosion will be at minimum and the removal of rock be a few tens of meters per glaciation cycle depending on the regional and local topographies.

However, one should consider possible impact of deeper glacial erosion of large discontinuities in the vicinity of the repository area. The major practically important impact on a repository is that deepening of existing large discontinuities, those of 1st order, which have a common spacing of some 50–100 km, can temporarily and permanently change the regional hydraulic gradients and hence also the water flow rate in the host rock if the repository is located within a few kilometres from the deep discontinuity.

Glacier-generated erosion of major fracture zones appears to be a real process in nature and it has been treated quite extensively in the literature. There are data that indicate possible erosion depths of many tens of meters and even 100 to 200 m, which can have a significant impact on the stress state and hydraulic performance of the rock.

The shape and thickness of the glaciers and the inclination of the base control the rate of glacier movement. The flow rate of the top surface of glaciers depends on the ratio of accumulation and melting and is on the order of 50–200 m per year in the Alps and up to 12 km per year in Himalaya. On Greenland, where the glacier thickness is up to a couple of kilometers, the flow rate has been reported to be 1–7 km per year. It is appreciably lower at the contact between the glacier and the bedrock than of the surface of the glacier but the numerous erosion scars on exposed rock demonstrate that the flowing ice exerted very high shear stresses on the underground.

Figure 6.15 shows a structural map of an island off the Swedish west coast with 50–100 m spacing of parallel valleys, emanating from erosion of 2nd and 3rd order discontinuities in crystalline rock. They extend downwards by many hundreds to thousands of meters but borings have shown that glaciation/deglaciation cycles have not rinsed the zones to larger depth than a few tens of meters, followed by filling the scars with eskers and sediments.

Major glacier-eroded valleys are of the type depicted in Fig. 6.16. It shows an area 15 km east of Gothenburg in southwestern Sweden with the Säveån river and the Aspen lake representing a large valley. The persistence is tens of kilometers and the width of its most narrow part one or a couple of hundred meters. It represents a discontinuity of 1st order or a major 2nd order weakness cleared by glacier movement from NE to ENE. Since the lake is 28 m deep and has bottom sediments of considerable thickness it is probable that glacial erosion of the fracture zone extended to more than 50 m depth. Erosive action to such depths is believed to have a practically important impact on the regional stress field and hydraulic performance of the rock mass.

Studies of deep canyons in Sweden, Finland and Norway have contributed to the understanding of glacial erosion of different types [24,25,26,27]. These studies have included erosion by glaciers and by meltwater flow in valleys of "subglacial" origin, i.e. formed by eroding glaciers and subglacial meltwater flow. The valleys, of which some are known to be very deep, like the Finnish Presidentinkuru valley (70 m), and the Norwegian Glomma valley (200 m) coincide with major discontinuities and are concluded to have been formed by glacial erosion.

Attempts to calculate the erosion by using numerical techniques have been made and we will look at one study of pedagogical value. Figure 6.17 shows a schematic picture of the loading case represented by an advancing ice front. It shows three parallel weak rock zones representing 1st, 2nd and 3rd order discontinuities that persist for at least one kilometer.

Applying the shape of a typical glacier, *Mer de Glace, Mont Blanc* (Fig. 6.18), the case considered for numerical modelling was that in Fig. 6.18. The glacier was assumed to be frozen to the ground to a varying distance from the edge and the depth

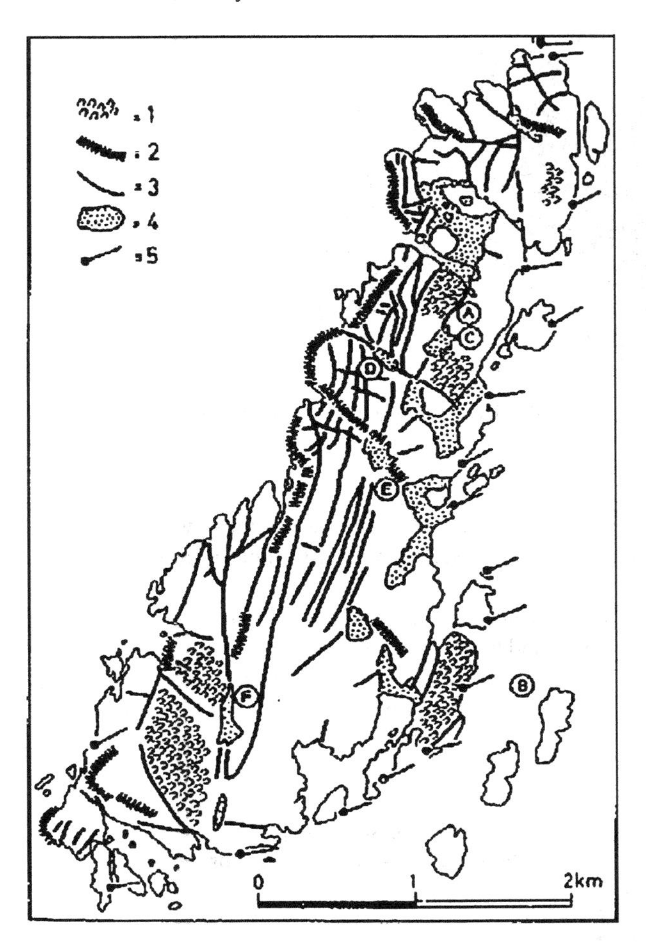

Fig. 6.15 Structure of Härmanö island [23]. (1) Giant stoss-sides, (2) Giant lee-sides, (3) "Micro-joint valleys", (4) Fine-grained sediments in rock basins, (5) Glacial striation.(North is up)

of erosion below the front of the glacier was also varied. It was further assumed that all material pushed up in front of the advancing ice front over the ground surface will be dispersed and moved ahead and to the sides and that it has no strength but a density of 1500 kg/m^3. The conceptual model implies that the glacier reaches deeper and deeper, approaching a maximum depth due to the increasing friction resistance (force) at the vertical boundaries.

The shear stresses that one likes to calculate are due to the creep of the ice mass and are obtained assuming different extensions of the adfreezing of the glacier. The ice is taken to behave as a viscous body with a viscosity of E13 Pas attacking a 1st order type discontinuity with typical material data. A friction angle of 35° was assumed for the glacier/underground contact considering the large amount of debris frozen in and a friction angle of 25° assumed for the contact between ice and eroded

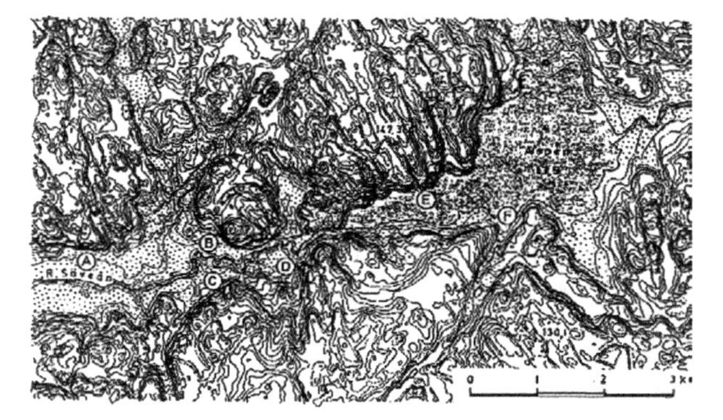

Fig. 6.16 Area east of Gothenburg with the river Säveån and the basin Lake Aspen [23]

material. The latter was considered as a Newtonian viscous mass with a viscosity of E10 Pas. Boundary element technique was employed for determining the erosive forces and penetration depth.

The ice was assumed to erode the surface and gradually penetrate the soft ground between the rock faces (Figs. 6.19 and 6.20). Because of friction between the ice and the rock and the mechanics of the material the flow velocity at the base of the ice will decrease as the depth increases until the point is reached at which the velocity

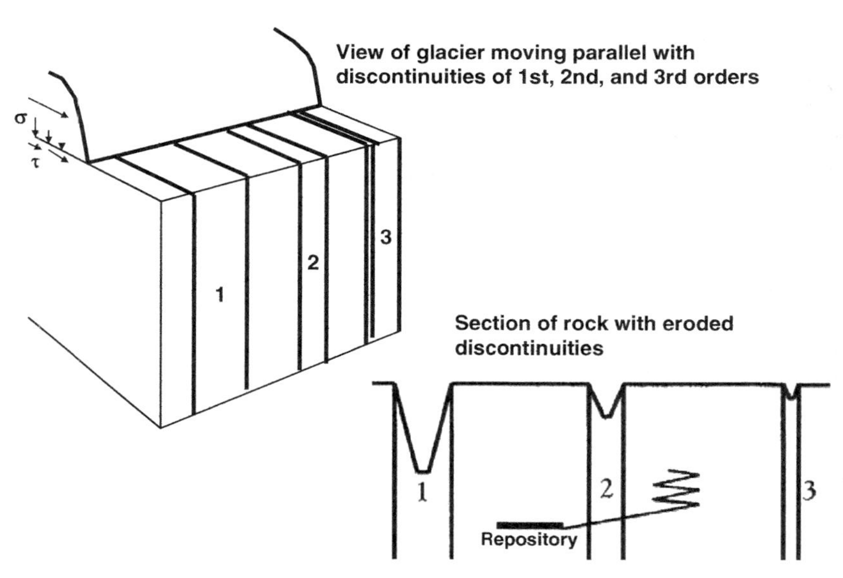

Fig. 6.17 Model for calculation of depth of erosion. *Right*: expected differences in erosion 1st, 2nd and 3rd order discontinuities

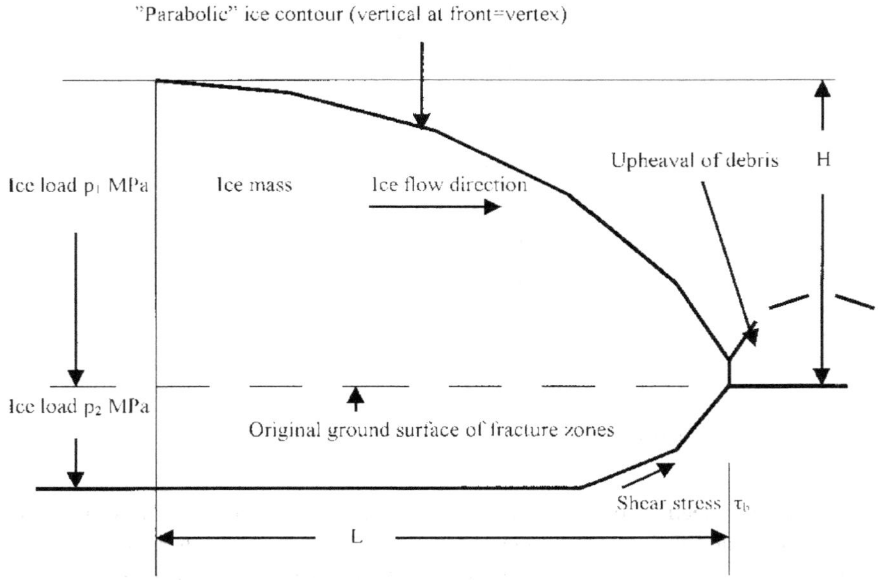

"Parabolic" ice contour (vertical at front=vertex)
Ice load p_1 MPa
Ice mass
Ice flow direction
Upheaval of debris
H
Ice load p_2 MPa
Original ground surface of fracture zones
Shear stress τ_b
L

Fig. 6.18 Conceptual model of an advancing glacier on land

will be zero or negligible at which erosion ceases. The strategy was to determine the depth at which this state is reached.

The applied velocity causes the glacier to slide over the soft material and along the rock walls. Friction was assumed at these surfaces and some velocity transmitted to the soft material in the discontinuity. The flow in the glacier is shown in Fig. 6.21 for the case where the glacier has penetrated the soft material to 180 m depth.

The flow rate distribution on the surface of the largest discontinuity in Fig. 6.17 ("1") was found to be about one tenth of the motion of the glacier front. Since the shearing rate at the erosion depth of 180 m was still not zero further erosion would be possible. However, a number of restricting factors, primarily the amount

5000m
1000m
500m
180m
Fracture Zone

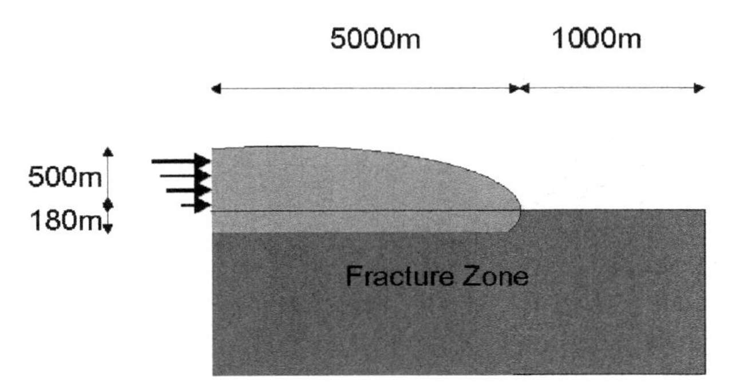

Fig. 6.19 Section of the model showing the penetration of the ice into the fracture zone

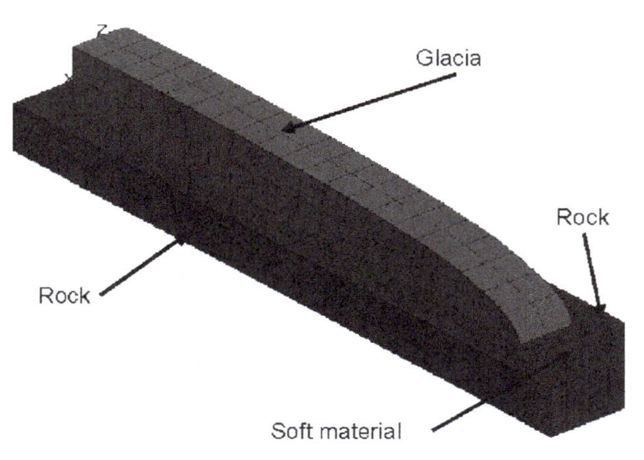

Fig. 6.20 View of the model showing the mesh and the key elements

of precipitation needed for maintaining the glacier shape, hindrance of discharge of eroded muck, and the general topography, would limit the erosion depth. It should be added that the deposition of debris contained in the ice and released when it ultimately melts, will ultimately fill up the deep scar.

The conclusion of all this is that glacier-generated erosion of major fracture zones appears to be a real process in nature and that cases described in the literature, supported by numerical modelling, suggest possible erosion depths of more than 100 m. This means that the impact of such erosion on the hydraulic and rock stress conditions in the adjacent rock hosting a repository need to be considered in the planning phase.

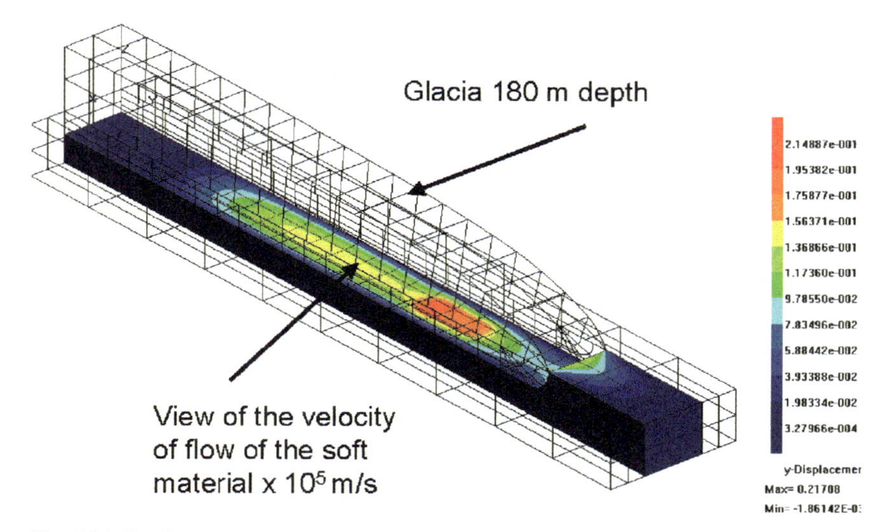

Fig. 6.21 Predicted velocity field in the glacier when it has reached down by 180 m

6.5 Salt and Argillaceous Rock

The softer rock types, salt and argillaceous rocks and clastic clay, are expected to undergo much more strain but less strong vibrations than crystalline rock for the same changes in regional stress fields and earthquake magnitudes, and creep is expected to smoothen boundaries between blocks and maintain coherence, hence causing less impact on percolation rates and radionuclide migration. Erosion by glaciers will be much more comprehensive than for crystalline rock. It should be added that strong metamorphic rock like gneiss is somewhat more ductile than granite and sustain strain with less comprehensive fracturing.

As concerns influence of glaciation NAGRA has conducted comprehensive studies of expected climatic changes and made preliminary estimates of general and glacial denudation rates for valleys corresponding to well over 100 m per glaciation cycle [28]. This certainly underlines the importance of considering glaciation in the site selection process.

6.6 Electrical Potentials

Electrical potentials are known to promote rapid corrosion of metal objects in boreholes and to have an impact on water migration in buffer clay. Corrosion of instruments in rock with electrical potentials generated by high voltage electrical transmission lines has been found to be extensive. It is therefore important to identify and assess electrical potentials in conjunction with site investigations and in the site selection process.

6.6.1 Natural Potentials

Measurements of natural potentials have been made by use of magnetotelluric equipment, and for recording potential differences, pairs of identical Pb-$PbCl_2$ electrodes have been used. The instruments are placed in boreholes and powered by batteries or solar panels. Measurements with commonly used equipment are usually successful in sedimentary rock while the high electrical resistance in crystalline rock make the results more uncertain. The most important conclusions from such investigations are:

- The amplitudes of the horizontal magnetic field components are very similar, whereas the amplitude of the vertical magnetic field is considerably lower, except during magnetic storms.
- Ongoing investigations by SKB indicate that electric fields are very similar in signature, but amplitude ratios indicate that the total electric field is strongly polarized due to lateral changes in the electrical conductivity close to the surface. Daily variations of a few nTesla may occur.

- DC voltages between deep boreholes can represent electric fields of about –250 mV/km in approximately N-S direction and 75 mV/km in the opposite direction (approximately E-W). With an approximate distance between the borehole points of about 250 m these fields correspond to voltage differences of about –60 and 20 mV, respectively.
- The inhomogeneous nature of the electric field contrasts to the homogeneity of the magnetic field.

6.6.2 Measurements in the Near-Field

Measurements in SKB's URL in granitic rock at Äspö where four full-scale copper canisters surrounded by dense smectite clay were located in 8 m deep deposition holes, have shown that two of the canisters have a potential that is higher than net-ground, while the others have potentials lower than net-ground. An obvious fact is that the potential was different in holes with no instrumentation in the buffer and in holes where the buffer was instrumented. Since the rock constitution and groundwater flow were nearly the same in all the holes it is obvious that the instrumentation caused the discrepancy.

6.6.3 Impact on Buffer

Electrical DC potentials are known to generate water and particle transport [29,30]. Water transport by applying electrical potentials, electro-osmosis, has been used on a large scale for dehydration and redirection of groundwater flow in order to stabilize natural and excavated slopes in clay. Particle transport in electrical fields has been used for sealing fractures in rock on an experimental scale, the principle being that clay particles have a negative charge and hence move towards the anode in DC fields [29]. On these grounds one can imagine that earth potentials can cause porewater migration in buffer clay and that water migration in the saturation phase of the buffer can generate electric potentials that may affect the rate of water saturation of initially relatively dry buffer materials, and the signals from instruments, and make the instruments corrode. Potentials of the magnitude that can exist in repository rock may change the rate of water migration in smectite-rich buffer by some 10% and would therefore not be of great practical importance but further research may show more impact. However, corrosion of copper canisters will be significant and be up to a few mm per 1000 years and hence be of practical importance.

6.6.4 Impact of Rock Structure

It is estimated on theoretical grounds that the electrical potentials in a rock mass can be high if the major discontinuities have a large spacing while they are expected to be lower where the rock contains numerous weaknesses of 3rd order.

The recommendation should be to select a region for locating a HLW repository where there are no transmission lines and where 3rd order discontinuities are frequent. Unfortunately, this is not favourable from the point of effectively using the deposition tunnels since frequent zones of this type means that many canister positions have to be omitted.

6.7 Practical Examples

6.7.1 Cystalline Rock

The principle of finding an optimum site for a HLW repository is primarily to identify the dominant discontinuities in the area by use of topographical and geological maps, and by visual inspection, followed by geophysical investigations, and gravimetric, magnetic, and electrical measurements. Once the major structural features have been identified and a preliminary regional structural model has been defined, borings are made, vertically as well as graded, for validation of the model and for getting information on petrology and structural details. By combination of boring for logging including primarily hydraulic measurements, sampling and analysis, and nuclear technologies, and more detailed geological and structural surveys the structural model is successively improved. The resolution can be high enough to define domains with certain intervals of fracture spacings or certain dominant petrological features.

A second or third examination of the region in question can result in generalized maps of 2nd and 3rd order discontinuities as they appear on the ground surface (Fig. 6.22). This figure refers to a study of topographical maps in different, equally large $3 \times 3\,\mathrm{km}^2$ areas in central Sweden, made for getting a first indication of the persistence and spacing of such discontinuities.

Comparison of the orientation of the major discontinuities in three areas gave the histograms in Fig. 6.23, indicating that the larger structural features are basically of the same type.

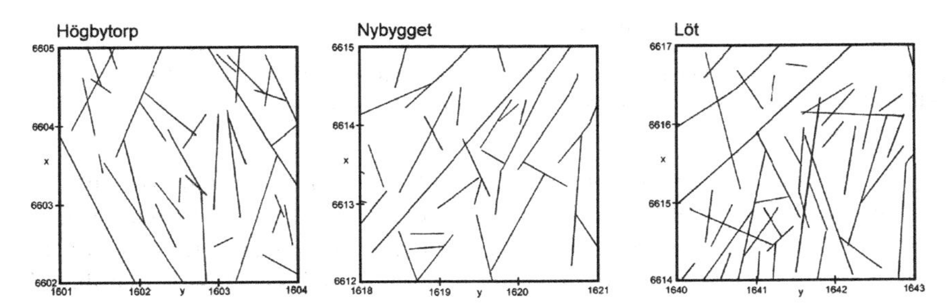

Fig. 6.22 Low-order discontinuities, mainly of 2nd order, in three areas in a geologically relatively homogeneous region north of Stockholm. There is a difference in orientation of the low-order discontinuities between the areas but statistical analyses showed obvious similarities in persistence and spacing. The edge length of the areas is 3 km

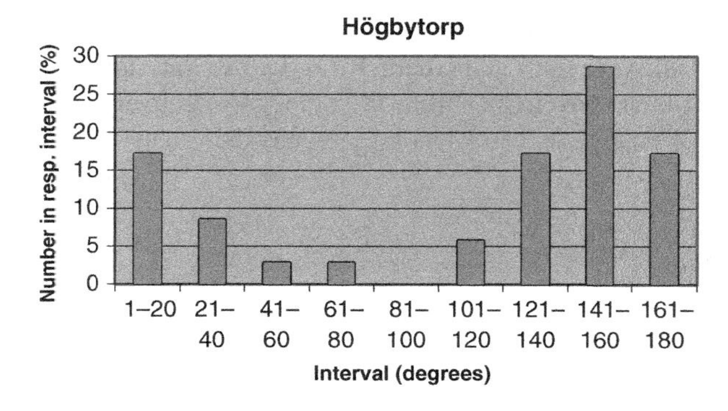

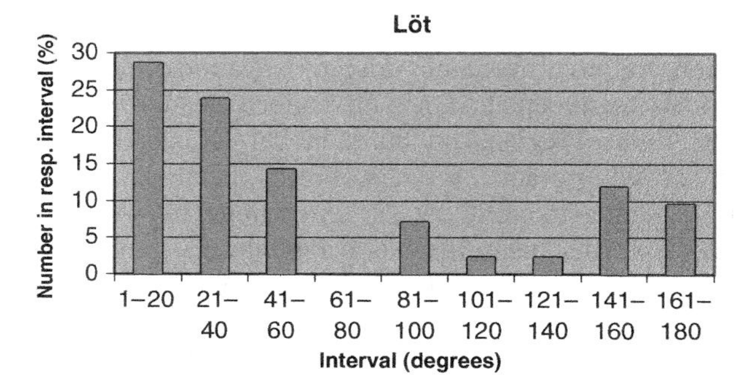

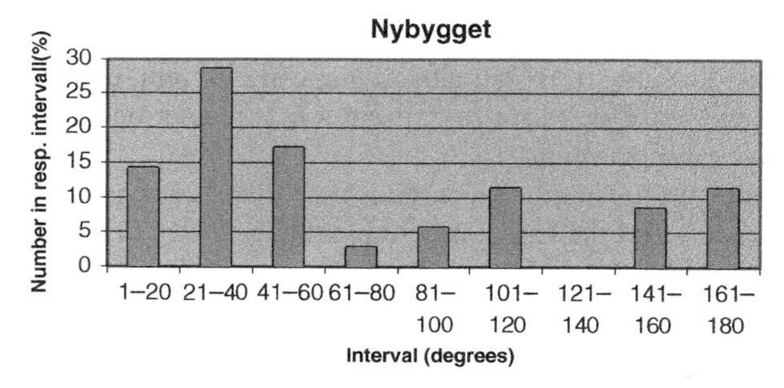

Fig. 6.23 Orientation of all major discontinuities in the respective area in Fig. 6.22. The histograms are sufficiently similar to suggest that the general structure is of the same type in the whole region

Figure 6.24 shows a schematic structure model of a candidate site in Sweden. The various elements represent discontinuities of 2nd and 3rd orders. The widest ones coloured blue, i.e. those representing 2nd order discontinuities, form a relatively regular pattern while the finer, 3rd order discontinuties are less regularly distributed and oriented.

At the stage when a structure model of this type has been worked out assessment has to be made of whether a repository can at all be located in the area with respect to the available space. This stage is illustrated by Fig. 6.25, which represents another

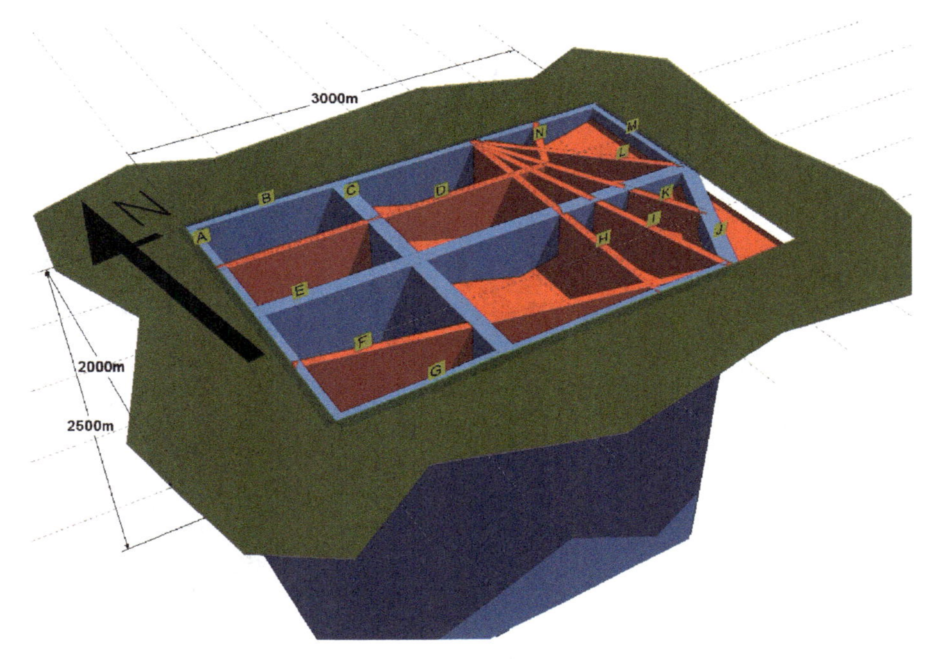

Fig. 6.24 Simplified and generalized model of candidate site (A). The green area represents the ground surface and the blue plates 2nd order discontinuities with 50–100 m width. The red plates are 2nd and 3rd order discontinuities with 10–50 m width. No repository panels have been marked

candidate site (B). The model shows a two-level repository for 6000 HLW canisters that does not fit in the largest space bounded by the 2nd order discontinuities D and F, without being intersected by the steep 2nd order discontinuity E. If the capacity has to be 12 000 canisters the repository would have to be split into many more parts but this would increase the cost substantially and instead suggest a three-level repository but this can lead to too high temperature in the rock and in the buffer clay. Construction of a repository with disposal at 400 and 450 m depths and possibly also at 500 m depth would be possible if the rock stress conditions are favourable and if the frequency and orientation of 3rd discontinuities are suitable.

Construction of ramps and shafts forming adits to a deep repository in crystalline rock will, by necessity, imply intersection of one or a few discontinuities of 2nd order, representing major fractures zones that contain strongly water-bearing stone/gravel/sand/silt parts, as well as very tight clay zones. This makes tunnel construction in such zones difficult because of strong water inflow and quick softening caused by sorption of water by the clay, which is often smectitic.

6.7.2 Argillaceous Rock, Salt Rock and Clastic Clay

The homogeneous character of domal salt, implying complete lack of low-order discontinuities makes other issues important in selecting a suitable position and size of

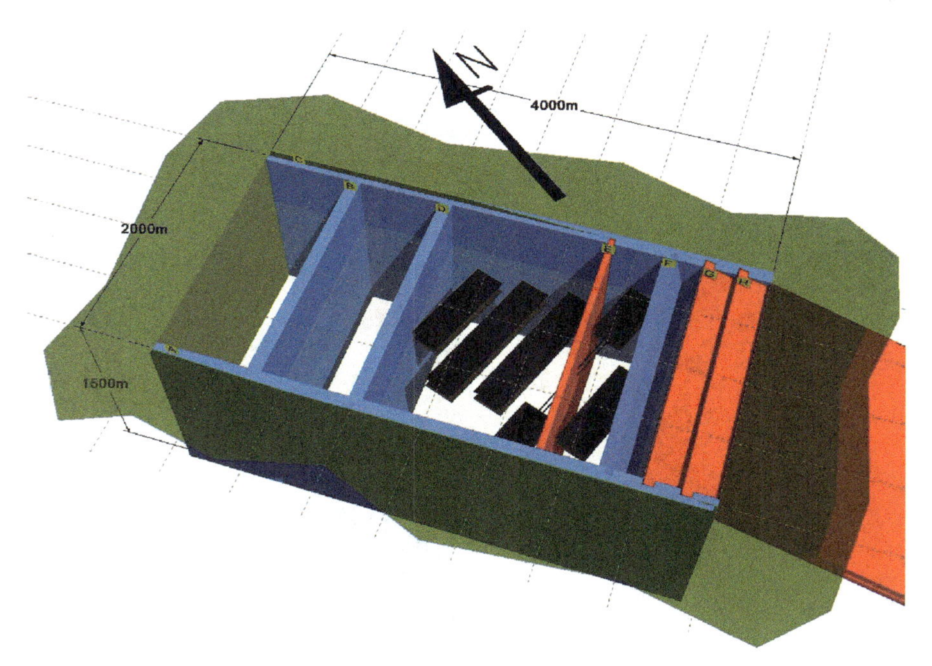

Fig. 6.25 Simplified and generalized model of an alternative site (B). The green area represents the ground surface and the blue plates 2nd order discontinuities with 50–100 m width. The red plates are 2nd and 3rd order discontinuities with 10–50 m width. Panels containing parallel rows of deposition tunnels are integrated at two depths

a repository, primarily the distance to the outer boundaries of the dome in question, and the presence of long-persisting soil layers. In bedded salt, however, such alient components can be frequent and homogeneous salt layers relatively thin. In certain areas, like in parts of Iraq, the total thickness of major salt beds is less than 50 m, which would be too little for hosting even a one-level repository considering the strong creep that can lead to accumulation of migrating canisters and creation of critical mass conditions.

For argillaceous rock and clastic clay, being formed as sediments, there are usually strong variations in mineralogy and physical properties but low-order discontinuities with significantly higher hydraulic conductivity and low mechanical strength are commonly much less abundant than in crystalline rock, meaning that finding a suitable site for a repository is not as demanding as for granite.

6.7.3 Electrical Fields

In crystalline rock numerous vertical and sub-vertical conductive fracture zones and mineralized zones can perturb current flow dramatically. If the current flows perpendicular to the zones electrical charges are set up at their boundaries, by which the electric field can become strongly enhanced or reduced and strong vertical currents generated. If the current flows parallel to the zones the electric field is more or less unperturbed, but the resulting current in such conductive zones may become magnified by the ratio of the electrical conductivity of the zone and of adjacent crystalline bedrock. Pipelines and cables with high electrical conductivity will act as conducting fracture zones, i.e. they can collect currents from the surroundings, resulting in enhanced corrosion.

It is obvious from these facts that the structural constitution of the repository rock is a major factor in the generation of natural electrical currents and potentials and that conductive features like cables can serve as short-circuiting conductors that can undergo corrosion and have a significant impact on data recorded by monitoring systems. As concluded from the earlier discussion of electrical potentials, inherent or induced by high voltage cables, areas with high electrical potentials should be avoided in the site selection process. The dependence of the potentials on the rock structure should be kept in mind. Thus, rock intersected by relatively closely spaced 2nd and 3rd order discontinuities is preferable since the electrical potential between them is lower than in rock with large distances between such discontinuities. From this point of view case A in Fig. 6.24 is more attractive than case B in Fig. 6.25.

6.8 The Ideal Location of a Repository

Is it possible to define the ideal location of a HLW repository? The answer would come naturally if the major desirable conditions listed below are at hand:

- Regional groundwater movement by hydraulic gradients should be nearly none.
- Crystalline rock should be chosen for safe construction and the site should be where only few strongly water-bearing low-order discontinuities are intersected.
- Earthquakes and other tectonic-related events should be of low magnitude.
- The rock stresses should be low or moderate. For a repository at 500 m depth the primary rock stresses should not exceed 30 MPa and the ratio of the major and minor principal stresses be in the interval 1–2.
- The repository should be located at least at 400 m depth for providing slow groundwater flow, sufficient margin to glacial erosion, and long migration paths to the biosphere for possibly released radionuclides.
- The repository should be at sea, which would serve as a recipient offering very strong dilution of possibly released radionuclides. The sea level should be approximately stable until the next glaciation occurs, i.e. at least for 10000 years.
- The crystalline bedrock hosting the repository should be covered by a few hundred meters of tight sedimentary rock, implying negligible water transport in vertical direction.

Cross section of island Gotland

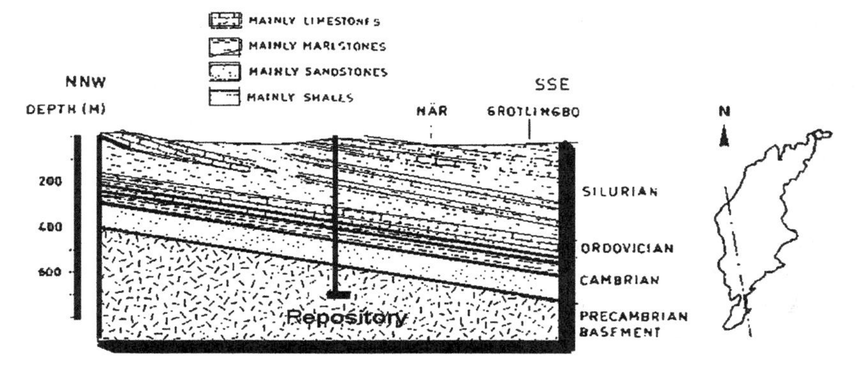

Fig. 6.26 Cross section of the island Gotland showing the series of sedimentary rock resting on the crystalline rock with the proposed repository

Some of these criteria are fulfilled by locating the repository on an island far from the mainland and all of them by the southernmost part of the island Gotland in the Baltic Sea (Fig. 6.26). It would hence be an ideal site. Thus, crystalline rock, i.e. gneiss, is reached at 400–500 m depth and here is where the repository should be constructed, preferably at 100 m below the upper boundary of this rock. It is covered by a few hundred meters of clay shales, marlstone, sandstone and limestone with about 50 bentonite horizons relatively evenly distributed in the profile [31,32]. The sedimentary rocks are mainly of Silurian age and have the typical features of sediments deposited in shallow sea between large coral reefs. They would serve as a perfectly tight lid of the system. There are traces of oil in the sandstone but the content of hydrocarbons is far from exploitable. The lack of hydraulic gradients in both the horizontal and vertical directions means that there is no driving force for water movement at all. Seismic and tectonic events are negligible and the rock stresses are probably moderate as indicated by the stability of 100 mm boreholes.

The question of how a repository in this area (Hoburgen) should be designed comes naturally and is readily answered. Thus, the deep-hole concept is not suitable while horizontal deposition tunnels of type KBS-H would be possible like concepts of types KBS-3 V. The version KBS-3i would be particularly suitable. The sealing of one or a couple of shafts is easily done by application of the same techniques as used in the HADES URL and deep German shafts.

6.9 Mine Repositories

6.9.1 General

One cannot avoid discussing the possibility of using abandoned mines for disposal of HLW. It is commonly thought that there is permanent veto against this concept because of the numerous objections raised. The major one is that no exploitable ore must be left so that the risk of future renewed mining is eliminated, and another being that the rock must be sufficiently tight and stable. The latter condition is difficult to achieve since mining aims at maximum fragmentation of the ore with no or only moderate interest in preserving the remaining rock, which would be the host rock of a repository. This depends, however, on the technique employed since certain methods can allow for reasonable preservation of the rock left after extraction of the ore.

Where subhorizontal ore bodies like metal-impregnated claystone beds have been mined, drifts for the ore extraction can be suitable for use as HLW disposal. Systems of such drifts extending from one or a few main tunnels forming fishbone-like patterns are particularly suitable since they resemble the deposition tunnels in the specially made HLW repositories. Big rooms, like those considered for disposal of hazardous chemical waste formed by extraction of steep ore bodies [32], are less suitable for practical reasons and because the HLW-generated temperature conditions may exceed commonly accepted levels.

A single-level mine geometry created by "room and pillar mining" of the type shown in Fig. 6.27 can offer very good conditions especially if the over- and underlying rocks consist of strong and tight argillaceous rock. This particular mine type was investigated in detail in the LRDT project [33] and consists of a regular array of pillars 7 meters long and wide and 5 meters high, repeated over a length of 150 meters.

The HLW canisters surrounded by buffer of highly compacted blocks of smectite-rich clay can be placed in "supercontainers" that are moved into well fitting tubes of the same metal as the containers. These are embedded in artificially prepared "argillaceous rock" compacted on site and composed so that the thermal conductivity is sufficient and the hydraulic conductivity low. Since water moving in from the surrounding rock will not reach the buffer until after hundreds to thousands of years, extensive corrosion of supercontainers and canisters desiccation can not jeopardize the isolation ability of these barriers. However, the very slow saturation of the buffer clay can cause permanent degradation of its hottest parts as in repositories in natural

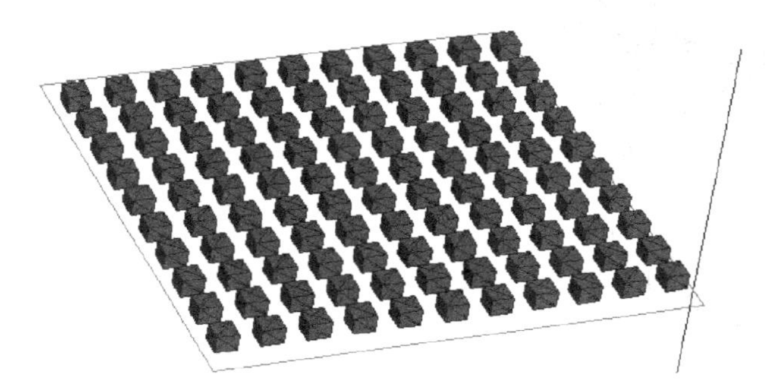

Fig. 6.27 Array configuration of a typical limestone room and pillar mine [33]

argillaceous rock. The initial degree of water saturation of the buffer must therefore be high, preferably more than 85%.

The individual components of the host rock, i.e. the pillars and contacting roof and base, are exposed to high stresses and the mechanical stability needs to be investigated in detail for certifying that collapse will not occur in the waste placement phase. The procedure for this has been outlined in various studies of which the LRDT project is particularly instructive. The stability models in this project were based on the Boundary Element Method, which has a number of advantages for modelling rock as only major discontinuities, and not the assumed homogeneous matrix between them, need to be described with elements.

Taking as an example one of the basic rock components, the pillars, the geometry is shown in Fig. 6.28 and the strain illustrated by Fig. 6.29 assuming the repository to be located at a depth of 400 m in sedimentary rock. The strong deformation of the roof and base for column spacings exceeding 3 m are caused by the low strength of the rock and the presence of an extensive EDZ.

The study verified the conclusion from several stability calculations of single deposition tunnels at a few hundred meter depth in argillaceous rock that the safety factor against collapse is very low. For the pillars in the rooms excavated in the same type of rock this factor is clearly even lower and construction of a pillared repository at this depth not possible. It would, however, be so if the depth were only about 200 m but the question would then be whether the distance to the ground surface is sufficient.

While most of the Organizations tend to consider the minimum distance from a HLW repository to the ground surface to be some 400 m it should be possible to locate it at a depth of only one or two hundred meters, applying the same performance criteria for the engineered barriers as for the deeper case. The basis of

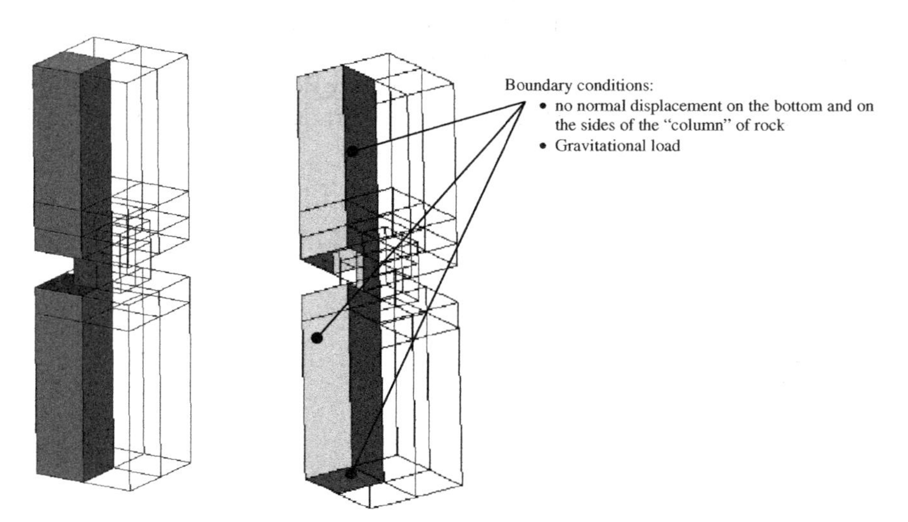

Fig. 6.28 Geometry of the columnar model

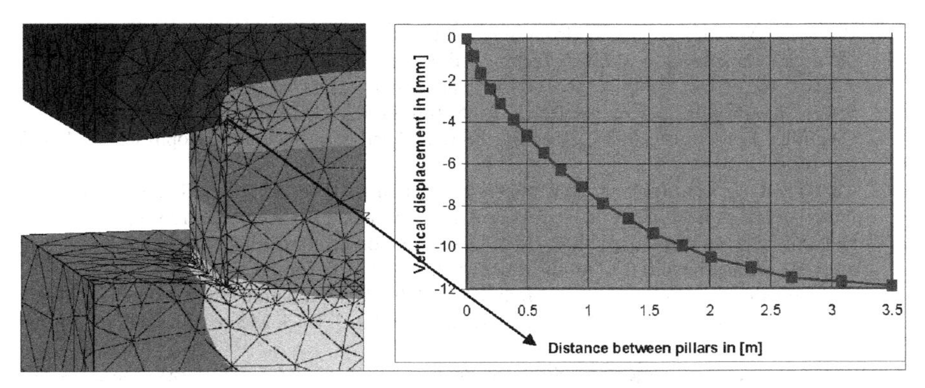

Fig. 6.29 Vertical (Z) displacement. Solid fill on a deformed mesh on the *left* and graphic on the *right*

this philosophy is that the performance of the rock with its spectrum of discontinuities will never be sufficiently well known to allow one to ascribe to it a certain isolation potential, but merely consider the rock to provide "mechanical support of the chemical apparatus", i.e. the repository. Taking the engineered barriers, i.e. the canisters and the buffer clay, to serve as isolation of the waste it would really not matter whether the depth is 200 or 400 m, keeping in mind that denudation including erosion by glaciers will not exceed a few tens of meters in a 100 000 year perspective except for where major, steep discontinuities are located. Naturally, the risk of deep glacial erosion of them would call for locating the repository at least a few kilometers from the nearest 1st order discontinuity.

6.9.2 Combined Mining and HLW Deposition

A special case worth considering is the possibility of continuing mining operations in one direction at one end of a suitable mine, leaving behind mined-out rooms that can be used for HLW disposal that proceeds at the same rate and in the same direction as the mining process. This idea, which requires application of logistics and detailed planning, has been proposed for storing low- and intermediate-level radioactive waste in a big copper mine in Kazahkstan. This concept may imply self-financing of disposal of HLW but naturally requires that the rock has enough strength and is sufficiently tight, and, of course, that the engineered barriers perform acceptably. These latter requirements would mean that no low-order discontinuities can be accepted to intersect the areas where the canisters are deposited.

References

1. Milnes AG, 1998. Crustal structure and regional tectonics of SE Sweden and the Baltic Sea. SKB Techn. Rep. TR-98-21. SKB, Stockholm.
2. Anderson EM, 1951. The dynamics of faulting. Hafner Publ. Co, New York.
3. Chinnery MA, 1966. Secondary faulting I. Theoretical aspect; Geological aspects. Can. J. Earth Sci., Vol. 3 (pp.163–190).

4. Bossart P, Mazurek M, 1991. Grimsel Test Site, Structural geology and water-flow-paths in the migration shear-zone. AGRA Technical Report 91-12. NAGRA, Wetiingen, Switzerland.
5. Zoback ML, 1992. First and second-order patterns of stress in the lithosphere: The World Stress Map Project. J. Geophys. Res., 97, B8, (pp. 11703–11728).
6. Cloetingh S, Wortel R, 1986. Stress in the Indo-Australian plate. Tectonophysics, Vol. 132 (pp. 46–67).
7. Coblentz DD, Sandiford M, Richardson RM, Zhou S, Hillis R, 1995. The origins of the intraplate stress field in continental Australia. Earth Planet. Sci. Lett. Vol. 133 (pp. 299–309).
8. Galybin AN, Mukhamediev ShA, 2004. Determination of elastic stresses from discrete data on stress orientations. Int. J. Solids Struct., Vol. 41 (18–19), (pp. 5125–5142).
9. Mukhamediev ShA, Galybin AN, Brady BHG, 2006. Determination of stress fields in elastic lithosphere by methods based on stress orientations. Int. J. Rock Mech. Min. Sci., 43, 1, (pp. 66-88).
10. Galybin AN, 2005. Project 10: Stress reconstruction. Annual ACcESS workshop 20/04/05, Brisbane, Australia.
11. Ammann M, Huber M, Mueller WH, 1992. Geologische Grundlagen fuer die hydrodynamische Modellierung des Kristallins der Nordschweiz. Dokumentierung der Arbeiten der Strukturgeologigruppe. NAGRA Intern. Ber. NAGRA, Wettingen.
12. ENRESA, 1994. Almacenamiento geologico profundo de resduos radioactivos de alta actividad (AGP). ENRESA Publ. Tecn. Num. 07/94.
13. Pusch R, 1997. Discontinuities in granite rocks. Suggested categorization for general description and rock mechanical calculations. SKB Report IPM D-97-03. SKB, Stockholm.
14. Pusch R, 1995. Rock Mechanics on a Geological Base. Developments in Engineering Geology, No.77. Elsevier Publ. Co (ISBN:0-444-89613-9).
15. Galybin AN, Dyskin AV, Brady BH, Mukhamediev ShA, 1999. Macroscopic interface sliding caused by accumulation of microzones of shear. Fracture and Damage Mechanics (Aliabadi MH, Ed) London, UK, (pp. 323–332).
16. Talbot C, Ramberg H, 1990. Some clarification of the tectonics of Äspö and its surroundings. SKB Swedish Hard Rock Laboratory, Progr. Report 25-90-15.
17. Schneider D, Brockmann E, Marti U, Schlatter A, Signer T, Wiget A, Wild U, 2002. National report of Switzerland: new developments in Swiss National Geodetic Surveying. EUREF, Publ. 11.
18. Alnaes H, 1994. Skiferens geologi – Fra brudd till Bruk. Byggsteinen No. 1/94, Steinforsk (Norw. Geotechnical Institute).
19. Pusch R, 1994. Waste disposal in rock. Developments in Geotechnical Engineering, 76. Elsevier Publ. Co. ISBN: 0-444-89449-7.
20. Pusch R, 1998. Practical visualization of rock structure. Eng. Geol., Vol. 49 (pp. 231–236).
21. Pusch R, Adey RA, 1998. Accurate computation of stress and strain od rock with discontinuities. Advances in Computational Structural Mechanics. Civil-Comp Ltd, Edinburgh, Scotland. (pp. 233–236).
22. Boulton GS, Kautsky U, Morén L, Wallroth T, 1999. Impact of long-term climate change on a deep geological repository for spent nuclear fuel. SKB Technical Report TR-99-05. SKB, Stockholm.
23. Rudberg S, 1973. Glacial erosion forms of medium size – a discussion based on four Swedish case studies. Z. Geomorph. N.F., Suppl. Bd. 17 (pp. 33–48).
24. Olvmo M, 1989. Meltwater canyons in Sweden. Diss. University of Göteborg. Department of Physical Geography.
25. Daniel, 1975. Glacialgeologi inom kartbladet Moskosel i mellersta Lappland. Sveriges Geologiska Unders. Ser. Ba nr 25.
26. Lundqvist J, 1969. Beskrivning till jordartskarta över Jämtlands län. Sveriges Geologiska Unders. Ser. Ca nr 45.
27. Dahl R, 1968. Glacial accumulations, drainage and ice recession in the Narvik-Skjomen district, Norway. Medd. Upsala Univ. Geogr. Inst., Ser. A, Nr. 232. (pp. 101–165).

28. Mueller WH, 1999. Geologische Entwicklung der Nordschweiz, Neotektonik und Langzeitzenarien Zuercher Weinland. NAGRA Technischer Bericht 99-08. NAGRA, Wettingen, Schweiz.
29. Pusch R, 1978. Rock sealing with bentonite by means of electrophoresis. Bull. Int. Ass. Eng. Geology, No. 18 (pp. 187–190).
30. Casagrande L, 1953. Review of past and current work on electro-osmotic stabilization of soils. Harvard Soil Mechanics, Ser. No. 45.
31. Pusch R, 1969. Geotechnical aspects of the interpretation of distorted strata in Silurian deposits. Diss. Acta Universitatis Stockholmiensis, Stockholm Contributions in Geology, Vol. XXI:2. Almqvist & Wiksell, Stockholm.
32. Snäll S, 1977. Silurian and Ordovician bentonites of Gotland (Sweden). Acta Universitatis Stockholmiensis, Stockholm Contributions in Geology, Vol. XXXI:1. Almqvist & Wiksell, Stockholm.
33. Popov V, Pusch R, (Eds), 2006. Disposal of hazardous waste in underground mines. WIT Press, Southampton, Boston. ISBN: 1-85312-750-7. ISSN: 1476-9581.

Chapter 7
Risk Assessment

7.1 General

Risk assessment associated with disposal of HLW is naturally a very important and comprehensive issue and even a very brief summary of the involved tasks would occupy dozens of books of this size. It therefore just touches on the subject and only refers to issues that have been dealt with in the preceding chapters, adding some general information of the risk of exposure to human beings by radiation, which underlines the necessity of practical and safe emplacement and retrieval of waste containers. The basis of the chapter is the generally accepted principle that isolation of radioactive waste is required for minimizing the risk of exposing the "biosphere" to radioactivity. In the present context, risk can be defined as the product of the probability of events that imply such exposure, and their health consequence. The first mentioned, which depends on the performance of the repository, is a fundamental issue for designers and licensing authorities, while the latter is of course a medical matter.

One should distinguish between risk related to external impact, like deliberate or accidental excitation of nuclear bombs or fall of supersized meteors, and risk by malfunctioning of the repository, taking all sorts of natural processes in the earth crust into consideration. We are considering here this latter issue and apply as a further restriction, the implicitely included policy of the Swedish Radiation Protection Institute (SSI), which agrees with what is stated by IAEA. Thus, following the general guidance of these authorities, the risk criteria concern repositories undisturbed by man, i.e. with the exclusion of intrusion into the repository and of human actions that disturb the immediate environment of the repository. SSI's regulation requires that the annual risk be less than E-6 and that it is valid in the first 100 000 years after closure. Before completing the construction of HLW repositories it is expected that their capability to isolate HLW beyond this period will be discussed with the authorities.

R. Pusch, *Geological Storage of Highly Radioactive Waste*,
DOI: 10.1007/978-3-540-77333-7_8, © Springer-Verlag Berlin Heidelberg 2008

7.2 Required Performance of the Repository

Performance requirements are defined by the appropriate laws and regulations in the different countries and they are not host rock specific. The national requirements are quite different in the different European countries. Safety assessment or performance assessment studies are carried out using pessimistic or conservative assumptions for determining whether or not these specific requirements are met. To account for unforeseen features, events and processes (FEPs), a large number of scenarios and parameter variations – impact analysis and sensitivity analysis – are investigated to assess the robustness of the proposed disposal system. Sufficient mechanical stability of repository drifts and rooms in the construction and waste application phases is a fundamental requirement and since argillaceous rock and clastic clay commonly have a rather low shear strength and exhibit significant anisotropy, while the rock stresses can be as high as in other geological media, this issue needs consideration [1]. Here, we will look a bit closer on the treatment of the risk issue in some of the EU-countries.

7.2.1 Sweden

Safety considerations form the basis of assessment of risk, primarily by identifying processes that can cause hazard, and by weighing their consequences. The processes of particular importance are those threatening the primary aim of a repository, i.e. to provide isolation of the waste, as well as the second aim, which is to delay migration of possibly released radionuclides, also termed as "attenuation of releases to the environment" [2]. An example of how one can proceed in a systematic study considering evaluation of FEPs is the following list of issues [3]:

- Basic scenario with outline of the evolution of radiation, temperature conditions, hydraulic performance, mechanical performance, and chemical constitution. The rock is assumed to maintain its structural constitution while the sea level will change. The EBS is assumed to retain its properties.
- Scenario of the formation of canister defects. The conditions are the same as in the basic scenario but a limited number of canisters are assumed to have defects emanating from the manufacturing.
- Scenario of climate changes. The conditions are the same as in the basic scenario but glaciation is assumed to take place. The performance of canisters is a number one issue.
- Scenario of tectonic events. The conditions are the same as in the basic scenario but seismic events, implying very moderate but quick shearing of deposition tunnels and holes, are assumed. The performance of canisters is a number one issue.

All the scenarios have an impact on the performance of the canisters, which are the most essential isolating components simply because canisters that remain intact will not give off radionuclides. Among the various related issues the following three have been in focus in formulating the risk concept:

- Radiation and release of radionuclides from spent fuel, handled, confined in canisters, and transported.
- Glaciation and its impact on deposition tunnels and holes and hence canisters.
- Degradation of engineered barriers by chemical processes and mechanical strain, which can all affect the canisters.

7.2.1.1 Spent Fuel

In the reactor, fission of the isotope U-235 and other fuel constituents yields new radioactive substances of which many are long-lived. Radionuclides in the spent fuel are fission and activation products, and actinides. Typical fission products are iodine (I-129, I-131), cesium (Cs-134, Cs-135, Cs-137) and strontium (Sr-90), while activation products are radioactive isotopes formed by uptake of neutrons in the atomic nuclei of the respective material. Actinides are remnants of the uranium and of new substances like plutonium and americium. Release of any of these elements to the biosphere and uptake by living species represents extreme hazard. Thus, the activity, expressed in terms of dose rate per hour, of the spent fuel is about 65 000 mSv/hour after 40 years of intermediate storage, the risk being illustrated by the fact that a person standing close to unshielded fuel will get a fatal dose (5000 mSv) in less than five minutes [4,5]. If the person steps back one hundred meters the dose rate is about 6.5 mSv per hour, meaning that he is exposed to the admissible normal annual dose in one hour. After 1000 years the dose rate has dropped to about 50 mSv/hour and after 10 000 years to about 5 mSv/hour and further to less than 1 mSv after 100 000 years. This demonstrates the need for effective and safe isolation for very long periods of time and, in particular, in the initial 50 years when the fuel is first stored under water and then placed in canisters, which are transported on public roads and within the repository area.

Retrievability of the waste, e.g. canisters, has been considered by SKB and techniques for it have been investigated and found feasible. However, it has been realized that extraction of canisters will be difficult because of the radiation early after the disposal and because of possible leakage of radionuclides after a longer period of time when the canisters can have been deformed or broken because of large rock strain. If the buffer has been significantly contaminated by radionuclides, retrieval would be very difficult and probably not possible. No detailed description of the practical handling and risk associated with canister retrieval has been worked out and there is no policy for assessing possibilities and limitations like in NAGRA's work.

7.2.1.2 Transport

A first risk moment is associated with transport of the waste. Spent fuel is in solid form and enclosed in tight containers that shield off radiation and protect the fuel from being damaged. For transport from the intermediate storage plant to the repository the spent fuel is enclosed in copper/iron canisters, which are in turn contained in transport cascs of 270 mm steel for total radiation shielding. These cascs are designed to withstand 2 MPa water pressure and 800°C for 30 minutes. The only realistic mishap that can possibly take place is if the transport vehicle is wrecked and the transport casc exposed to fire so that the waste becomes heated [4,5]. The amount of released cesium will accumulate on the ground surface and the contamination be similar to what struck northern Sweden by the Tjernobyl accident.

7.2.1.3 Impact of Glaciation and Performance and Evolution of the Repository

Glaciation is expected but not in the nearest 10 000 years although preceding permafrost could start already in a few thousand years. Applying models for the growth of large glaciers that are based on data deduced from preceding Quaternary glaciations, the shape and height of forthcoming glaciers have been predicted and also the associated evolution of stresses on the repository level. This has been shown to lead to overstressing of the near-field rock and comprehensive breakage of deposition holes and tunnels. The performance of the canisters due to the very significant rock strain has not yet been determined and the risk of release of radionuclides has not been ultimately defined [3].

Taking the formation of permafrost and growth of big glaciers to occur much more rapidly than presently assumed, the canisters may be in a critical condition already after 10 000 years from now. The dose rate would then be a few mSv/hour, but released radionuclides will be prevented from reaching up to the ground surface since the rock would be frozen to a depth of more than 100 meters. Melting of the glacier would require at least another 10 000 years, at which time the dose rate is even lower and radionuclides appearing at the ground surface being washed out in the arctic sea at negligible concentration.

While this scenario would obviously not be critical, another process has recently been proposed to cause an important but not yet defined risk. This process is proposed to be dispersion of the buffer clay in the low-electrolyte melt water that will occupy the voids of the rock mass in the course of the ice retreat. The dispersion of the clay is proposed to be followed by transport of the released clay particles out from the near-field where the amount of canister-embedding buffer could drop to a critically low value. The loss of buffer would of course be disastrous but there is presently no complete conceptual model for large-scale loss of buffer nor any unanimous estimate of the rate and comprehension of this process.

7.2.2 Switzerland

In Switzerland, the requirements are outlined in the regulatory guideline R21 from the authorities [1] stating that a dose rate of 0.1 mSv/year must not be exceeded at any time. The allowed dose rate is very low in comparison to the natural dose rates, which depend very much on local conditions and range from 1 to about 20 mSv/year.

According to the present plans the radioactive waste will be disposed in a geological repository at a certain depth in argillaceous rock, the tightness of which is deemed to be sufficient to prevent radionuclides to reach the ground surface in nearly any time perspective. This conclusion has been reached taking exogenic processes as well as tectonics and seismic events into consideration [2]. A major issue that has emerged concerns retrieval of canisters, which may become necessary and which can represent considerable risk [1]. The reason for retrieval of the waste can be:

- Social or political considerations.
- Monitoring or new research results indicating that the repository is not safe.
- Part of the "waste", especially spent fuel, being needed as a resource for further use.

In all cases, the waste has to be retrieved without exposing the workers, people living in the vicinity of the repository, or the environment, to danger.

Clays in the form of argillaceous rock may cause more difficulties in developing a retrievability plan than hard rocks. Liners or any other support that keep tunnels open are likely to degrade with time due to alteration of concrete and corrosion of steel. In addition, supports may fail due to the additional thermal stress induced during the thermal pulse. Therefore, significant technical effort could be necessary to retrieve the waste. Depending on the appropriate regulations, different levels of emphasis are given to retrievability or reversibility. If it has a high priority, the layout of the emplacement tunnels and holes favours smaller units, which means rather than several hundred meter long emplacement tunnels containing a large number of containers, shorter emplacement tunnels or boreholes with few containers.

In principle, retrievability of waste canisters in clay, regardless of the type, is feasible at nearly any time but becomes more and more difficult as re-saturation of the repository progresses. Therefore, a staged approach is proposed, implying that retrievability is always possible but has different technical implications. The different stages could be defined in the following simplified way:

7.2.2.1 Stage 1: Operational Stage

At this stage, all access, operational and construction tunnels are open and accessible, trained crews are available and the re-saturation of the buffer is at a very early stage because of the very low hydraulic conductivity of the clayey host rock and the very limited frequency of water-conducting features. This means the buffer can be fairly easily retrieved and the waste emplacement be reversed.

7.2.2.2 Stage 2: Monitoring Stage

The emplacement is then finished, and part of the tunnel network backfilled and sealed, while the access tunnels and ramps are still open. Limited trained manpower is available but all necessary information on the construction and emplacement work is available. The reopening of the backfilled access and operational tunnel systems is standard mining work and no radiological shielding is necessary here. However, the retrieval of the waste from the emplacement unit will be more difficult because the buffer material is probably partly saturated and has built up a significant swelling pressure, and temperature will have increased significantly. The retrieval of the buffer associated with required reinforcement of the tunnel support is time consuming but feasible, as has in principle been shown in the URLs. Short emplacement units clearly have an advantage in such a situation.

7.2.2.3 Stage 3: Post-Closure Stage

At this stage all underground structures are backfilled and the repository not accessible anymore. Detailed studies have to be made of how to re-access the closed repository, whether by opening old access ramp or shafts or by constructing new ones. Retrievability during this stage is very time consuming and expensive but still feasible. Depending on the different programs, additional stages could be necessary or appropriate.

Monitoring starts with the initial characterisation of the repository site. The changes of the baseline conditions due to construction, operation and closure can then be followed. The monitoring program in tight rocks like clays will probably be significantly different from that in hard and brittle rocks as changes in water levels or hydraulic heads at some distance from the underground tunnels can hardly be detected.

From technical and scientific points of view, long-term monitoring is not mandatory for a repository that is designed to be passively safe, but it is requested by government or public in most countries. Such monitoring has to be designed to ensure that safety of the repository is not compromised by the monitoring system. This means that sensors should not be placed too close to the waste and cables should be avoided (hydraulic short-circuit). To avoid undesirable effects due to monitoring, a pilot repository close to, but hydraulically isolated from the main repository, is preferably constructed. This pilot repository would contain a small but representative amount of waste in a configuration that mimics the conditions in the main repository as planned in Switzerland. This pilot repository could then be monitored during a special extended monitoring phase and closed at a later stage. Monitoring programs are not defined at the moment; research work is conducted in existing and upcoming projects to define monitoring strategies in more detail.

7.2.3 Belgium

The regulating authority in Belgium has not yet determined any dose or risk constraints for the deep disposal of conditioned high-level and long-lived waste [1]. Belgian legislation and regulations normally conform, or are being brought into line, with ICRP recommendations as included in the basic standards of the IAEA and the EU. A maximum limit for the dose constraint used for exposures that are expected to occur, has been defined at 0.3 mSv/year. Also a risk constraint used for potential exposures, i.e. that may or may not occur, has been defined as the product of the probability of exposure to a certain dose and the likelihood of death as a result of that dose. A factor of about 5E-2 per Sv is used by ICRP for this likelihood, and for a dose constraint of 0.3 mSv/year, a risk constraint of 1 to 2E-5 per year is assumed.

7.3 Current Repository Design Principles

The design of repositories is based on the concept of multi-barrier systems and regulatory requirements, which differ from country to country. All concepts for the disposal of HLW including spent fuel must provide safety by a combination of EBS and the natural geological barriers. No single barrier must jeopardize the function of other barriers. The host rock should provide a stable, protected environment for the EBS, ensuring their longevity. Crystalline rock is very good in this respect and salt and clastic clay less good but the two last mentioned provide better retardation of radionuclides that may escape from the EBS provided that no continuous EDZ is formed.

7.3.1 Repositories in Crystalline and Argillaceous Rock

Crystalline and strong argillaceous rock provide very good mechanical stability to repositories meaning that very little or moderately comprehensive support in the form of bolting, shotcreting, grouting or casting of bulkheads is required in the construction and waste placement phases. The frequent low-order discontinuities in crystalline rock makes it more difficult to find space for deposition tunnels and holes than in argillaceous and other types of rock. Argillaceous rock is normally weaker than crystalline rock, which makes it necessary to locate repositories closer to the ground surface than in crystalline rock. Access to repositories is possible via ramps and shafts. Ramps have a much stronger impact on the groundwater pressure and flow than shafts.

EBS consist of metal canisters, the most lasting ones being copper in reducing environment, dense smectite clay for serving as buffer, and smectitic backfill in deposition tunnels, from which buffer-equipped deposition holes extend, and in other places where tightness is required. The clay buffer and backfills are chemically compatible with the rock.

Risk is associated with construction work in weak parts of the rock, with degradation of buffers and backfills, and especially with breakage and leakage of canisters.

7.3.2 Repositories in Salt Rock

Construction of repositories in salt is simpler, quicker and safer than in brittle rock and can be made deeper than in the other rocks. The lack of discontinuities in domal salt gives very good opportunities for locating rooms etc in rational patterns but the strong creep potential requires frequent maintenance. Access to repositories is only via shafts. Bedded salt contains layers or lenses of sediments that at least partly determine the location and size of deposition rooms [1].

EBS consists of metal containers or canisters that can be weak and thin, or strong and heavy, depending of the type of waste, and backfills consist of crushed salt. The lack of water minimizes corrosion and hence implies long-lived EBS.

Risk is associated with construction of wide rooms that can collapse, and, possibly, with very high gas pressures that can jeopardize the stability of the host rock. Also, the presence of brine inclusions can cause problems in excavation of tunnels and rooms and cause risk of collapse.

7.3.3 Repositories in Clastic Clay

The repository design should minimize perturbation of the barrier performance of the clay host rock. The specific characteristics of the clay must not be irreversibly disturbed by mechanical operations. This may require use of specialized excavation and construction techniques to prepare galleries with limited dimensions [1].

Construction of repositories in clastic clay is much more difficult and has to be made closer to the ground surface than in the other rocks because of the low mechanical strength. The lack of low-order discontinuities in suitable deposits of clastic clay offers good possibilities for rational utilization of the rock, the problem being that the thickness of accessible homogeneous clay sediments is limited. The poor stability requires very quick construction of support of excavations for which concrete or steel elements are used. This may cause problems with chemically induced dissolution of cement and clay and production of hydrogen gas. Since dense clay, of the type that can be considered for hosting a repository, has a very low gas conductivity, high gas pressures can evolve if the amount of steel support is big. For a repository at a depth of 300 m it may rise to 10 MPa and cause large deformation of the EBS. Higher gas pressure may yield disastrous outburst of gas from the repository via a channel formed all the way up the ground surface.

Risk is associated with construction because of the low strength of the clay mass and collapse is a common threat. In a longer perspective, heating can increase the porewater pressure and reduce the effective pressure so that instability arises. Gas production can theoretically result in large strain and collapse and assessment of the risk associated with these two events would be required.

7.4 Design Requirements Related to Safety

7.4.1 EBS

The safety-related requirements are based on general design principles and safety strategies [1]. The system must provide physical containment by limitation of water influx into the repository during construction and in the thermal phase, and delay and disperse the escape of radionuclides from the waste matrix through the disposal system. It must also dilute the concentration of radionuclides escaped from the disposal system, and limit access to the disposed waste. More specifically, the EBS should fulfil the following:

1. The EBS has been attributed the role in the disposal system of ensuring containment of the radionuclides during the operational and thermal phases of the repository. The required time periods may be derived from the consideration that complex interactions between components of the engineered barriers and also radionuclide migration under a significant temperature gradient should be avoided during the thermal phase. Both phenomena are generally considered as potentially enhancing radionuclide release. The duration of the period during the operational phase in which the waste containers and their immediate environment are exposed to the corrosive effects of air, should be minimised. The reliability of the materials used and operations applied should be ensured by a Quality Assurance and Quality Control (QA/QC) Program elaborated for each component (e.g. for weldings).
2. The different components of the engineered barriers should be conceived and chosen in such a way that they do not adversely affect the performance of the host rock. This implies that Portland cement should not be used in concrete abutments in argillaceous or clastic clay because of the degrading impact of the high pH, and that steel should not be extensively used because of its gas-producing potential and its impact on clay by forming cementing components. Still, concrete based on the use of high-pH cement may still be considered provided that the products from its reaction with the host rock and other barriers can serve acceptably.
3. The temperature increase in the near field should be limited for avoiding boiling in the thermal period. A maximum temperature requirement of 100°C at the interface between the canisters and the surrounding material has been proposed for the Belgian, Swedish and Finnish concepts. This requirement should minimize the temperature impact on the performance of the engineered barriers and on the function of the host rock.

7.4.2 Design Requirements Related to the Assessment of Long-Term Radiological Safety

The importance of long-term radiological safety is of greatest importance and primarily puts a demand on the EBS [1]. One of the related issues is the location of repositories, for which the following recommendations have been proposed:

1. In argillaceous and clastic clay the horizontal disposal galleries should be located in the middle part of the clay layer in order to maximize the effective thickness of the layer of clay around the waste.
2. The disposal system should possess robustness in order to limit the uncertainties related to the assessment of its long-term performance, i.e. the characteristics of its components should exhibit reasonable predictability. It is important to avoid as much as possible any occurrence of chemical or physical interactions between the waste, the engineered barriers, and the host rock.
3. As the decision to suspend reprocessing has been taken rather recently in some countries, design studies for the disposal of spent fuel (SF) have not always

progressed as far as the vitrified waste studies. However, as a general rule, the different waste types including different forms of HLW and long-lived ILW will be disposed of separately (compartmentalisation).

7.4.3 Design Requirements Related to Safety During the Operational Phase

1. The repository design should provide radiological protection to the personnel operating the repository facility. The national legislations with respect to the maximum allowed doses have to be respected.
2. Clays, even overconsolidated claystones, are low-strength materials and the current disposal concepts favour keeping emplacement units open for a short time only. Therefore, concurrent construction and emplacement work is planned, which could lead to conflicts in operational radiological safety. In such cases, it is foreseen to work in separate areas, with separate access and infrastructure – i.e. regulated and non-regulated areas. This concerns the Belgian concept and the proposed use of mines discussed in Chap. 6.

7.4.4 Design Requirements Related to Criticality

All the conditions of normal and contingency operation and all the conditions of long-term containment of the waste must be such that they take adequate account of the risks of criticality. This is fulfilled by the proposed repository concepts that are planned to be realized in crystalline, salt, and argillaceous rock, as well as in clastic clay.

7.4.5 Design Requirements Related to Non-radiological Environmental Impact

Depending on the geological conditions at a given site special design requirements can be specified to minimize the effect of the repository on the environment, respecting for instance temperature increase in nearby aquifers, stockpiling of excavated rock and soil and storing used material on the ground surface.

7.4.6 Design Requirements Related to Flexibility

The design concepts are based on the "observational method" principle, that is, the design should be flexible enough to adaptate to improved knowledge of the host clay formation and of the barrier materials gained in the course of excavation. As a general rule, the development and implementation of a disposal system should follow a step-by-step approach. The concerned time period covers many decades, during which new insights may be acquired or new technologies developed and implemented. It should therefore remain possible to reverse or to defer certain decisions on technical solutions.

7.4.7 Design Requirements Related to Retrievability of the Waste

Retrievability or reversibility of the waste emplacement may become a legal obligation in countries that are not favouring disposal in salt rock or in clastic clay in which identifying of the position of heavy canisters and bringing them up the biosphere would be extremely difficult or impossible. For all other rock types retrievability should be considered for each of the lifetime phases of the repository. However, design options introduced to facilitate retrievability should under no circumstances adversely affect the operational or long-term safety of the repository.

7.4.8 Design Requirements Related to Technical Feasibility

The repository design must be technically feasible. Therefore, the design of the repository components and transportation equipment should be based on well-known systems and mechanisms that have proven their effectiveness in the industry.

References

1. Svemar Ch, 2005. Cluster Repository Project (CROP). Final Report of European Commission Contract FIR1-CT-2000-2003, Brussels, Belgium.
2. NAGRA, 2002. Project Opalinus Clay, Safety Report. NAGRA, Wettingen, Switzerland.
3. SKB, 2003. Planning Report for the Safety Assessment SR-Can. SKB Technical Report TR-03-08. SKB, Stockholm.
4. Wilson CK, Shaw KB, Hughes JS, 1995. Accidents and Incidents in the transport of radioactive material – an analysis of 37 years of experience. Proc. PATRAM'95, The 11th Int. Conf. on the Packaging and Transportation of Radioactive Materials.
5. Pettersson L, Ringi M, 1997. Använt kärnbränsle – Transporter. SKB Report R-97-22. SKB, Stockholm.

Index

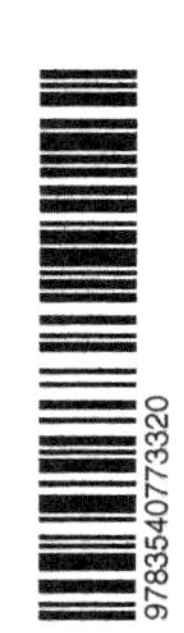
9783540773320